Pandas1.x 实例精解

[美] 马特·哈里森 等著

刘 鹏 译

清华大学出版社

北 京

内 容 简 介

本书详细阐述了与 Pandas 相关的基本解决方案，主要包括 Pandas 基础、DataFrame 基本操作、创建和保留 DataFrame，开始数据分析，探索性数据分析，选择数据子集，过滤行，对齐索引，分组以进行聚合、过滤和转换，将数据重组为规整形式，组合 Pandas 对象，时间序列分析，使用 Matplotlib、Pandas 和 Seaborn 进行可视化，调试和测试等内容。此外，本书还提供了相应的示例、代码，以帮助读者进一步理解相关方案的实现过程。

本书适合作为高等院校计算机及相关专业的教材和教学参考书，也可作为相关开发人员的自学用书和参考手册。

北京市版权局著作权合同登记号 图字：01-2021-6445

Copyright © Packt Publishing 2020.First published in the English language under the title
Pandas 1.x Cookbook.
Simplified Chinese-language edition © 2022 by Tsinghua University Press.All rights reserved.

本书中文简体字版由 Packt Publishing 授权清华大学出版社独家出版。未经出版者书面许可，不得以任何方式复制或抄袭本书内容。

本书封面贴有清华大学出版社防伪标签，无标签者不得销售。
版权所有，侵权必究。举报：010-62782989，beiqinquan@tup.tsinghua.edu.cn。

图书在版编目（CIP）数据

Pandas1.x 实例精解 /（美）马特•哈里森，等著；刘鹏译. —北京：清华大学出版社，2022.6
书名原文：Pandas 1.x Cookbook
ISBN 978-7-302-60960-5

Ⅰ. ①P… Ⅱ. ①马… ②刘… Ⅲ. ①软件工具—程序设计 Ⅳ. ①TP311.561

中国版本图书馆 CIP 数据核字（2022）第 089015 号

责任编辑：贾小红
封面设计：刘　超
版式设计：文森时代
责任校对：马军令
责任印制：丛怀宇

出版发行：清华大学出版社
网　　址：http://www.tup.com.cn，http://www.wqbook.com
地　　址：北京清华大学学研大厦 A 座　　邮　编：100084
社 总 机：010-83470000　　邮　购：010-62786544
投稿与读者服务：010-62776969，c-service@tup.tsinghua.edu.cn
质量反馈：010-62772015，zhiliang@tup.tsinghua.edu.cn

印 装 者：定州启航印刷有限公司
经　　销：全国新华书店
开　　本：185mm×230mm　　印　张：36.25　　字　数：726 千字
版　　次：2022 年 6 月第 1 版　　印　次：2022 年 6 月第 1 次印刷
定　　价：159.00 元

产品编号：090911-01

译　者　序

如果说数据是信息时代的石油，那么数据分析人员就是石油工人，他们负责从原始数据中提炼出见解，为成功的预测和决策提供科学可信的佐证。可以说，他们就是真正的"神算子"，而且这种神算既不靠夜观天象的玄学，也不靠脑袋一热的灵感，而是靠海量的源于真实世界的数据采集、整理、提取、计算、建模、训练、预测和验证过程。

一般认为，数据分析人员（或数据科学家）应该具有数据分析、统计学和机器学习等多方面的核心能力。Pandas 是基于 NumPy 的一种工具，学习和掌握它可以为解决数据分析任务提供极大的帮助。Pandas 的名称来自面板数据（panel data）和 Python 数据分析（data analysis），它纳入了大量库和一些标准的数据模型，提供了高效操作数据集所需的工具。Pandas 最初是作为金融数据分析工具而开发出来的，但现在它和 Python 结合，已经可以构成强大而高效的数据分析环境。

本书是学习 Pandas 方面的实用教程，它以秘笈的方式，提供了大量的 Pandas 基础操作和高级技巧。其中包括基本的 Series 和 DataFrame 操作（了解数据类型、选择列、调用 Series 方法、了解 Series 的操作、使用 Series 方法链、重命名列名、创建和删除列、选择多个 DataFrame 列、使用方法选择列、排序列名称、统计 DataFrame 摘要信息、比较缺失值、转置 DataFrame 操作的方向）；基础数据分析方法（开发数据分析例程、通过更改数据类型减少内存使用量、通过排序选择每个组中的最大值）；探索性数据分析（使用分类数据和连续数据、跨越分类比较连续值、比较两个连续列、使用 Pandas 分析库）；数据子集选择技巧（同时选择 DataFrame 行和列、使用整数和标签选择数据、按字典序切片）；数据过滤（构造多个布尔条件、用布尔数组过滤、比较行过滤和索引过滤、使用唯一索引和排序索引进行选择、转换 SQL WHERE 子句）；索引对齐（检查索引对象、生成笛卡儿积、了解索引暴增现象、给不相等的索引填充值、添加来自不同 DataFrames 的列、突出显示每一列的最大值）；分组和聚合、数据重组和规整（使用 stack 将变量值规整为列名称、使用 melt 将变量值规整为列名称、同时堆叠多组变量、反转已堆叠的数据、在 groupby 聚合之后取消堆叠、使用 groupby 聚合复制.pivot_table 方法的功能、重命名轴的级别以方便数据重塑）；时间序列分析（智能分割时间序列、用时间数据过滤列、使用仅适用于 DatetimeIndex 的方法）；使用 Matplotlib 和 Pandas 等进行可视化以及测试 Pandas 代码等。

在翻译本书的过程中，为了更好地帮助读者理解和学习，本书以中英文对照的形式保留了大量的原文术语，这样的安排不但方便读者理解书中的代码，而且也有助于读者通过网络查找和利用相关资源。

本书由刘鹏翻译，此外黄进青也参与了部分翻译工作。由于译者水平有限，难免有疏漏和不妥之处，在此诚挚欢迎读者提出任何意见和建议。

译　者

前　　言

　　Pandas 是一个使用 Python 创建和处理结构化数据的库。什么是结构化？其实就是按行和列组织的表格数据，就像你在 Excel 电子表格或 SQL 数据库中找到的那样。数据科学家、分析师、程序员和工程师等都将利用它来提取所需的数据。

　　Pandas 仅限于"小数据"（这里说的"小"是指数据可以容纳在单台机器的内存中）。但是，其语法和操作已被其他项目采用或启发了它们的应用，这些项目包括 PySpark、Dask、Modin、cuDF、Baloo、Dexplo、Tabel、StaticFrame 等。这些项目有不同的目标，但其中一些将扩展到大数据。因此，Pandas 的功能使它逐渐成为与结构化数据进行交互的事实上的 API，在这种情况下，了解 Pandas 的工作原理是很有必要也很有价值的。

　　本书作者之一 Matt Harrison 经营着一家负责企业培训的公司，其主营业务是为想要提高 Python 和数据处理技能水平的大型公司提供人员培训服务。因此，这些年来，他已经教会了成千上万的人使用 Python 和 Pandas。编写本书的目的是帮助许多人更好地理解和应用 Pandas，破解他们的迷惑。尽管 Pandas 优点不少，但也有一些难解或令人困惑的地方。本书将详细介绍有关 Pandas 中的各种操作，指导读者了解可能遇到的一些难点，以便能够真正掌握和处理它们。

本书读者

　　本书以操作秘笈的形式编写，包含近 100 个秘笈，从非常简单的应用到高级操作技巧都有涵盖。所有秘笈力求以清晰、简洁、现代的惯用 Pandas 代码编写。"实战操作"部分详细介绍各个秘笈的操作步骤，"原理解释"部分对秘笈的每一步都进行非常详细的阐释。绝大多数秘笈还提供"扩展知识"部分，使读者能够举一反三，发展出自己的操作技巧。本书包含大量的 Pandas 代码，并提供了配套的源数据文件，以便读者跟随操作和对照学习。

　　概括而言，前 7 章中的秘笈比后面 7 章中的 Pandas 操作更简单，并更着重于 Pandas 的基础应用，而后 7 章的重点则是更高级的操作技巧，并且更多地以项目为导向。由于本书涵盖的范围和难度都较广，因此对新手和日常用户都有用。根据我们的经验，即使是经常使用 Pandas 的人也可能并未掌握其很多编码技巧。这也和 Pandas 本身的特性有关，因为几乎总是有多种方法可以完成相同的操作。但是，如果读者不熟悉 Pandas 的话，那么采用的方法可能是效率最低的。对于同一个问题，两个 Pandas 解决方案之间在性能上相差一个数量级,这种情况并不罕见。

阅读本书需要掌握一定的 Python 基础知识。我们假定读者熟悉 Python 中所有常见的内置数据容器,如列表、集合、字典和元组。

内容介绍

本书共包含 14 章,具体内容如下。

第 1 章"Pandas 基础",详细介绍 Pandas 的两个数据结构,即 Series 和 DataFrame。此外,还解释它们的组成部分和相关术语。数据的每一列必须仅具有一种数据类型,并且每种数据类型都被涵盖。对此,本章详细讨论每种数据类型,并介绍如何使用方法链等操作。

第 2 章"DataFrame 基本操作",重点介绍数据分析人员在数据分析期间执行的最关键和最典型的操作。

第 3 章"创建和保留 DataFrame",讨论提取数据和创建 DataFrame 的各种方法,包括读取 CSV 文件、Excel 电子表格、JSON 格式数据和 HTML 表格等。

第 4 章"开始数据分析",介绍在读入数据之后应该开始执行的操作,例如通过更改数据类型减少内存使用量、从最大中选择最小、通过排序选择每个组中的最大值和计算追踪止损单价格等,这些都是比较实用的技巧。

第 5 章"探索性数据分析",介绍用于比较数字数据和分类数据的基本分析技术。本章还演示常见的可视化技术。

第 6 章"选择数据子集",介绍选择数据的不同子集的多种方法,包括选择 Series 数据、选择 DataFrame 行、同时选择 DataFrame 行和列、使用整数和标签选择数据、按字典序切片等,这些操作包含一定的技巧,粗心的用户可能会感到困惑。

第 7 章"过滤行",介绍查询数据以基于布尔条件选择数据子集的过程,包括构造多个布尔条件、用布尔数组过滤、使用查询方法提高布尔索引的可读性,以及使用布尔值、整数位置和标签进行选择等。

第 8 章"对齐索引",主要讨论非常重要但却经常被误解的索引对象。错误使用索引会导致许多错误的结果,本章中的秘笈演示如何正确使用索引来提供有力的结果。

第 9 章"分组以进行聚合、过滤和转换"介绍强大的分组功能,这些功能在数据分析期间总是必需的。你可以构建自定义函数以应用于分组。

第 10 章"将数据重组为规整形式",阐释规整数据的定义及其重要性,并演示如何将许多不同形式的杂乱数据集转换为规整数据集。

第 11 章"组合 Pandas 对象",介绍许多可用于垂直或水平组合 DataFrame 和 Series

的方法，包括将新行追加到 DataFrame、将多个 DataFrame 连接在一起以及连接到 SQL 数据库等操作。此外，还详细阐释 concat、join 和 merge 方法之间的区别。

第 12 章 "时间序列分析"，讨论时间序列的强大功能，它使得分析人员可以按任何时间维度进行数据剖析。

第 13 章 "使用 Matplotlib、Pandas 和 Seaborn 进行可视化"，本章主要介绍 Matplotlib 库，该库负责 Pandas 中的所有可视化绘图。此外，还介绍 Pandas 绘图方法以及 Seaborn 库，Seaborn 库能够产生 Pandas 中无法直接获得的美观的可视化效果。

第 14 章 "调试和测试"，探讨测试 DataFrame 和 Pandas 代码的机制。如果你打算在生产环境中部署 Pandas，那么本章将帮助你建立对代码的信心。本章介绍的具体操作包括转换数据、测试.apply 方法的性能、使用 Dask、Pandarell 和 Swifter 等提高.apply 方法的性能、检查代码、在 Jupyter 中进行调试、管理数据的完整性、结合使用 pytest 和 Pandas 以及使用 Hypothesis 库生成测试等。

充分利用本书

要充分利用本书，你也许需要执行以下操作。

首先，也是最重要的，你应该下载本书所有代码，这些代码都被存储在 Jupyter Notebook 中。阅读每个秘笈时，请在 Notebook 中运行代码的每个步骤。在运行代码时，请确保自己进行更多的探索。

其次，在浏览器中打开 Pandas 官方说明文档，其网址如下。

http://pandas.pydata.org/pandas-docs/stable/

Pandas 说明文档是一个很好的资源，其中包含超过 1000 页的材料。在文档中有大多数操作 Pandas 的示例，通常可以从 See also（另请参阅）部分中直接链接它们。当然，它的缺陷是，虽然涵盖了大多数基础操作，但示例采用的却是虚拟数据，这些虚拟数据并不能反映你在分析现实世界中的数据集时可能遇到的情况。

本书需要的软件包

Pandas 是用于 Python 编程语言的第三方程序包，在出版本书时，它的版本为 1.0.1（目前，Python 的版本为 3.8）。本书中的示例在 Python 3.6 及更高版本中都应该可以正常工作。

你可以通过多种方式在计算机上安装 Pandas 和本书提到的其余库，但是最简单的方法是安装 Anaconda 发行版。该版本由 Anaconda 创建，将所有流行的用于科学计算的库打包到一个可下载的文件中，该文件可在 Windows、macOS 和 Linux 上使用。你可以访问以下页面以获取 Anaconda 发行版。

https://www.anaconda.com/distribution

除了所有科学计算库外，Anaconda 发行版还附带 Jupyter Notebook，这是一个基于浏览器的程序，可使用 Python 和其他多种语言进行开发。本书的所有秘笈都是在 Jupyter Notebook 内部开发的，所有代码都已提供。

当然，不使用 Anaconda 发行版也可以安装本书所需的所有库。感兴趣的读者可访问 Pandas 安装页面，其网址如下。

http://pandas.pydata.org/pandas-docs/stable/install.html

下载示例代码文件

读者可以从 www.packtpub.com 中下载本书的示例代码文件。具体步骤如下。

（1）登录或注册 www.packtpub.com。

（2）在 Search（搜索）框中输入本书名称 *Pandas 1.x Cookbook* 的一部分（不分区大小写，并且不必输入完全），即可看到本书出现在推荐下拉菜单中，如图 P-1 所示。

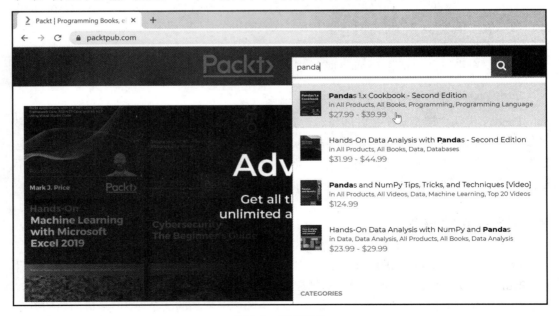

图 P-1　搜索书名

（3）选择 Pandas 1.x Cookbook 一书，并在其详细信息页面中单击 Download code from GitHub（从 GitHub 上下载代码）按钮，如图 P-2 所示。需要说明的是，你需要登录此网站才能看到该下载按钮（注册账号是免费的）。

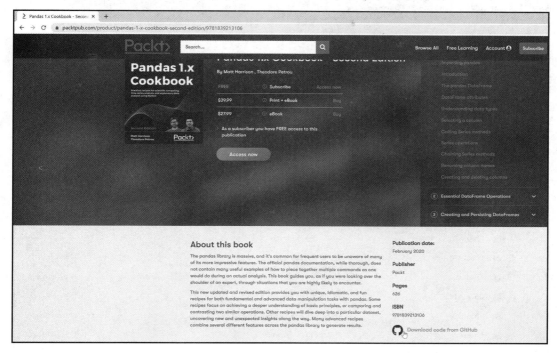

图 P-2　选择本书书名

本书代码包在 GitHub 上的托管地址如下。

https://github.com/PacktPublishing/Pandas-Cookbook-Second-Edition

在下载页面上，单击 Code（代码）按钮，然后选择 Download ZIP 即可下载本书代码包，如图 P-3 所示。

如果代码有更新，则也会在现有 GitHub 存储库上更新。

下载文件后，请确保使用最新版本软件解压或析取文件夹。

- ❏　WinRAR/7-Zip（Windows 系统）。
- ❏　Zipeg/iZip/UnRarX（Mac 系统）。
- ❏　7-Zip/PeaZip（Linux 系统）。

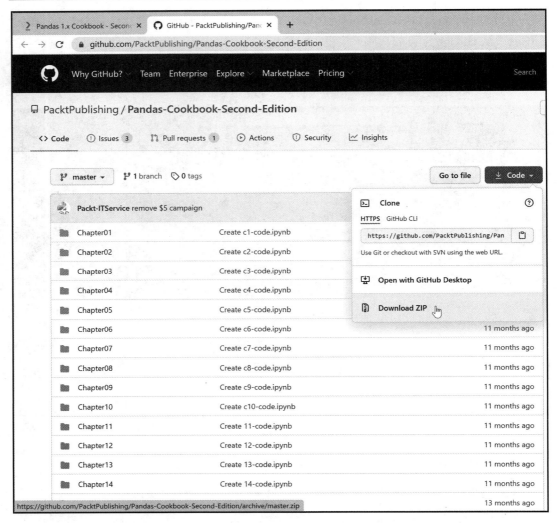

图 P-3　下载本书代码包

运行 Jupyter Notebook

要跟随本书秘笈进行操作，建议启动并运行 Jupyter Notebook，以便读者可以在阅读秘笈的同时运行代码。与仅阅读本书相比，在计算机上练习操作可以使读者自己进行探索并能够更深入地了解与本书秘笈相关的更多内容。

假设你已经在计算机上安装了 Anaconda 发行版，则可以从 Anaconda 图形用户界面或命令行中启动 Jupyter Notebook。两个选项任选其一，我们强烈建议你使用命令行。如果你打算使用 Python 做很多事情，那么需要从这一刻就开始适应命令行。

安装 Anaconda 之后，打开命令提示符（在 Windows 的搜索栏中输入 cmd，或在 Mac 或 Linux 上打开终端），然后输入以下命令。

```
$ jupyter-notebook
```

你不必从主目录而是可以从任何位置处运行上述命令，浏览器中的内容将反映该位置。

尽管现在已经启动了 Jupyter Notebook 程序，但是实际上我们还没有启动可以在 Python 中进行开发的单个 Notebook。对此，你可以单击页面右侧的 New（新建）按钮，这将下拉列出所有可能使用的内核的列表。如果你刚刚下载了 Anaconda，则只有一个可用的内核（Python 3）。选择 Python 3 内核后，将在浏览器中打开一个新选项卡，你可以在其中开始编写 Python 代码。

当然，你也可以打开以前创建的 Notebook，而不用开始一个新的。为此，可在 Jupyter Notebook 浏览器主页提供的文件系统中导航，然后选择要打开的 Notebook。所有 Jupyter Notebook 文件均以.ipynb 作为扩展名。

或者，你也可以使用云提供商的 Notebook 环境。Google 和 Microsoft 都提供了已预装 Pandas 的免费 Notebook 环境。

下载彩色图像

我们还提供了一个 PDF 文件，其中包含本书中使用的屏幕截图/图表的彩色图像。你可以通过以下网址下载。

https://static.packt-cdn.com/downloads/9781839213106_ColorImages.pdf

本书约定

本书中使用了许多文本约定。

（1）有关代码块的设置形式如下。

```
import pandas as pd
import numpy as np
```

```
movies = pd.read_csv("data/movie.csv")
movies
```

（2）当要强调代码块的特定部分时，相关行或项目以粗体显示。示例如下。

```
import pandas as pd
import numpy as np
movies = pd.read_csv("data/movie.csv")
movies
```

（3）任何命令行输入或输出都采用如下所示的粗体代码形式。

```
>>> employee = pd.read_csv('data/employee.csv')
>>> max_dept_salary = employee.groupby('DEPARTMENT')['BASE_SALARY'].max()
```

（4）术语或重要单词采用中英文对照形式，在括号内保留其英文原文。示例如下。

可以在单个 Figure 中绘制折线图（line plot）、散点图（scatter plot）和条形图（bar plot）。散点图是唯一需要为 x 和 y 值指定列的图形。

（5）对于界面词汇则保留其英文原文，在后面使用括号添加其中文翻译。示例如下。

可以看到，前两个目的地机场都在夏威夷，即 HNL（火奴鲁鲁国际机场）和 OGG（卡胡鲁伊机场），考虑到美国的地理情况，这一点不足为奇。

（6）本书使用了以下两个图标。

表示警告或重要的注意事项。

表示提示或小技巧。

每个秘笈的假设

应该假定在每个秘笈的开头，将 NumPy、Matplotlib 和 Pandas 都导入了名称空间。为了将绘图直接嵌入 Notebook 中，还必须运行魔术命令%matplotlib inline。

此外，所有数据都被存储在 data 目录中，并且通常被存储为 CSV 文件，以便可以使用 read_csv 函数直接读取。

```
>>> %matplotlib inline
>>> import numpy as np
>>> import matplotlib.pyplot as plt
>>> import pandas as pd
>>> my_dataframe = pd.read_csv('data/dataset_name.csv')
```

数据集说明

本书共使用了大约 20 个数据集。在完成秘笈中的操作步骤时，了解每个数据集的背景信息可能会非常有帮助。有关每个数据集的详细说明，请访问以下网址中提供的 dataset_description Jupyter Notebook。

https://github.com/PacktPublishing/Pandas-Cookbook-Second-Edition

每个数据集都有一个列的列表，并提供了每一列的信息以及如何获取数据的注释。

编写体例

本书大多数章节是采用秘笈形式编写的，每节就是一个秘笈，每个秘笈中又分别包括"实战操作""原理解释""扩展知识"小节，使你既能学习 Pandas 实用操作，又能了解其相关的知识和原理，从而真正掌握和领会 Pandas 应用技巧。

关于作者

Matt Harrison 从 2000 年开始使用 Python。他经营着 MetaSnake 公司,这是一家提供 Python 和数据科学方面企业培训的公司。

Matt Harrison 还是 *Machine Learning Pocket Reference*(《机器学习口袋宝典》)、*Illustrated Guide to Python 3*(《Python 3 绘本指南》)和 *Learning the Pandas Library*(《学习 Pandas 库》)以及其他畅销书籍的作者。

Theodore Petrou 是数据科学家和 Dunder Data 公司的创始人,后者是一家致力于探索性数据分析的专业教育公司。他还是 Houston Data Science(休斯顿数据科学)线下聚会的牵头人,Houston Data Science 是一个有 2000 多名成员的聚会组织,其主要目标是将本地数据分析爱好者聚集在一起,以探讨数据科学实践。在创建 Dunder Data 公司之前,Theodore Petrou 是一家大型石油服务公司 Schlumberger 的数据科学家,他在该公司的绝大部分工作就是研究和探索数据。

Theodore Petrou 的一些项目包括使用针对性的情感分析来从工程师文本中发现故障的根本原因,开发定制的客户端/服务器仪表板应用程序以及实时 Web 服务以避免对销售商品的错误定价。Theodore Petrou 拥有莱斯大学(位于美国休斯顿市郊)的统计硕士学位。在成为数据科学家之前,他常利用自己的分析技能来玩扑克游戏和教授数学。他还是通过实践学习理念的坚定支持者,常在 Stack Overflow 技术问答网站上回答有关 Pandas 的问题。

关于审稿人

Simon Hawkins 拥有伦敦帝国学院的航空工程硕士学位。他职业生涯的早期在国防和核能领域担任技术分析师,专注于针对高完整性设备的各种建模功能和仿真技术。后来,他过渡到电子商务世界,重点转向数据分析。如今,他对万物分析数据科学感兴趣,并且是 Pandas 核心开发团队的成员。

目 录

第 1 章 Pandas 基础 ... 1
1.1 导入 Pandas ... 1
1.2 介绍 Pandas ... 1
1.3 关于 Pandas DataFrame ... 2
1.3.1 实战操作 ... 2
1.3.2 原理解释 ... 3
1.4 了解 DataFrame 属性 ... 4
1.4.1 实战操作 ... 4
1.4.2 原理解释 ... 5
1.4.3 扩展知识 ... 6
1.5 了解数据类型 ... 6
1.5.1 实战操作 ... 7
1.5.2 原理解释 ... 8
1.5.3 扩展知识 ... 9
1.6 选择列 ... 9
1.6.1 实战操作 ... 10
1.6.2 原理解释 ... 13
1.6.3 扩展知识 ... 13
1.7 调用 Series 方法 ... 14
1.7.1 实战操作 ... 14
1.7.2 原理解释 ... 19
1.7.3 扩展知识 ... 19
1.8 了解 Series 的操作 ... 20
1.8.1 实战操作 ... 21
1.8.2 原理解释 ... 23
1.8.3 扩展知识 ... 23
1.9 使用 Series 方法链 ... 26

    1.9.1 实战操作 26
    1.9.2 原理解释 27
    1.9.3 扩展知识 28
1.10 重命名列名 31
    1.10.1 实战操作 31
    1.10.2 原理解释 31
    1.10.3 扩展知识 32
1.11 创建和删除列 34
    1.11.1 实战操作 34
    1.11.2 原理解释 40
    1.11.3 扩展知识 40

第 2 章 DataFrame 基本操作

2.1 介绍 43
2.2 选择多个 DataFrame 列 43
    2.2.1 实战操作 43
    2.2.2 原理解释 44
    2.2.3 扩展知识 45
2.3 使用方法选择列 45
    2.3.1 实战操作 46
    2.3.2 原理解释 48
    2.3.3 扩展知识 48
2.4 排序列名称 49
    2.4.1 实战操作 50
    2.4.2 原理解释 52
    2.4.3 扩展知识 52
2.5 统计 DataFrame 摘要信息 52
    2.5.1 实战操作 53
    2.5.2 原理解释 55
    2.5.3 扩展知识 55
2.6 使用 DataFrame 方法链 56
    2.6.1 实战操作 56

2.6.2　原理解释 57
　　　2.6.3　扩展知识 57
　2.7　了解 DataFrame 的操作 58
　　　2.7.1　实战操作 59
　　　2.7.2　原理解释 62
　　　2.7.3　扩展知识 63
　2.8　比较缺失值 63
　　　2.8.1　做好准备 63
　　　2.8.2　实战操作 64
　　　2.8.3　原理解释 66
　　　2.8.4　扩展知识 66
　2.9　转置 DataFrame 操作的方向 67
　　　2.9.1　实战操作 67
　　　2.9.2　原理解释 69
　　　2.9.3　扩展知识 69
　2.10　确定大学校园的多样性 70
　　　2.10.1　实战操作 70
　　　2.10.2　原理解释 74
　　　2.10.3　扩展知识 74
第 3 章　创建和保留 DataFrame 77
　3.1　介绍 77
　3.2　从头开始创建 DataFrame 77
　　　3.2.1　实战操作 77
　　　3.2.2　原理解释 78
　　　3.2.3　扩展知识 78
　3.3　编写 CSV 80
　　　3.3.1　实战操作 80
　　　3.3.2　扩展知识 81
　3.4　读取大型 CSV 文件 82
　　　3.4.1　实战操作 82
　　　3.4.2　原理解释 88

3.4.3 扩展知识 89
3.5 使用 Excel 文件 90
  3.5.1 实战操作 90
  3.5.2 原理解释 91
  3.5.3 扩展知识 91
3.6 使用 ZIP 文件 92
  3.6.1 实战操作 92
  3.6.2 原理解释 95
  3.6.3 扩展知识 95
3.7 与数据库协同工作 95
  3.7.1 实战操作 95
  3.7.2 原理解释 97
3.8 读取 JSON 97
  3.8.1 实战操作 97
  3.8.2 原理解释 100
  3.8.3 扩展知识 100
3.9 读取 HTML 表格 100
  3.9.1 实战操作 101
  3.9.2 原理解释 105
  3.9.3 扩展知识 106

第 4 章 开始数据分析 107
4.1 介绍 107
4.2 开发数据分析例程 107
  4.2.1 实战操作 108
  4.2.2 原理解释 110
  4.2.3 扩展知识 110
4.3 数据字典 111
4.4 通过更改数据类型减少内存使用量 112
  4.4.1 实战操作 112
  4.4.2 原理解释 115
  4.4.3 扩展知识 116

4.5 从最大中选择最小 ... 117
4.5.1 实战操作 ... 118
4.5.2 原理解释 ... 119
4.5.3 扩展知识 ... 119
4.6 通过排序选择每组中的最大值 ... 119
4.6.1 实战操作 ... 119
4.6.2 原理解释 ... 121
4.6.3 扩展知识 ... 122
4.7 使用 sort_values 复制 nlargest ... 123
4.7.1 实战操作 ... 123
4.7.2 原理解释 ... 125
4.8 计算追踪止损单价格 ... 126
4.8.1 实战操作 ... 126
4.8.2 原理解释 ... 128
4.8.3 扩展知识 ... 128

第5章 探索性数据分析 ... 129
5.1 介绍 ... 129
5.2 摘要统计 ... 129
5.2.1 实战操作 ... 130
5.2.2 原理解释 ... 132
5.2.3 扩展知识 ... 132
5.3 查看列类型 ... 132
5.3.1 实战操作 ... 132
5.3.2 原理解释 ... 133
5.3.3 扩展知识 ... 134
5.4 分类数据 ... 137
5.4.1 实战操作 ... 137
5.4.2 原理解释 ... 140
5.4.3 扩展知识 ... 141
5.5 连续数据 ... 145
5.5.1 实战操作 ... 145

5.5.2 原理解释 148
5.5.3 扩展知识 149
5.6 跨越分类比较连续值 151
- 5.6.1 实战操作 151
- 5.6.2 原理解释 153
- 5.6.3 扩展知识 153
5.7 比较两个连续列 157
- 5.7.1 实战操作 157
- 5.7.2 原理解释 162
- 5.7.3 扩展知识 163
5.8 使用分类值比较分类值 165
- 5.8.1 实战操作 165
- 5.8.2 原理解释 171
5.9 使用 Pandas 分析库 171
- 5.9.1 实战操作 172
- 5.9.2 原理解释 173

第 6 章 选择数据子集 175
6.1 介绍 175
6.2 选择 Series 数据 175
- 6.2.1 实战操作 176
- 6.2.2 原理解释 179
- 6.2.3 扩展知识 180
6.3 选择 DataFrame 行 182
- 6.3.1 实战操作 182
- 6.3.2 原理解释 184
- 6.3.3 扩展知识 185
6.4 同时选择 DataFrame 行和列 185
- 6.4.1 实战操作 185
- 6.4.2 原理解释 187
- 6.4.3 扩展知识 188
6.5 使用整数和标签选择数据 188

　　　　　目　　录　　　　　·XXIII·

   6.5.1 实战操作 188
   6.5.2 原理解释 189
   6.5.3 扩展知识 189
 6.6 按字典序切片 190
   6.6.1 实战操作 190
   6.6.2 原理解释 192
   6.6.3 扩展知识 192

第 7 章　过滤行 193
 7.1 介绍 193
 7.2 计算布尔统计信息 193
   7.2.1 实战操作 194
   7.2.2 原理解释 195
   7.2.3 扩展知识 196
 7.3 构造多个布尔条件 196
   7.3.1 实战操作 197
   7.3.2 原理解释 197
   7.3.3 扩展知识 198
 7.4 用布尔数组过滤 199
   7.4.1 实战操作 199
   7.4.2 原理解释 201
   7.4.3 扩展知识 202
 7.5 比较行过滤和索引过滤 202
   7.5.1 实战操作 203
   7.5.2 原理解释 203
   7.5.3 扩展知识 204
 7.6 使用唯一索引和排序索引进行选择 205
   7.6.1 实战操作 205
   7.6.2 原理解释 207
   7.6.3 扩展知识 207
 7.7 转换 SQL WHERE 子句 208
   7.7.1 实战操作 209

      7.7.2 原理解释 210
      7.7.3 扩展知识 211
  7.8 使用查询方法提高布尔索引的可读性 212
      7.8.1 实战操作 212
      7.8.2 原理解释 213
      7.8.3 扩展知识 213
  7.9 使用.where 方法保留 Series 大小 214
      7.9.1 实战操作 214
      7.9.2 原理解释 218
      7.9.3 扩展知识 218
  7.10 屏蔽 DataFrame 行 218
      7.10.1 实战操作 218
      7.10.2 原理解释 220
      7.10.3 扩展知识 221
  7.11 使用布尔值、整数位置和标签进行选择 221
      7.11.1 实战操作 221
      7.11.2 原理解释 224

第 8 章 对齐索引 225
  8.1 介绍 225
  8.2 检查 Index 对象 225
      8.2.1 实战操作 225
      8.2.2 原理解释 227
      8.2.3 扩展知识 227
  8.3 生成笛卡儿积 228
      8.3.1 实战操作 228
      8.3.2 原理解释 229
      8.3.3 扩展知识 229
  8.4 了解索引暴增现象 231
      8.4.1 实战操作 231
      8.4.2 原理解释 233
      8.4.3 扩展知识 233

8.5 给不相等的索引填充值 234
8.5.1 实战操作 234
8.5.2 原理解释 236
8.5.3 扩展知识 237
8.6 添加来自不同 DataFrames 中的列 239
8.6.1 实战操作 239
8.6.2 原理解释 242
8.6.3 扩展知识 242
8.7 突出显示每列的最大值 244
8.7.1 实战操作 245
8.7.2 原理解释 250
8.7.3 扩展知识 251
8.8 使用方法链复制 .idxmax 252
8.8.1 实战操作 252
8.8.2 原理解释 256
8.8.3 扩展知识 257
8.9 查找最常见的列的最大值 258
8.9.1 实战操作 258
8.9.2 原理解释 259
8.9.3 扩展知识 259

第 9 章 分组以进行聚合、过滤和转换 **261**
9.1 介绍 261
9.2 定义聚合 262
9.2.1 实战操作 262
9.2.2 原理解释 264
9.2.3 扩展知识 265
9.3 使用多个列和函数进行分组和聚合 265
9.3.1 实战操作 266
9.3.2 原理解释 268
9.3.3 扩展知识 269
9.4 分组后删除多重索引 271

9.4.1 实战操作 271
9.4.2 原理解释 274
9.4.3 扩展知识 274
9.5 使用自定义聚合函数进行分组 275
9.5.1 实战操作 275
9.5.2 原理解释 277
9.5.3 扩展知识 277
9.6 使用*args 和**kwargs 自定义聚合函数 279
9.6.1 实战操作 279
9.6.2 原理解释 281
9.6.3 扩展知识 281
9.7 检查 groupby 对象 282
9.7.1 实战操作 282
9.7.2 原理解释 285
9.7.3 扩展知识 286
9.8 筛选少数族裔占多数的州 286
9.8.1 实战操作 287
9.8.2 原理解释 288
9.8.3 扩展知识 288
9.9 通过减肥赌注做出改变 289
9.9.1 实战操作 289
9.9.2 原理解释 294
9.9.3 扩展知识 295
9.10 计算每个州的 SAT 加权平均成绩 296
9.10.1 实战操作 297
9.10.2 原理解释 299
9.10.3 扩展知识 300
9.11 按连续变量分组 301
9.11.1 实战操作 302
9.11.2 原理解释 303
9.11.3 扩展知识 304
9.12 计算城市之间的航班总数 305

	9.12.1 实战操作	305
	9.12.2 原理解释	308
	9.12.3 扩展知识	309
9.13	寻找最长的准点航班连续记录	310
	9.13.1 实战操作	310
	9.13.2 原理解释	314
	9.13.3 扩展知识	316

第 10 章 将数据重组为规整形式 ... 319

10.1	介绍	319
10.2	使用 stack 将变量值规整为列名称	321
	10.2.1 实战操作	322
	10.2.2 原理解释	324
	10.2.3 扩展知识	324
10.3	使用 melt 将变量值规整为列名称	326
	10.3.1 实战操作	326
	10.3.2 原理解释	327
	10.3.3 扩展知识	327
10.4	同时堆叠多组变量	328
	10.4.1 实战操作	329
	10.4.2 原理解释	330
	10.4.3 扩展知识	330
10.5	反转已堆叠的数据	331
	10.5.1 实战操作	332
	10.5.2 原理解释	335
	10.5.3 扩展知识	335
10.6	在 groupby 聚合之后取消堆叠	336
	10.6.1 实战操作	337
	10.6.2 原理解释	338
	10.6.3 扩展知识	339
10.7	使用 groupby 聚合复制 .pivot_table 方法的功能	340
	10.7.1 实战操作	340

10.7.2 原理解释 342
10.7.3 扩展知识 342
10.8 重命名轴的级别以方便数据的重塑 344
10.8.1 实战操作 344
10.8.2 原理解释 348
10.8.3 扩展知识 349
10.9 对多个变量存储为列名称的情况进行规整 350
10.9.1 实战操作 350
10.9.2 原理解释 354
10.9.3 扩展知识 354
10.10 对多个变量存储为单个列的情况进行规整 356
10.10.1 实战操作 356
10.10.2 原理解释 359
10.10.3 扩展知识 360
10.11 对多个值存储在同一单元格中的情况进行规整 360
10.11.1 实战操作 361
10.11.2 原理解释 362
10.11.3 扩展知识 362
10.12 对变量存储在列名称和值中的情况进行规整 363
10.12.1 实战操作 364
10.12.2 原理解释 365
10.12.3 扩展知识 365

第 11 章 组合 Pandas 对象 367
11.1 介绍 367
11.2 将新行追加到 DataFrame 367
11.2.1 实战操作 367
11.2.2 原理解释 372
11.2.3 扩展知识 372
11.3 将多个 DataFrame 连接在一起 373
11.3.1 实战操作 374
11.3.2 原理解释 376

11.3.3 扩展知识 376
11.4 了解 concat 函数、.join 和.merge 方法之间的区别 377
  11.4.1 实战操作 378
  11.4.2 原理解释 383
  11.4.3 扩展知识 384
11.5 连接到 SQL 数据库 385
  11.5.1 实战操作 386
  11.5.2 原理解释 389
  11.5.3 扩展知识 390

第 12 章 时间序列分析 393
12.1 介绍 393
12.2 了解 Python 和 Pandas 日期工具之间的区别 393
  12.2.1 实战操作 394
  12.2.2 原理解释 398
12.3 智能分割时间序列 399
  12.3.1 实战操作 400
  12.3.2 原理解释 403
  12.3.3 扩展知识 404
12.4 用时间数据过滤列 404
  12.4.1 实战操作 404
  12.4.2 原理解释 407
  12.4.3 扩展知识 408
12.5 使用仅适用于 DatetimeIndex 的方法 408
  12.5.1 实战操作 409
  12.5.2 原理解释 414
  12.5.3 扩展知识 415
12.6 计算每周犯罪数 415
  12.6.1 实战操作 416
  12.6.2 原理解释 418
  12.6.3 扩展知识 418
12.7 分别汇总每周犯罪和交通事故 419

12.7.1 实战操作 420
12.7.2 原理解释 422
12.7.3 扩展知识 424
12.8 按星期和年份衡量犯罪情况 425
12.8.1 实战操作 425
12.8.2 原理解释 432
12.8.3 扩展知识 434
12.9 使用匿名函数进行分组 435
12.9.1 实战操作 435
12.9.2 原理解释 438
12.10 按 Timestamp 和其他列分组 438
12.10.1 实战操作 439
12.10.2 原理解释 442
12.10.3 扩展知识 443

第 13 章 使用 Matplotlib、Pandas 和 Seaborn 进行可视化 445
13.1 介绍 445
13.2 Matplotlib 入门 446
13.3 Matplotlib 的面向对象指南 447
13.3.1 实战操作 450
13.3.2 原理解释 454
13.3.3 扩展知识 458
13.4 使用 Matplotlib 可视化数据 458
13.4.1 实战操作 458
13.4.2 原理解释 462
13.4.3 扩展知识 463
13.5 Pandas 绘图基础 466
13.5.1 实战操作 467
13.5.2 原理解释 469
13.5.3 扩展知识 470
13.6 可视化航班数据集 471
13.6.1 实战操作 471

13.6.2　原理解释 482
13.7　使用堆积面积图发现新兴趋势 484
　　　13.7.1　实战操作 484
　　　13.7.2　原理解释 488
13.8　了解 Seaborn 和 Pandas 之间的区别 489
　　　13.8.1　实战操作 489
　　　13.8.2　原理解释 495
13.9　使用 Seaborn 网格进行多变量分析 496
　　　13.9.1　实战操作 496
　　　13.9.2　原理解释 499
　　　13.9.3　扩展知识 500
13.10　使用 Seaborn 在钻石数据集中发现辛普森悖论 502
　　　13.10.1　实战操作 503
　　　13.10.2　原理解释 507
　　　13.10.3　扩展知识 507

第 14 章　调试和测试 509
14.1　转换数据 509
　　　14.1.1　实战操作 509
　　　14.1.2　原理解释 513
14.2　测试.apply 方法的性能 514
　　　14.2.1　实战操作 514
　　　14.2.2　原理解释 515
　　　14.2.3　扩展知识 515
14.3　使用 Dask、Pandarell 和 Swifter 等提高 .apply 方法的性能 516
　　　14.3.1　实战操作 517
　　　14.3.2　原理解释 518
14.4　检查代码 519
　　　14.4.1　实战操作 520
　　　14.4.2　原理解释 523
　　　14.4.3　扩展知识 523
14.5　在 Jupyter 中进行调试 523

| 14.5.1　实战操作 524
| 14.5.2　原理解释 526
| 14.5.3　扩展知识 526
| 14.6　管理数据的完整性 527
| 14.6.1　实战操作 527
| 14.6.2　原理解释 534
| 14.7　结合使用 pytest 和 Pandas 535
| 14.7.1　实战操作 535
| 14.7.2　原理解释 539
| 14.7.3　扩展知识 539
| 14.8　使用 Hypothesis 库生成测试 540
| 14.8.1　实战操作 540
| 14.8.2　原理解释 545

第 1 章 Pandas 基础

本章包含以下秘笈。
- 导入 Pandas。
- 关于 Pandas DataFrame。
- 了解 DataFrame 属性。
- 了解数据类型。
- 选择列。
- 调用 Series 方法。
- 了解 Series 的操作。
- 使用 Series 方法链。
- 重命名列名。
- 创建和删除列。

1.1 导入 Pandas

Pandas 库的大多数用户都愿意使用导入别名，将其称为 pd。通常情况下，本书不会重复显示 Pandas 和 NumPy 的导入操作，所以在此统一做介绍。其语句如下。

```
>>> import pandas as pd
>>> import numpy as np
```

1.2 介绍 Pandas

本章的目的是通过全面阐释 Series 和 DataFrame 数据结构来介绍 Pandas 的应用基础。对于 Pandas 用户来说，了解 Series 和 DataFrame 之间的区别非常重要。

Pandas 库对于处理结构化（structured）数据很有用。什么是结构化数据？存储在表中的数据（如 CSV 文件、Excel 电子表格或数据库表等）都是结构化的。非结构化（unstructured）数据由自由格式的文本、图像、声音或视频组成。如果你发现自己要处理结构化数据，那么 Pandas 将对你很有用。

本章将介绍如何从 DataFrame（二维数据集）中选择一个数据列，该数据列将作为

Series（一维数据集）返回。处理这些一维对象可以轻松显示不同方法和运算符的工作方式。许多 Series 方法都会返回另一个 Series 作为输出，这导致有可能连续调用其他方法，这称为方法链（method chaining）。

Series 和 DataFrame 的 Index 组件是将 Pandas 与其他大多数数据分析库区分开的组件，并且是了解有多少种操作有效的关键。当我们将其用作 Series 值的有意义的标签时，即可明白这个对象的强大。

本章最后两个秘笈包含在数据分析期间经常要执行的任务。

1.3 关于 Pandas DataFrame

在深入研究 Pandas 之前，有必要先了解 DataFrame 的组件。

1.3.1 实战操作

从视觉上来说，Pandas DataFrame 的输出显示（在 Jupyter Notebook 中）看起来只不过是由行和列组成的普通数据表。隐藏在 DataFrame 表面之下的是 3 个组件，即索引（index）、列（column）和数据（data），如图 1-1 所示。分析人员必须了解这 3 个组件，以最大程度地发挥 DataFrame 的全部潜能。

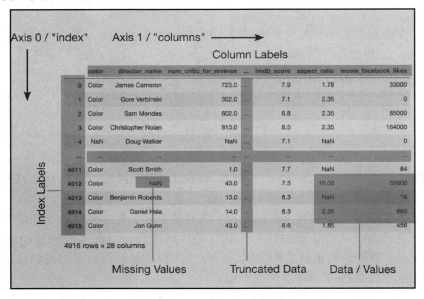

图 1-1　DataFrame 剖析

原 文	译 文
Axis 0/"index"	轴 0/索引
Axis 1/"columns"	轴 1/列
Column Labels	列标签
Index Labels	索引标签
Missing Values	缺失值
Truncated Data	截断的数据
Data/Values	数据/值

图 1-1 是将 movie（电影）数据集读入 Pandas DataFrame 中的结果，并提供了其所有主要组件的标签示意图。读入数据集的代码如下。

```
>>> movies = pd.read_csv("data/movie.csv")
>>> movies
       color         direc/_name   ...   aspec/ratio   movie/likes
0      Color         James Cameron  ...          1.78         33000
1      Color         Gore Verbinski ...          2.35             0
2      Color         Sam Mendes     ...          2.35         85000
3      Color     Christopher Nolan  ...          2.35        164000
4        NaN         Doug Walker    ...           NaN             0
...      ...                 ...   ...           ...           ...
4911   Color         Scott Smith    ...           NaN            84
4912   Color                 NaN   ...         16.00         32000
4913   Color     Benjamin Roberds  ...           NaN            16
4914   Color         Daniel Hsia    ...          2.35           660
4915   Color           Jon Gunn     ...          1.85           456
```

1.3.2 原理解释

Pandas 首先使用 read_csv 函数将数据从磁盘读取到内存中，然后读取到 DataFrame 中。按照约定，术语索引标签（index label）和列名称（column name）分别是指索引和列的各个成员。术语索引（index）是从整体意义上指称所有索引标签，而术语列（column）则是从整体意义上指称所有列名称。

索引和列名称中的标签允许根据索引和列名称提取数据。稍后将显示这一点。索引也可用于对齐（alignment）。当多个 Series 或 DataFrame 被组合在一起时，在进行任何计算之前，索引首先需要对齐，稍后的秘笈也会显示这一点。

列和索引统称为轴（axes）。更具体地说，索引是轴 0，列是轴 1。

Pandas 使用 NaN（not a number，不是数字）表示缺失值。请注意，即使 color（彩色）

列包含的是字符串值，它也使用 NaN 表示缺失值。

列中间的 3 个连续点（...）表示至少存在一列，但由于列数超过了预定义的显示限制，因此未显示。默认情况下，Pandas 显示 60 行和 20 列，但是由于书本的版面有限制，因此数据只能通过部分截断的方式纳入页面中。

.head 方法接收一个可选参数 n，该参数控制显示的行数。n 的默认值为 5。类似地，.tail 方法返回的是最后 n 行。在图 1-1 中即显示了前 5 行和后 5 行的数据。

1.4 了解 DataFrame 属性

可以从 DataFrame 访问 3 个 DataFrame 组件（索引、列和数据）中的每一个。你可能要对单个组件而不是对整个 DataFrame 进行操作。一般来说，尽管我们可以将数据提取到 NumPy 数组中，但是除非所有列都是数字列；否则，我们通常将其保留在 DataFrame 中。DataFrame 是管理异类数据列的理想选择，而 NumPy 数组与其相形之下则逊色不少。

此秘笈将 DataFrame 的索引、列和数据提取到它们自己的变量中，然后演示如何从同一对象继承列和索引。

1.4.1 实战操作

（1）使用 DataFrame 属性 columns、index 和值（to_numpy）分别将列、索引和数据分配给它们自己对应的变量（columns、index 和 data）。

```
>>> movies = pd.read_csv("data/movie.csv")
>>> columns = movies.columns
>>> index = movies.index
>>> data = movies.to_numpy()
```

（2）显示每个变量组件的值。

```
>>> columns
Index(['color', 'director_name', 'num_critic_for_reviews', 'duration',
       'director_facebook_likes', 'actor_3_facebook_likes',
'actor_2_name',
       'actor_1_facebook_likes', 'gross', 'genres', 'actor_1_name',
       'movie_title', 'num_voted_users', 'cast_total_facebook_likes',
       'actor_3_name', 'facenumber_in_poster', 'plot_keywords',
       'movie_imdb_link', 'num_user_for_reviews', 'language','country',
       'content_rating', 'budget', 'title_year', 'actor_2_
```

```
facebook_likes',
        'imdb_score', 'aspect_ratio', 'movie_facebook_likes'],
        dtype='object')
>>> index
RangeIndex(start=0, stop=4916, step=1)
>>> data
array([ ['Color', 'James Cameron', 723.0, ..., 7.9, 1.78, 33000],
        ['Color', 'Gore Verbinski', 302.0, ..., 7.1, 2.35, 0],
        ['Color', 'Sam Mendes', 602.0, ..., 6.8, 2.35, 85000],
        ...,
        ['Color', 'Benjamin Roberds', 13.0, ..., 6.3, nan, 16],
        ['Color', 'Daniel Hsia', 14.0, ..., 6.3, 2.35, 660],
        ['Color', 'Jon Gunn', 43.0, ..., 6.6, 1.85, 456]],
dtype=object)
```

（3）输出每个 DataFrame 组件的 Python 类型（也就是输出结果中的最后一个小点后面的单词）。

```
>>> type(index)
<class 'pandas.core.indexes.range.RangeIndex'>
>>> type(columns)
<class 'pandas.core.indexes.base.Index'>
>>> type(data)
<class 'numpy.ndarray'>
```

（4）索引和列密切相关，并且都是 Index 的子类。这使得分析人员可以对索引和列执行类似的操作，示例如下。

```
>>> issubclass(pd.RangeIndex, pd.Index)
True
>>> issubclass(columns.__class__, pd.Index)
True
```

1.4.2 原理解释

索引和列其实代表的是相同的事物，只是沿着的轴不同。它们有时也分别被称为行索引（row index）和列索引（column index）。

Pandas 中有许多类型的索引对象。如果未指定索引类型，则 Pandas 将使用 RangeIndex。RangeIndex 是 Index 的子类，类似于 Python 的 range 对象。由于 RangeIndex 的整个值序列都不会被加载到内存中（除非必须这样做），因此节省了内存。另外，RangeIndex 的定义包括其开始值、停止值和步长值。

1.4.3 扩展知识

在可能的情况下，可以使用哈希表来实现 Index 对象，该哈希表允许快速选择和数据对齐。Index 对象与 Python 集合相似，因为它们支持诸如相交（intersection）和并集（union）之类的操作，但是由于它们是有序的并且可以有重复的条目，因此它们其实是不同的。

请注意 DataFrame 属性.values 返回 NumPy n 维数组或 ndarray 的方式。大多数 Pandas 对象都严重依赖 ndarray。在索引、列和数据之下就是 NumPy ndarray。可以将这些 Pandas 对象视为构建许多其他对象的 Pandas 的基本对象。要理解这一点，可以研究索引和列的值：

```
>>> index.to_numpy()
array([ 0, 1, 2, ..., 4913, 4914, 4915], dtype=int64))
>>> columns.to_numpy()
array(['color', 'director_name', 'num_critic_for_reviews', 'duration',
       'director_facebook_likes', 'actor_3_facebook_likes',
       'actor_2_name', 'actor_1_facebook_likes', 'gross', 'genres',
       'actor_1_name', 'movie_title', 'num_voted_users',
       'cast_total_facebook_likes', 'actor_3_name',
       'facenumber_in_poster', 'plot_keywords', 'movie_imdb_link',
       'num_user_for_reviews', 'language', 'country', 'content_rating',
       'budget', 'title_year', 'actor_2_facebook_likes', 'imdb_score',
       'aspect_ratio', 'movie_facebook_likes'], dtype=object)
```

综上所述，我们通常不会访问底层的 NumPy 对象，而是倾向于将对象保留为 Pandas 对象，并使用 Pandas 操作。当然，分析人员经常会将 NumPy 函数应用于 Pandas 对象。

1.5 了解数据类型

从广义上来划分，数据可以被分为连续数据和分类数据。连续数据（continuous data）始终是数字的，代表某种度量，如身高、重量或薪水，因此连续数据具有无限的数字上的可能性；而分类数据（categorical data）则代表离散的有限数量的值，如汽车颜色、扑克牌花色或谷物品牌等。

Pandas 并未从广义上将数据分为连续数据和分类数据。相反，Pandas 对许多不同的数据类型都有精确的技术定义。下面介绍常见的 Pandas 数据类型。

❑ float：NumPy 浮点值类型，它支持缺失值。

- int:NumPy 整数值类型,它不支持缺失值。
- 'Int64':Pandas 整数值类型,它可为 None 值。
- object:用于存储字符串(和混合类型)的 NumPy 类型。
- 'category':Pandas 分类值类型,它支持缺失值。
- bool:NumPy 布尔值类型,它不支持缺失值(None 值变为 False,np.nan 变为 True)。
- 'boolean':Pandas 布尔值类型,它可为 None 值。
- datetime64 [ns]:NumPy 日期值类型,它支持缺失值(NaT)。

在此秘笈中,我们将显示 DataFrame 中每一列的数据类型。在提取数据后,至关重要的是要知道每列中保存的数据类型,因为它会从根本上改变可能进行的操作的类型。

1.5.1 实战操作

(1)使用.dtypes 属性显示每一列的名称及其数据类型。

```
>>> movies = pd.read_csv("data/movie.csv")
>>> movies.dtypes
color                       object
director_name               object
num_critic_for_reviews      float64
duration                    float64
director_facebook_likes     float64
                             ...
title_year                  float64
actor_2_facebook_likes      float64
imdb_score                  float64
aspect_ratio                float64
movie_facebook_likes        int64
Length: 28, dtype: object
```

(2)使用.value_counts 方法返回每种数据类型的计数。

```
>>> movies.dtypes.value_counts()
float64    13
int64       3
object     12
dtype: int64
```

(3)使用.info 方法了解数据。

```
>>> movies.info()
<class 'pandas.core.frame.DataFrame'>
RangeIndex: 4916 entries, 0 to 4915
Data columns (total 28 columns):
color                       4897 non-null object
director_name               4814 non-null object
num_critic_for_reviews      4867 non-null float64
duration                    4901 non-null float64
director_facebook_likes     4814 non-null float64
actor_3_facebook_likes      4893 non-null float64
actor_2_name                4903 non-null object
actor_1_facebook_likes      4909 non-null float64
gross                       4054 non-null float64
genres                      4916 non-null object
actor_1_name                4909 non-null object
movie_title                 4916 non-null object
num_voted_users             4916 non-null int64
cast_total_facebook_likes   4916 non-null int64
actor_3_name                4893 non-null object
facenumber_in_poster        4903 non-null float64
plot_keywords               4764 non-null object
movie_imdb_link             4916 non-null object
num_user_for_reviews        4895 non-null float64
language                    4904 non-null object
country                     4911 non-null object
content_rating              4616 non-null object
budget                      4432 non-null float64
title_year                  4810 non-null float64
actor_2_facebook_likes      4903 non-null float64
imdb_score                  4916 non-null float64
aspect_ratio                4590 non-null float64
movie_facebook_likes        4916 non-null int64
dtypes: float64(13), int64(3), object(12)
memory usage: 1.1+ MB
```

1.5.2 原理解释

每个DataFrame列均列出一种类型。例如，aspect_ratio（电影宽高比）列中的每个值都是64位浮点数，而movie_facebook_likes（影片在Facebook上的喜好值）列中的每个值都是64位整数。Pandas默认使用其核心数字类型（64位浮点数和64位整数），而不考虑所有数据容纳在内存中所需的大小问题。即使列完全由整数值 0 组成，其数据类型仍将为 int64（64 位整数）。

当调用 .dtypes 属性时，.value_counts 方法将返回 DataFrame 中所有数据类型的计数。

object 数据类型是一种有别于其他类型的数据类型。object 数据类型的列可以包含任何有效 Python 对象的值。一般来说，当列属于 object 数据类型时，它表示整个列都是字符串。当你加载 CSV 文件且字符串列包含缺失值时，Pandas 会在该单元格中标记一个 NaN（float）值。因此，该列中可能同时包含 object 和浮点（缺失）值。

.dtypes 属性会将该列显示为 object（在 Series 中显示为 O），而非混合类型的列（即包含字符串和浮点数的列）。

在 Series 中的示例如下。

```
>>> pd.Series(["Paul", np.nan, "George"]).dtype
dtype('O')
```

.info 方法除了输出非空值的计数外，还会输出数据类型信息。此外，.info 方法还列出了 DataFrame 使用的内存量。这是很有用的信息，但是仅输出在屏幕上。如果需要使用数据，.dtypes 属性将返回一个 Pandas Series。

1.5.3 扩展知识

几乎所有的 Pandas 数据类型都是从 NumPy 中构建的。这种紧密的集成使用户可以更轻松地集成 Pandas 和 NumPy 操作。随着 Pandas 变得越来越大和越来越流行，事实证明 object 数据类型对于包含字符串值的所有列都太通用了。Pandas 创建了自己的分类数据类型，以处理具有固定数量的可能值的字符串（或数字）列。

1.6 选 择 列

从 DataFrame 中选择单个列将返回一个 Series（其索引与 DataFrame 相同）。它是数据的一个单一维度，仅由索引和数据组成。分析人员也可以在不使用 DataFrame 的情况下自己创建一个 Series，但更常见的是将它们从 DataFrame 中提取出来。

本秘笈研究了两种不同的语法，以选择单列数据（即一个 Series）。一种语法使用索引操作符（index operator）；另一种语法则使用属性访问（attribute access），也称为点表示法（dot notation）。

1.6.1 实战操作

(1) 将列名称作为字符串传递给索引操作符以选择一个数据 Series。

```
>>> movies = pd.read_csv("data/movie.csv")
>>> movies["director_name"]
0          James Cameron
1         Gore Verbinski
2             Sam Mendes
3       Christopher Nolan
4            Doug Walker
              ...
4911          Scott Smith
4912                  NaN
4913      Benjamin Roberds
4914           Daniel Hsia
4915             Jon Gunn
Name: director_name, Length: 4916, dtype: object
```

(2) 也可以使用属性访问的方式来完成相同的任务。

```
>>> movies.director_name
0          James Cameron
1         Gore Verbinski
2             Sam Mendes
3       Christopher Nolan
4            Doug Walker
              ...
4911          Scott Smith
4912                  NaN
4913      Benjamin Roberds
4914           Daniel Hsia
4915             Jon Gunn
Name: director_name, Length: 4916, dtype: object
```

(3) 还可以从 .loc 和 .iloc 属性的索引中提取出 Series。前者允许分析人员按列名提取，而后者则可以按位置提取。这些方式在 Pandas 说明文档中分别被称为基于标签（label-based）和基于位置（positional-based）。

使用 .loc 属性时，需要指定行和列的选择器，由逗号分隔。其中：行选择器是一个切片对象（slice object），它是一个没有开始或结束名称的冒号（:），这意味着选择该维的所有值；列选择器则是要提取出 Series 数据的列的名称，如 director_name（导演姓名）列。

使用.iloc 属性时，同样需要指定行和列的选择器。其中：行选择器是一个切片对象，它是一个没有开始或结束名称的冒号（:），这意味着选择该维的所有值；列选择器为 1 则表示要提取出第二列的数据作为 Series（请记住，Python 的索引计数是从 0 开始的）。

.loc 和.iloc 属性的应用示例如下。

```
>>> movies.loc[:, "director_name"]
0            James Cameron
1           Gore Verbinski
2              Sam Mendes
3         Christopher Nolan
4              Doug Walker
               ...
4911           Scott Smith
4912                   NaN
4913       Benjamin Roberds
4914           Daniel Hsia
4915              Jon Gunn
Name: director_name, Length: 4916, dtype: object

>>> movies.iloc[:, 1]
0            James Cameron
1           Gore Verbinski
2              Sam Mendes
3         Christopher Nolan
4              Doug Walker
               ...
4911           Scott Smith
4912                   NaN
4913       Benjamin Roberds
4914           Daniel Hsia
4915              Jon Gunn
Name: director_name, Length: 4916, dtype: object
```

（4）Jupyter 将以等宽字体显示 Series，并显示该 Series 的索引、类型、长度和名称。此外，Jupyter 还将根据 Pandas 配置设置截断数据，如图 1-2 所示。

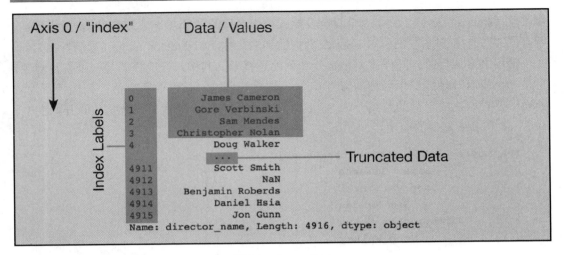

图 1-2 Series 显示

原　　文	译　　文
Axis 0/"index"	轴 0/索引
Data/Values	数据/值
Index Labels	索引标签
Truncated Data	截断的数据

还可以使用适当的属性查看该 Series 的索引、类型、长度和名称。

```
>>> movies["director_name"].index
RangeIndex(start=0, stop=4916, step=1)

>>> movies["director_name"].dtype
dtype('O')

>>> movies["director_name"].size
4196

>>> movies["director_name"].name
'director_name'
```

（5）验证输出是否为 Series。

```
>>> type(movies["director_name"])
<class 'pandas.core.series.Series'>
```

（6）请注意，即使将类型报告为 object，但由于缺少值，因此 Series 中也会同时包含浮点数和字符串。我们可以将 .apply 方法与 type 函数一起使用，以获取 Series 中每个成员所包含的数据类型。这里我们不必查看整个 Series 的结果，而是将 .unique 方法链接到结果，即可查看 director_name 列中找到的唯一数据类型。

```
>>> movies["director_name"].apply(type).unique()
array([<class 'str'>, <class 'float'>], dtype=object)
```

1.6.2 原理解释

Pandas DataFrame 通常包含多列（当然，它也可以只有一列）。这些列中的每一列都可以被提取出来并被视为一个 Series。

有很多机制可以从 DataFrame 中提取列。一般来说，最简单的方法是尝试将其作为 DataFrame 的属性进行访问。属性访问可通过点表示法完成。这样做有以下好处。

- 较少的输入量。
- Jupyter 将提供列名称的自动补全功能。
- Jupyter 将提供 Series 属性的自动补全功能。

当然，该方法也有以下缺陷。

- 仅适用于包含有效 Python 属性名称且与现有 DataFrame 属性不冲突的列。
- 无法创建新列，只能更新现有列。

什么是有效的 Python 属性？即以字符开头并包含下画线的字母数字序列。一般来说，它们都是小写字母，以遵循标准的 Python 命名约定。这意味着带有空格或特殊字符的列名称将不适用于属性访问方法。

使用索引操作符（[）选择列名则没有这方面的限制，任何列名都可以使用，并且也可以使用索引操作符创建和更新列。当使用索引操作符时，Jupyter 也会提供列名称的自动补全功能，但遗憾的是，在随后的 Series 属性上则不会自动补全。

很多分析人员常用的是属性访问方式，因为 Series 属性能够自动补全的功能确实非常方便。但是，在使用属性访问方式时，分析人员需要确保列名称是有效的 Python 属性名称，并且与现有的 DataFrame 属性不冲突。

此外，分析人员也可以尝试不使用属性访问或索引操作符的方式进行更新，而是改为使用 .assign 方法进行更新。在本书中将看到许多使用 .assign 的示例。

1.6.3 扩展知识

要在 Jupyter 中使用自动补全功能，可以在输入小点后按 Tab 键，也可以在索引访问中输入开始字符后按 Tab 键。Jupyter 将弹出一个自动补全列表，你可以使用向上或向下箭头键选择其中一个，然后按 Enter 键自动补全。

1.7 调用 Series 方法

Pandas 的典型工作流程允许在 Series 和 DataFrame 的执行语句之间进行切换。调用 Series 方法是使用 Series 提供的功能的主要方式。

Series 和 DataFrame 都具有强大的功能。分析人员可以使用内置的 dir 函数来查看 Series 的所有属性和方法。在下面的代码中，还显示了 Series 和 DataFrame 共有的属性和方法的数量。它们共享绝大多数的属性和方法名称。

```
>>> s_attr_methods = set(dir(pd.Series))
>>> len(s_attr_methods)
471
>>> df_attr_methods = set(dir(pd.DataFrame))
>>> len(df_attr_methods)
458
>>> len(s_attr_methods & df_attr_methods)
400
```

可以看到，这两个对象都有很多功能。当然，初学者也不必为此感到担心，因为大多数 Pandas 用户只会使用到其功能的一部分，并且其工作良好。

本秘笈涵盖了最常见、功能最强大的 Series 方法和属性。而对于 DataFrame 来说，许多方法的作用几乎是等效的。

1.7.1 实战操作

（1）在读取 movie（电影）数据集之后，选择两个具有不同数据类型的 Series。例如，director_name（导演姓名）列包含字符串（Pandas 称其为 object 或 O 数据类型），而 actor_1_facebook_likes（演员 1——主演在 Facebook 上的喜欢/点赞值）列则包含数值数据（正式的数据类型为 float64）。

```
>>> movies = pd.read_csv("data/movie.csv")
>>> director = movies["director_name"]
>>> fb_likes = movies["actor_1_facebook_likes"]

>>> director.dtype
```

```
dtype('O')

>>> fb_likes.dtype
dtype('float64')
```

(2).head 方法列出了 Series 的前 5 个条目。也可以提供一个可选参数来更改返回的条目数。另一个选择是使用.sample 方法查看一些数据。根据数据集的情况，这可能有助于更好地了解数据，因为第一行可能与后续行迥然不同。

```
>>> director.head()
0         James Cameron
1        Gore Verbinski
2           Sam Mendes
3      Christopher Nolan
4          Doug Walker
Name: director_name, dtype: object

>>> director.sample(n=5, random_state=42)
2347       Brian Percival
4687         Lucio Fulci
691         Phillip Noyce
3911       Sam Peckinpah
2488     Rowdy Herrington
Name: director_name, dtype: object

>>> fb_likes.head()
0     1000.0
1    40000.0
2    11000.0
3    27000.0
4      131.0
Name: actor_1_facebook_likes, dtype: float64
```

(3) Series 的数据类型通常决定了哪种方法最有用。例如，用于 object 数据类型 Series 的最有用的方法之一是.value_counts，该方法可以计算频率。

```
>>> director.value_counts()
Steven Spielberg    26
Woody Allen         22
Clint Eastwood      20
Martin Scorsese     20
Ridley Scott        16
                    ..
```

```
Eric England          1
Moustapha Akkad       1
Jay Oliva             1
Scott Speer           1
Leon Ford             1
Name: director_name, Length: 2397, dtype: int64
```

（4）.value_counts 方法通常对于包含 object 数据类型的 Series 更为有用，但有时也可以提供对数值类型 Series 的深入了解。例如，与 fb_likes 列一起使用时，可以看到数据已经将较高的数字四舍五入到最接近的千位，因为极为不可能有如此多的电影收到的喜欢/点赞值恰好是 1000。

```
>>> fb_likes.value_counts()
1000.0     436
11000.0    206
2000.0     189
3000.0     150
12000.0    131
            ...
362.0        1
216.0        1
859.0        1
225.0        1
334.0        1
Name: actor_1_facebook_likes, Length: 877, dtype: int64
```

（5）可以使用.size 或.shape 属性或内置的 len 函数对 Series 中的元素数进行计数。此外，.unique 方法将返回具有唯一值的 NumPy 数组。

```
>>> director.size
4916
>>> director.shape
(4916,)
>>> len(director)
4916
>>> director.unique()
array([ 'James Cameron', 'Gore Verbinski', 'Sam Mendes', ...,
        'Scott Smith', 'Benjamin Roberds', 'Daniel Hsia'],
dtype=object)
```

（6）另外，还有一个.count 方法，该方法不返回项目数，而是返回非缺失值的数量。

```
>>> director.count()
```

```
4814

>>> fb_likes.count()
4909
```

(7) 使用.min（最小值）、.max（最大值）、.mean（平均值）、.median（中值）和.std（标准差）方法可以获得基本摘要统计信息。

```
>>> fb_likes.min()
0.0

>>> fb_likes.max()
640000.0

>>> fb_likes.mean()
6494.488490527602

>>> fb_likes.median()
982.0

>>> fb_likes.std()
15106.986883848309
```

(8) 为了简化步骤（7），也可以使用.describe 方法一次性返回汇总统计信息和一些分位数。当.describe 与 object 数据类型列一起使用时，将返回完全不同的输出。

```
>>> fb_likes.describe()
count      4909.000000
mean       6494.488491
std       15106.986884
min           0.000000
25%         607.000000
50%         982.000000
75%       11000.000000
max      640000.000000
Name: actor_1_facebook_likes, dtype: float64

>>> director.describe()
count              4814
unique             2397
top     Steven Spielberg
freq                 26
Name: director_name, dtype: object
```

（9）使用.quantile 方法可以计算数字数据的分位数。请注意，如果传递的是单个值，则获得标量输出；如果传递的是一个列表，则输出为 Pandas Series。

```
>>> fb_likes.quantile(0.2)
510.0
>>> fb_likes.quantile(
...     [0.1, 0.2, 0.3, 0.4, 0.5, 0.6, 0.7, 0.8, 0.9]
... )
0.1      240.0
0.2      510.0
0.3      694.0
0.4      854.0
0.5      982.0
0.6     1000.0
0.7     8000.0
0.8    13000.0
0.9    18000.0
Name: actor_1_facebook_likes, dtype: float64
```

（10）由于步骤（6）中的.count 方法返回的值小于在步骤（5）中找到的 Series 元素的总数，因此我们知道每个 Series 中都有缺失值。.isna 方法可用于确定每个单独的值是否缺失。该结果是一个 Series。你可以将它视为一个布尔数组（它是一个包含布尔值并且与原始 Series 具有相同索引和长度的 Series）。

```
>>> director.isna()
0       False
1       False
2       False
3       False
4       False
        ...
4911    False
4912    True
4913    False
4914    False
4915    False
Name: director_name, Length: 4916, dtype: bool
```

（11）可以使用.fillna 方法替换 Series 中的所有缺失值。

```
>>> fb_likes_filled = fb_likes.fillna(0)
>>> fb_likes_filled.count()
4916
```

（12）要删除 Series 元素中包含缺失值的条目，可以使用.dropna 方法。

```
>>> fb_likes_dropped = fb_likes.dropna()
>>> fb_likes_dropped.size
4909
```

1.7.2 原理解释

本秘笈中选择介绍这些方法的原因在于，它们在数据分析中使用的频率很高。

本秘笈中的步骤返回不同类型的对象。

在步骤（1）中，.head 方法的结果是另一个 Series。.value_counts 方法也会生成一个 Series，但具有原始 Series 的唯一值作为索引，而计数则作为其值。

在步骤（5）和步骤（6）中，.size 属性和.count 方法都将返回标量值，而.shape 属性则返回的是一个单项元组。这是从 NumPy 借来的约定，该约定允许使用任意维度的数组。

在步骤（7）中，每个单独的方法都将返回一个标量值。

在步骤（8）中，.describe 方法返回一个 Series，其所有摘要统计信息名称均作为索引，而统计信息则作为其值。

在步骤（9）中，.quantile 方法是灵活的，当传递单个值时，该方法返回一个标量值，但是在给定一个列表时则返回一个 Series。

在步骤（10）～步骤（12）中，分别对应的.isna、.fillna 和.dropna 方法都将返回一个 Series。

1.7.3 扩展知识

.value_counts 方法是信息量最大的 Series 方法之一，在探索性分析（尤其是分类列）中被大量使用。.value_counts 方法默认返回计数，但是通过将 normalize（归一化）参数设置为 True，也可以返回相对频率，从而提供了另一种分布视图，示例如下。

```
>>> director.value_counts(normalize=True)
Steven Spielberg      0.005401
Woody Allen           0.004570
Clint Eastwood        0.004155
Martin Scorsese       0.004155
Ridley Scott          0.003324
                        ...
Eric England          0.000208
```

```
Moustapha Akkad    0.000208
Jay Oliva          0.000208
Scott Speer        0.000208
Leon Ford          0.000208
Name: director_name, Length: 2397, dtype: float64
```

在此秘笈中，我们还观察到.count 方法的结果与.size 属性不匹配，并以此确定该 Series 中包含缺失值。其实，还有一种更直接的方法则是检查.hasnans 属性。

```
>>> director.hasnans
True
```

对于.isna 方法，还有一个对应的.notna 方法。.notna 方法将为所有非缺失值返回 True。

```
>>> director.notna()
0       True
1       True
2       True
3       True
4       True
        ...
4911    True
4912    False
4913    True
4914    True
4915    True
Name: director_name, Length: 4916, dtype: bool
```

另外还有一个.isnull 方法，该方法是.isna 方法的别名。.isna 方法更常用，这可能是因为分析人员们都比较懒，.isna 方法需要输入的字母更少？当然，这只是一个玩笑，其实，Pandas 在很多地方都使用 NaN，这是真的。在 Pandas 或 Python 世界的任何地方，我们几乎从未见过 NULL，这可能才是分析人员更喜欢.isna 方法而不是.isnull 方法的原因。

1.8　了解 Series 的操作

Python 中存在大量用于操作对象的运算符。例如，将加号运算符放在两个整数之间时，Python 会将这两个整数加在一起。

```
>>> 5 + 9    # 加号运算符示例，5 和 9 相加
14
```

Series 和 DataFrame 支持许多 Python 运算符/操作符。一般来说，使用运算符时会返回新的 Series 或 DataFrame。

在此秘笈中，我们将会对不同的 Series 对象应用各种运算符，以产生具有完全不同值的新 Series。

1.8.1 实战操作

（1）选择 imdb_score（互联网电影资料库评分）列作为一个 Series。

```
>>> movies = pd.read_csv("data/movie.csv")
>>> imdb_score = movies["imdb_score"]
>>> imdb_score
0       7.9
1       7.1
2       6.8
3       8.5
4       7.1
       ...
4911    7.7
4912    7.5
4913    6.3
4914    6.3
4915    6.6
Name: imdb_score, Length: 4916, dtype: float64
```

（2）使用加号运算符向每个 Series 元素中均添加 1。

```
>>> imdb_score + 1
0       8.9
1       8.1
2       7.8
3       9.5
4       8.1
       ...
4911    8.7
4912    8.5
4913    7.3
4914    7.3
4915    7.6
Name: imdb_score, Length: 4916, dtype: float64
```

（3）其他基本算术运算符还包括减号（-）、乘法（*）、除法（/）和乘幂（**），这些运算符使用标量值时工作方式是相似的。例如，我们可以将每个 Series 元素均乘以 2.5。

```
>>> imdb_score * 2.5
0        19.75
1        17.75
2        17.00
3        21.25
4        17.75
          ...
4911     19.25
4912     18.75
4913     15.75
4914     15.75
4915     16.50
Name: imdb_score, Length: 4916, dtype: float64
```

（4）Python 使用双斜杠（//）表示向下取整的除法。向下取整的除法运算符会截断除法结果。百分号（%）是求余运算符，它将返回除法运算之后的余数。Series 实例还支持以下运算。

```
>>> imdb_score // 7
0        1.0
1        1.0
2        0.0
3        1.0
4        1.0
          ...
4911     1.0
4912     1.0
4913     0.0
4914     0.0
4915     0.0
Name: imdb_score, Length: 4916, dtype: float64
```

（5）Series 支持 6 个比较运算符，分别是大于（>）、小于（<）、大于或等于（>=）、小于或等于（<=）、等于（==）和不等于（!=）。每个比较运算符都会根据条件的比较结果将 Series 中的每个值转换为 True 或 False。其结果是一个布尔数组，这对于后面秘笈中的过滤操作非常有用。

```
>>> imdb_score > 7
0        True
```

```
1          True
2          False
3          True
4          True
           ...
4911       True
4912       True
4913       False
4914       False
4915       False
Name: imdb_score, Length: 4916, dtype: bool
>>> director = movies["director_name"]
>>> director == "James Cameron"
0          True
1          False
2          False
3          False
4          False
           ...
4911       False
4912       False
4913       False
4914       False
4915       False
Name: director_name, Length: 4916, dtype: bool
```

1.8.2 原理解释

此秘笈中使用的所有运算符/操作符都可以将相同的操作应用于 Series 中的每个元素。在原生 Python 中，这将需要一个 for 循环在应用该操作之前遍历序列中的每个项目。Pandas 在很大程度上依赖于 NumPy 库，该库允许进行向量化计算，也可以对整个数据序列进行操作而无须显式编写 for 循环。每个操作都会返回一个具有相同索引但包含新值的新 Series。

1.8.3 扩展知识

此秘笈中使用的所有运算符都具有等效的方法，这些方法可产生完全相同的结果。例如，在前面步骤（2）中的 imdb_score+1 运算就可以使用.add 方法来达到同样的效果。

当我们将方法链接在一起时，使用方法而非运算符十分有用。

示例如下。

```
>>> imdb_score.add(1)     # 和 imdb_score + 1 是等效的
0       8.9
1       8.1
2       7.8
3       9.5
4       8.1
        ...
4911    8.7
4912    8.5
4913    7.3
4914    7.3
4915    7.6
Name: imdb_score, Length: 4916, dtype: float64


>>> imdb_score.gt(7)      # 和 imdb_score > 7 是等效的
0       True
1       True
2       False
3       True
4       True
        ...
4911    True
4912    True
4913    False
4914    False
4915    False
Name: imdb_score, Length: 4916, dtype: bool
```

为什么 Pandas 要提供与这些运算符等效的方法？就其性质而言：一方面，运算符/操作符仅以一种方式进行操作；另一方面，方法却可以包含允许更改其默认功能的参数。

后面的秘笈还将进一步探讨这一点，不过这里可以先提供一个小示例进行说明。以 .sub 方法为例，该方法可以对 Series 执行减法。使用减号（-）运算符进行减法运算时，缺失值将被忽略。但是，在使用 .sub 方法时，则允许指定 fill_value 参数以代替缺失值。

```
>>> money = pd.Series([100, 20, None])
>>> money - 15
0    85.0
1     5.0
2     NaN
```

```
dtype: float64

>>> money.sub(15, fill_value=0)
0     85.0
1      5.0
2    -15.0
dtype: float64
```

表 1-1 列出了运算符和对应的方法。

表 1-1 运算符和对应的方法

运算符分组	运算符	Series 方法名称
算术运算符	+	.add
	-	.sub
	*	.mul
	/	.div
	//	.floordiv
	%	.mod
	**	.pow
比较运算符	<	.lt
	>	.gt
	<=	.le
	>=	.ge
	==	.eq
	!=	.ne

你可能对 Python Series 对象或与此相关的任何对象在遇到运算符时知道该怎么处理感到好奇。例如，表达式 imdb_score * 2.5 如何知道将 Series 中的每个元素乘以 2.5？其实，Python 具有一种内置的标准化方法，使对象可以使用特殊方法与运算符/操作符进行通信。

这些特殊方法就是对象在遇到运算符时在内部调用的方法。特殊方法始终分别以两个下画线开头和结尾（如.__mul__）。因此，它们也被称为 dunder 方法，因为实现运算符的方法被双下画线所包围。dunder 是某个懒惰的分析人员凭空造出来的词汇，因为他真正想说的是 double underscore（双下画线）。

每当使用乘法运算符时，都会调用上面举例的特殊方法.__mul__。Python 将 imdb_score * 2.5 表达式解释为 imdb_score.__mul__(2.5)。

使用特殊方法和使用运算符没有什么区别，因为它们在做完全相同的事情。运算符只是特殊方法的语法糖（syntactic sugar）。当然，调用.mul 方法与调用.__mul__方法则是不同的，这和.mul 方法与乘法（*）运算符的区别一样。

1.9 使用 Series 方法链

在 Python 中，每个变量都指向一个对象，许多属性和方法都返回新对象。这允许使用属性访问的方式顺序调用方法，这被称为方法链（method chaining）或流编程（flow programming）。Pandas 是一个非常适合方法链的库，因为许多 Series 和 DataFrame 方法返回更多的 Series 和 DataFrame，因此可以调用更多方法。

为了更好地理解方法链，我们可以用一个句子来将事件链转换为方法链。考虑以下句子。

某人开车去商店买食物，然后开车回家，准备食材，做饭，上菜，吃饭，最后洗碗。

该句子的 Python 版本可能采用以下形式。

```
(person .drive('store')
.buy('food')
.drive('home')
.prepare('food')
.cook('food')
.serve('food')
.eat('food')
.cleanup('dishes')
)
```

在上面的代码中，person 是调用方法的对象（或类的实例）。每个方法都返回另一个实例，允许发生调用链。传递给每个方法的参数将指定方法的操作方式。

尽管可以在单行中写入整个方法链，但更可取的是每行写入一个方法。由于 Python 通常不允许单个表达式按多行编写，因此我们有两个选择：第一种方式是将所有内容括在括号中；第二种方式是在每行末尾加反斜杠（\），以指示该行在下一行继续。为了进一步提高代码的可读性，还可以垂直对齐方法调用。

此秘笈显示了使用 Pandas Series 的类似方法链。

1.9.1 实战操作

（1）加载 movie 数据集，并从其中抽出两列。

```
>>> movies = pd.read_csv("data/movie.csv")
>>> fb_likes = movies["actor_1_facebook_likes"]
>>> director = movies["director_name"]
```

（2）附加到链条末尾的两种最常见方法是.head 或.sample 方法。这抑制了长输出。如果结果 DataFrame 非常宽，则还可以使用.T 属性转置结果。对于较短的方法链，不需要将每个方法都放在不同的行上，示例如下。

```
>>> director.value_counts().head(3)
Steven Spielberg      26
Woody Allen           22
Clint Eastwood        20
Name: director_name, dtype: int64
```

（3）计算缺失值数量的常见方法是在调用.isna 方法之后链接.sum 方法。

```
>>> fb_likes.isna().sum()
7
```

（4）fb_likes 的所有非缺失值应为整数，因为在 Facebook 平台上表达对影片或演员的喜好时，应为喜欢或不喜欢，不可能有小数值。在大多数 Pandas 版本中，任何包含缺失值的数字列都必须具有其数据类型为 float（Pandas 0.24 引入了 Int64 类型，该类型支持缺失值，但默认情况下不使用）。如果用 0 值填充了 fb_likes 的缺失值，则可以使用.astype 方法将其转换为整数类型。

```
>>> fb_likes.dtype
dtype('float64')
>>> (fb_likes.fillna(0).astype(int).head())
0     1000
1    40000
2    11000
3    27000
4      131
Name: actor_1_facebook_likes, dtype: int64
```

1.9.2 原理解释

在步骤（2）中，首先使用.value_counts 方法返回 Series，然后链接.head 方法以选择前 3 个元素。最后返回的对象是一个 Series，也可以在其上链接更多方法。

在步骤（3）中，.isna 方法创建了一个布尔数组。Pandas 分别将 False 和 True 视为 0 和 1，因此.sum 方法将返回缺失值的数量。

步骤（4）中的3个链接方法中的每一个都返回一个Series。它看起来似乎不太直观，但是.astype方法将返回具有不同数据类型的全新Series。

1.9.3 扩展知识

方法链的一个潜在缺点是调试变得困难。因为在方法调用期间创建的中间对象都未被存储在变量中，所以方法可能很难在链中跟踪其发生的精确位置。

将每个调用单独放在一行，这是一个不错的主意，因为它有助于调试更复杂的命令。在构建方法链时，笔者通常一次只建立一种方法，但偶尔也需要返回以前的代码中或对其稍作调整。

为了调试这种代码，首先可以注释掉除第一个命令以外的所有命令。然后取消对第一个链的注释，确保它能正常工作，接着继续进行下一个。

例如，假设要调试前面步骤（4）中的代码，则可以注释掉最后两个方法调用，并确保.fillna方法的工作是正常的。

```
>>> (
...     fb_likes.fillna(0)
...     # .astype(int)
...     # .head()
... )
0         1000.0
1        40000.0
2        11000.0
3        27000.0
4          131.0
           ...
4911       637.0
4912       841.0
4913         0.0
4914       946.0
4915        86.0
Name: actor_1_facebook_likes, Length: 4916, dtype: float64
```

然后取消对下一个方法的注释，并确保该方法正常工作。

```
>>> (
...     fb_likes.fillna(0).astype(int)
...     # .head()
... )
0         1000
1        40000
```

```
2         11000
3         27000
4           131
            ...
4911        637
4912        841
4913          0
4914        946
4915         86
Name: actor_1_facebook_likes, Length: 4916, dtype: int64
```

调试链的另一种方法是调用 .pipe 方法以显示中间值。需要给 Series 上的 .pipe 方法传递一个函数，该函数接收 Series 作为输入并可以返回任何内容（但如果要在方法链中使用 .pipe 方法，那么我们希望返回 Series）。

以下函数 debug_ser 将输出中间结果的值。

```
>>> def debug_ser(ser):
...     print("BEFORE")
...     print(ser)
...     print("AFTER")
...     return ser

>>> (fb_likes.fillna(0).pipe(debug_ser).astype(int).head())
BEFORE
0         1000.0
1        40000.0
2        11000.0
3        27000.0
4          131.0
            ...
4911       637.0
4912       841.0
4913         0.0
4914       946.0
4915        86.0
Name: actor_1_facebook_likes, Length: 4916, dtype: float64
AFTER
0         1000
1        40000
2        11000
3        27000
4          131
Name: actor_1_facebook_likes, dtype: int64
```

如果要创建一个全局变量来存储中间值，也可以使用.pipe方法。

```
>>> intermediate = None
>>> def get_intermediate(ser):
...     global intermediate
...     intermediate = ser
...     return ser

>>> res = (
...    fb_likes.fillna(0)
...    .pipe(get_intermediate)
...    .astype(int)
...    .head()
... )

>>> intermediate
0        1000.0
1       40000.0
2       11000.0
3       27000.0
4         131.0
          ...
4911      637.0
4912      841.0
4913        0.0
4914      946.0
4915       86.0
Name: actor_1_facebook_likes, Length: 4916, dtype: float64
```

如本秘笈开头所述，可以将反斜杠用于多行代码。因此，也可以将前面的步骤（4）重写为以下形式。

```
>>> fb_likes.fillna(0)\
...    .astype(int)\
...    .head()
0     1000
1    40000
2    11000
3    27000
4      131
Name: actor_1_facebook_likes, dtype: int64
```

当然，有些分析人员更喜欢在方法链上使用括号，因为在向链中添加方法时，不断添加尾部反斜杠可能会令有些人讨厌。

1.10 重命名列名

DataFrame 上最常见的操作之一是重命名（rename）列名称。分析人员重命名列名称的动机之一是确保这些列名称是有效的 Python 属性名称。这意味着列名称不能以数字开头，而是带下画线的小写字母数字。好的列名称还应该是描述性的，言简意赅，并且不应与现有的 DataFrame 或 Series 属性冲突。

在此秘笈中，我们将重命名列名称。重命名的动机是使代码更易于理解，并让你的环境对你有所帮助。前面我们介绍过，如果你使用点表示法访问 Series，则 Jupyter 将允许你自动补全 Series 方法（但不允许在索引访问时自动补全方法）。

1.10.1 实战操作

（1）读取 movie 数据集。

```
>>> movies = pd.read_csv("data/movie.csv")
```

（2）DataFrame 的重命名方法接收将旧值映射到新值的字典。可以为这些列创建一个字典，如下所示。

```
>>> col_map = {
...     "director_name": "director",
...     "num_critic_for_reviews": "critic_reviews",
... }
```

（3）将字典传递给重命名方法，并将结果分配给新变量。

```
>>> movies.rename(columns=col_map).head()
   color            director  ... aspec/ratio  movie/likes
0  Color       James Cameron  ...        1.78        33000
1  Color      Gore Verbinski  ...        2.35            0
2  Color         Sam Mendes  ...        2.35        85000
3  Color   Christopher Nolan  ...        2.35       164000
4    NaN         Doug Walker  ...         NaN            0
```

1.10.2 原理解释

DataFrame 上的.rename 方法允许重命名列标签。可以通过给列属性赋值来重命名列。1.10.3 节将显示如何通过赋值给.column 属性进行重命名。

1.10.3 扩展知识

在此秘笈中，我们更改了列名称。如果需要，还可以使用.rename 方法重命名索引。如果列是字符串值，则更有意义。

因此，我们可以将索引设置为 movie_title（电影片名）列，然后将这些值映射为新值。

```
>>> idx_map = {
...     "Avatar": "Ratava",
...     "Spectre": "Ertceps",
...     "Pirates of the Caribbean: At World's End": "POC",
... }
>>> col_map = {
...     "aspect_ratio": "aspect",
...     "movie_facebook_likes": "fblikes",
... }
>>> (
...     movies.set_index("movie_title")
...     .rename(index=idx_map, columns=col_map)
...     .head(3)
... )
             color   director_name   ...  aspect   fblikes
movie_title                          ...
Ratava       Color   James Cameron   ...  1.78     33000
POC          Color   Gore Verbinski  ...  2.35         0
Ertceps      Color   Sam Mendes      ...  2.35     85000
```

重命名行标签和列标签有多种方法。可以将 Python 列表赋值给索引和列属性。当列表具有与行和列标签相同数量的元素时，此赋值有效。

以下代码就显示了这样一个示例。我们将从 CSV 文件中读取数据，并使用 index_col 参数告诉 Pandas 将 movie_title 列用作索引。然后，在每个 Index 对象上使用.to_list 方法来创建 Python 标签列表。接着，在每个列表中修改 3 个值，然后将这 3 个值重新赋值给.index 和.column 属性。

```
>>> movies = pd.read_csv(
...     "data/movie.csv", index_col="movie_title"
... )
>>> ids = movies.index.to_list()
>>> columns = movies.columns.to_list()
# 使用列表赋值重命名行和列标签
```

```
>>> ids[0] = "Ratava"
>>> ids[1] = "POC"
>>> ids[2] = "Ertceps"
>>> columns[1] = "director"
>>> columns[-2] = "aspect"
>>> columns[-1] = "fblikes"
>>> movies.index = ids
>>> movies.columns = columns
>>> movies.head(3)
          color        director  ...  aspect  fblikes
Ratava    Color   James Cameron  ...    1.78    33000
POC       Color  Gore Verbinski  ...    2.35        0
Ertceps   Color     Sam Mendes  ...    2.35    85000
```

另一种选择是将一个函数传递给.rename方法。该函数接收一个列名称并返回一个新名称。假设列中有空格和大写字母，则此代码将清除它们。

```
>>> def to_clean(val):
...     return val.strip().lower().replace(" ", "_")

>>> movies.rename(columns=to_clean).head(3)
          color        director  ...  aspect  fblikes
Ratava    Color   James Cameron  ...    1.78    33000
POC       Color  Gore Verbinski  ...    2.35        0
Ertceps   Color     Sam Mendes  ...    2.35    85000
```

在某些Pandas代码中，还可以看到用于清除列名的列表推导式。使用新的清除列表，可以将结果重新赋值给.columns属性。假设列中有空格和大写字母，此代码将清除它们。

```
>>> cols = [
...     col.strip().lower().replace(" ", "_")
...     for col in movies.columns
... ]
>>> movies.columns = cols
>>> movies.head(3)
          color        director  ...  aspect  fblikes
Ratava    Color   James Cameron  ...    1.78    33000
POC       Color  Gore Verbinski  ...    2.35        0
Ertceps   Color     Sam Mendes  ...    2.35    85000
```

因为此代码更改了原始DataFrame，所以可以考虑使用.rename方法。

1.11 创建和删除列

在数据分析过程中，可能需要创建新列来表示新变量。一般来说，这些新列将从数据集的已有的列中被创建。Pandas 有若干种不同的方法可以向 DataFrame 中添加新列。

在此秘笈中，我们学习使用.assign 方法在 movie 数据集中创建新列，然后使用.drop 方法删除列。

1.11.1 实战操作

（1）创建新列的方法之一是进行索引赋值。请注意，这不会返回新的 DataFrame，而是会更改现有的 DataFrame。如果给列赋予标量值，那么该标量值将对该列中的每个单元格使用该值。例如，我们可以在 movie 数据集中创建一个 has_seen 列，以指示我们是否看过该电影。我们将为每个值赋值为 0。默认情况下，新列将被追加到末尾。

```
>>> movies = pd.read_csv("data/movie.csv")
>>> movies["has_seen"] = 0
```

（2）尽管上述方法有效且很常见，但是我们也可以使用方法链和.assign 方法。这将返回带有新列的新 DataFrame（注意，这和步骤（1）中的方法不同，步骤（1）中的方法是更改现有的 DataFrame）。因为.assign 方法使用参数名称作为列名称，所以该列名称必须是有效的参数名称。

```
>>> movies = pd.read_csv("data/movie.csv")
>>> idx_map = {
...     "Avatar": "Ratava",
...     "Spectre": "Ertceps",
...     "Pirates of the Caribbean: At World's End": "POC",
... }
>>> col_map = {
...     "aspect_ratio": "aspect",
...     "movie_facebook_likes": "fblikes",
... }
>>> (
...     movies.rename(
...         index=idx_map, columns=col_map
...     ).assign(has_seen=0)
... )
```

```
        color       director_name    ...   fblikes   has_seen
0       Color       James Cameron    ...     33000          0
1       Color      Gore Verbinski    ...         0          0
2       Color         Sam Mendes    ...     85000          0
3       Color    Christopher Nolan   ...    164000          0
4         NaN        Doug Walker    ...         0          0
...       ...               ...     ...       ...        ...
4911    Color        Scott Smith    ...        84          0
4912    Color                NaN    ...     32000          0
4913    Color    Benjamin Roberds    ...        16          0
4914    Color        Daniel Hsia    ...       660          0
4915    Color          Jon Gunn    ...       456          0
```

（3）在 movie 数据集中，有若干列包含有关 Facebook 平台上给予电影点赞次数的数据。我们可以将 Facebook 上所有对演员和导演表示喜欢的列加起来，并将结果赋值给 total_likes 列。可以通过两种方法做到这一点。

第一种方法是直接使用运算符将这些表示喜欢的列加在一起。

```
>>> total = (
...     movies["actor_1_facebook_likes"]
...     + movies["actor_2_facebook_likes"]
...     + movies["actor_3_facebook_likes"]
...     + movies["director_facebook_likes"]
... )
```

```
>>> total.head(5)
0     2791.0
1    46563.0
2    11554.0
3    95000.0
4        NaN
dtype: float64
```

第二种方法是使用方法链，因此这里需要调用 .sum 方法。我们可以传递一个列的列表以选择 .loc，并且仅提取想要求和的那些列。

```
>>> cols = [
...     "actor_1_facebook_likes",
...     "actor_2_facebook_likes",
...     "actor_3_facebook_likes",
...     "director_facebook_likes",
```

```
...    ]
>>> sum_col = movies.loc[:, cols].sum(axis="columns")
>>> sum_col.head(5)
0     2791.0
1    46563.0
2    11554.0
3    95000.0
4      274.0
dtype: float64
```

然后，可以将此 Series 赋值给新列。请注意：当使用加号（+）运算符时，结果中会出现缺失值（NaN）；而使用 .sum 方法在默认情况下会忽略缺失值，因此使用此方法会得到一个不同的结果。

```
>>> movies.assign(total_likes=sum_col).head(5)
   color        direc/_name  ...  movie/likes  total/likes
0  Color      James Cameron  ...        33000       2791.0
1  Color     Gore Verbinski  ...            0      46563.0
2  Color         Sam Mendes  ...        85000      11554.0
3  Color   Christopher Nolan ...       164000      95000.0
4    NaN        Doug Walker  ...            0        274.0
```

另外，可将函数作为参数传递给 .assign 方法。此函数接收 DataFrame 作为输入，并应返回一个 Series。

```
>>> def sum_likes(df):
...     return df[
...         [
...             c
...             for c in df.columns
...             if "like" in c
...             and ("actor" in c or "director" in c)
...         ]
...     ].sum(axis=1)

>>> movies.assign(total_likes=sum_likes).head(5)
   color        direc/_name  ...  movie/likes  total/likes
0  Color      James Cameron  ...        33000       2791.0
1  Color     Gore Verbinski  ...            0      46563.0
2  Color         Sam Mendes  ...        85000      11554.0
3  Color   Christopher Nolan ...       164000      95000.0
4    NaN        Doug Walker  ...            0        274.0
```

（4）1.7 节 "调用 Series 方法"中已经介绍过，movie 数据集包含缺失值。如步骤（3）中使用加号（+）运算符将数字列彼此相加时，如果包含缺失值，则结果为 NaN；但是，如果使用.sum 方法，则会将 NaN 转换为 0。

现在可以来检查，这两种方法产生的新列中是否包含缺失值。

```
>>> (
...     movies.assign(total_likes=sum_col)["total_likes"]
...     .isna()
...     .sum()
... )
0

>>> (
...     movies.assign(total_likes=total)["total_likes"]
...     .isna()
...     .sum()
... )
122
```

也可以用 0 填充缺失值。

```
>>> (
...     movies.assign(total_likes=total.fillna(0))[
...         "total_likes"
...     ]
...     .isna()
...     .sum()
... )
0
```

（5）movie 数据集还有一列名为 cast_total_facebook_likes（影片的全体演员在 Facebook 平台上的总喜欢/点赞值）。

如果使用 cast_total_facebook_likes 列和新创建的列 total_likes 计算出百分比值将会很有趣。不过，在创建百分比列之前，我们还需要先进行一些基本数据验证。我们将确保 cast_total_facebook_likes 大于或等于 total_likes。

```
>>> def cast_like_gt_actor(df):
...     return (
...         df["cast_total_facebook_likes"]
...         >= df["total_likes"]
...     )
```

```
>>> df2 = movies.assign(
...     total_likes=total,
...     is_cast_likes_more=cast_like_gt_actor,
... )
```

（6）is_cast_likes_more 列现在是一个布尔数组，里面包含的全部是布尔值。可以使用 .all 方法检查此列的所有值是否均为 True。

```
>>> df2["is_cast_likes_more"].all()
False
```

（7）事实证明，至少有一部电影的 total_likes 数量多于 cast_total_facebook_likes。这可能是因为导演在 Facebook 平台上收获的喜欢/点赞值未计入全体演员的总喜欢值。因此，我们可以回溯并删除 total_likes 列。这可以在列参数中使用 .drop 方法来实现。

```
>>> df2 = df2.drop(columns="total_likes")
```

（8）重新创建一个仅包含演员收获的喜欢/点赞值总数的 Series。

```
>>> actor_sum = movies[
...     [
...         c
...         for c in movies.columns
...         if "actor_" in c and "_likes" in c
...     ]
... ].sum(axis="columns")

>>> actor_sum.head(5)
0     2791.0
1    46000.0
2    11554.0
3    73000.0
4      143.0
dtype: float64
```

（9）现在再次检查 cast_total_facebook_likes 中的所有值是否都大于或等于 actor_sum。可以使用大于或等于（>=）运算符或 .ge 方法来执行此操作。

```
>>> movies["cast_total_facebook_likes"] >= actor_sum
0    True
1    True
2    True
3    True
4    True
```

```
                ...
4911            True
4912            True
4913            True
4914            True
4915            True
Length: 4916, dtype: bool

>>> movies["cast_total_facebook_likes"].ge(actor_sum)
0               True
1               True
2               True
3               True
4               True
                ...
4911            True
4912            True
4913            True
4914            True
4915            True
Length: 4916, dtype: bool

>>> movies["cast_total_facebook_likes"].ge(actor_sum).all()
True
```

（10）计算 actor_sum 和 cast_total_facebook_likes 喜欢值的百分比。

```
>>> pct_like = actor_sum.div(
...     movies["cast_total_facebook_likes"]
... ).mul(100)
```

（11）验证该 Series 的最小值和最大值是否为 0～100。

```
>>> pct_like.describe()
count    4883.000000
mean       83.327889
std        14.056578
min        30.076696
25%        73.528368
50%        86.928884
75%        95.477440
max       100.000000
dtype: float64
```

（12）可以使用 movie_title（电影片名）列作为索引来创建一个 Series。Series 构造函数允许同时传递值和索引。

```
>>> pd.Series(
...     pct_like.to_numpy(), index=movies["movie_title"]
... ).head()
movie_title
Avatar                                      57.736864
Pirates of the Caribbean: At World's End    95.139607
Spectre                                     98.752137
The Dark Knight Rises                       68.378310
Star Wars: Episode VII - The Force Awakens  100.000000
dtype: float64
```

1.11.2 原理解释

许多 Pandas 操作都是很灵活的，列的创建就是其中之一。该秘笈既赋给了一个标量值——如步骤（1）所示，又赋给了一个 Series 值——如步骤（2）所示，它们都可以创建新列。

步骤（3）使用了加号（+）运算符和.sum 方法将 4 个不同的 Series 加总在一起。步骤（4）使用了方法链来查找和填充缺失值。步骤（5）使用了大于或等于比较运算符返回布尔 Series，然后在步骤（6）中使用了.all 方法检查其每个值是否为 True。

.drop 方法接收要删除的行或列的名称。默认情况下将按索引名称删除行。要删除列，必须将 axis 参数设置为 1 或'columns'。也就是说，axis 的默认值为 0 或'index'。

步骤（8）和步骤（9）在没有 total_likes 列的情况下重新执行了步骤（3）～步骤（6）的操作。步骤（10）最终计算出自步骤（4）以来我们想要的百分比列。步骤（11）则验证了百分比值为 0～100。

1.11.3 扩展知识

可以使用.insert 方法将新列插入 DataFrame 的特定位置处。.insert 方法接收新列的整数位置作为其第一个参数，接收新列的名称作为其第二个参数，接收新列的值作为其第三个参数。对此，需要使用.get_loc Index 方法查找列名称的整数位置。

.insert 方法可就地修改调用 DataFrame，因此不会有赋值语句。此外，.insert 方法还将返回 None。因此，很多分析人员更喜欢使用.assign 方法创建新列。如果需要排序，则可以将列的有序列表传递给索引操作符（或传递给.loc 属性）。

例如，要计算每部电影的收益，可以从票房总收入 gross 中减去预算 budget，然后将新列插入 gross 列之后。其代码如下。

```
>>> profit_index = movies.columns.get_loc("gross") + 1
>>> profit_index
9
>>> movies.insert(
...     loc=profit_index,
...     column="profit",
...     value=movies["gross"] - movies["budget"],
... )
```

除了使用.drop 方法删除列，还有一种方法是使用 del 语句，但是这种方法也不会返回一个新的 DataFrame，因此最好还是使用.drop 方法。

```
>>> del movies["director_name"]
```

第 2 章 DataFrame 基本操作

本章包含以下秘笈。
- 选择多个 DataFrame 列。
- 使用方法选择列。
- 排序列名称。
- 统计 DataFrame 摘要信息。
- 使用 DataFrame 方法链。
- 了解 DataFrame 的操作。
- 比较缺失值。
- 转置 DataFrame 操作的方向。
- 确定大学校园的多样性。

2.1 介 绍

本章介绍 DataFrame 的许多基本操作。许多秘笈与第 1 章"Pandas 基础"中的秘笈相似,只不过第 1 章主要讨论的是 Series 的操作。

2.2 选择多个 DataFrame 列

可以通过将列名称传递给 DataFrame 的索引操作符来选择单个列。本书 1.6 节"选择列"对此已有介绍。一般来说,分析人员需要关注和处理的是当前工作数据集的一个子集,而这正是通过选择多个列来获得的。

在此秘笈中,我们将从 movie 数据集中选择所有的 actor(演员)和 director(导演)相关数据列。

2.2.1 实战操作

(1)读取 movie 数据集,并将所需列的列表传递给索引操作符。

```
>>> import pandas as pd
>>> import numpy as np
>>> movies = pd.read_csv("data/movie.csv")
>>> movie_actor_director = movies[
...     [
...         "actor_1_name",
...         "actor_2_name",
...         "actor_3_name",
...         "director_name",
...     ]
... ]
>>> movie_actor_director.head()
    actor_1_name    actor_2_name    actor_3_name   director_name
0    CCH Pounder      Joel Dav...      Wes Studi     James Ca...
1    Johnny Depp      Orlando ...      Jack Dav...   Gore Ver...
2    Christop...      Rory Kin...      Stephani...   Sam Mendes
3     Tom Hardy       Christia...      Joseph G...   Christop...
4    Doug Walker      Rob Walker             NaN    Doug Walker
```

（2）在某些情况下，我们仅需要选择 DataFrame 的一列，此时使用索引操作可以返回一个 Series 或一个 DataFrame。如果我们传递一个包含单个项目的列表，则将返回一个 DataFrame；如果只传递一个包含列名称的字符串，则将返回一个 Series。

```
>>> type(movies[["director_name"]])
<class 'pandas.core.frame.DataFrame'>

>>> type(movies["director_name"])
<class 'pandas.core.series.Series'>
```

（3）我们也可以使用 .loc 属性按名称提取出一列。因为此索引操作要求首先传递一个行选择器，所以我们将使用冒号（:）表示一个选择所有行的切片。这也可以返回一个 DataFrame 或 Series。

```
>>> type(movies.loc[:, ["director_name"]])
<class 'pandas.core.frame.DataFrame'>

>>> type(movies.loc[:, "director_name"])
<class 'pandas.core.series.Series'>
```

2.2.2 原理解释

DataFrame 的索引操作符非常灵活，可以接收许多不同的对象。如果传递的是一个字符串，那么它将返回一维的 Series；如果将列表传递给索引操作符，那么它将以指定顺序返回列表中所有列的 DataFrame。

步骤（2）显示了如何选择单个列作为 DataFrame 和 Series。一般来说，可以使用字符串选择单个列，从而得到一个 Series。如果需要一个 DataFrame，则可以将列名称放在一个单元素的列表中。

步骤（3）显示了如何使用 .loc 属性提取 Series 或 DataFrame。

2.2.3 扩展知识

在索引操作符内部传递长列表可能会导致可读性问题。为了解决这个问题，你可以先将所有列名称保存到一个列表变量中。以下代码可获得与步骤（1）相同的结果。

```
>>> cols = [
...     "actor_1_name",
...     "actor_2_name",
...     "actor_3_name",
...     "director_name",
... ]
>>> movie_actor_director = movies[cols]
```

使用 Pandas 时，最常见的异常之一是 KeyError。该问题主要是由于错误地输入了列或索引名称。每当尝试不使用列表进行多列选择时，都很容易引发相同的错误。

```
>>> movies[
...     "actor_1_name",
...     "actor_2_name",
...     "actor_3_name",
...     "director_name",
... ]
Traceback (most recent call last):
    ...
KeyError: ('actor_1_name', 'actor_2_name', 'actor_3_name', 'director_name')
```

2.3 使用方法选择列

尽管列选择通常是使用索引操作符完成的，但仍有一些 DataFrame 方法可以按其他方式进行选择。例如，.select_dtypes 和 .filter 就是两个很有用的可以选择列的方法。

如果要按类型选择，则需要熟悉 Pandas 数据类型。在 1.5 节"了解数据类型"中对 Pandas 数据类型进行了详细说明。

2.3.1 实战操作

（1）读取 movie 数据集，然后缩短要显示的列名，最后使用.get_dtype_counts 方法输出包含每种特定数据类型的列数。

```
>>> movies = pd.read_csv("data/movie.csv")
>>> def shorten(col):
...     return (
...         str(col)
...         .replace("facebook_likes", "fb")
...         .replace("_for_reviews", "")
...     )
>>> movies = movies.rename(columns=shorten)
>>> movies.get_dtype_counts()
float64    13
int64       3
object     12
dtype: int64
```

（2）使用.select_dtypes 方法仅选择数据类型为整数的列。

```
>>> movies.select_dtypes(include="int").head()
   num_voted_users  cast_total_fb  movie_fb
0           886204           4834     33000
1           471220          48350         0
2           275868          11700     85000
3          1144337         106759    164000
4                8            143         0
```

（3）如果要选择所有数字列，则可以将字符串 number 传递给 include 参数。

```
>>> movies.select_dtypes(include="number").head()
   num_critics  duration  ...  aspect_ratio  movie_fb
0        723.0     178.0  ...          1.78     33000
1        302.0     169.0  ...          2.35         0
2        602.0     148.0  ...          2.35     85000
3        813.0     164.0  ...          2.35    164000
4          NaN       NaN  ...           NaN         0
```

（4）如果需要整数和字符串列，则可以执行以下操作。

```
>>> movies.select_dtypes(include=["int", "object"]).head()
    color        direc/_name  ...  conte/ating  movie_fb
0   Color      James Cameron  ...        PG-13     33000
1   Color     Gore Verbinski  ...        PG-13         0
2   Color         Sam Mendes  ...        PG-13     85000
3   Color   Christopher Nolan ...        PG-13    164000
4     NaN        Doug Walker  ...          NaN         0
```

（5）要排除仅包含浮点数字的列，可执行以下操作。

```
>>> movies.select_dtypes(exclude="float").head()
    color     director_name  ...  content_rating  movie_fb
0   Color       James Ca...  ...           PG-13     33000
1   Color       Gore Ver...  ...           PG-13         0
2   Color        Sam Mendes  ...           PG-13     85000
3   Color        Christop...  ...          PG-13    164000
4     NaN       Doug Walker  ...             NaN         0
```

（6）选择列的另一种方法是使用.filter方法。此方法很灵活，可以根据使用的参数搜索列名（或索引标签）。在这里，我们将使用like参数搜索所有Facebook列或包含确切字符串fb的列名称。like参数将检查列名中的子字符串。

```
>>> movies.filter(like="fb").head()
   director_fb  actor_3_fb  ...  actor_2_fb  movie_fb
0          0.0       855.0  ...       936.0     33000
1        563.0      1000.0  ...      5000.0         0
2          0.0       161.0  ...       393.0     85000
3      22000.0     23000.0  ...     23000.0    164000
4        131.0         NaN  ...        12.0         0
```

（7）.filter方法有很多相关技巧（或参数）。例如，如果使用items参数，则可以传入一个列名称的列表。

```
>>> cols = [
...     "actor_1_name",
...     "actor_2_name",
...     "actor_3_name",
...     "director_name",
... ]
>>> movies.filter(items=cols).head()
      actor_1_name  ...    director_name
0      CCH Pounder  ...    James Cameron
1      Johnny Depp  ...   Gore Verbinski
2   Christoph Waltz  ...       Sam Mendes
```

```
3          Tom Hardy     ...    Christopher Nolan
4        Doug Walker     ...         Doug Walker
```

（8）.filter 方法还允许使用正则表达式（regular expression），通过 regex 参数搜索列。例如，我们可以搜索列名称中某处有数字的所有列。

```
>>> movies.filter(regex=r"\d").head()
    actor_3_fb   actor_2_name   ...    actor_3_name   actor_2_fb
0        855.0     Joel Dav...  ...       Wes Studi       936.0
1       1000.0     Orlando ...  ...      Jack Dav...    5000.0
2        161.0     Rory Kin...  ...      Stephani...     393.0
3      23000.0     Christia...  ...      Joseph G...   23000.0
4          NaN     Rob Walker   ...             NaN      12.0
```

2.3.2 原理解释

步骤（1）列出了所有不同数据类型的频率。或者也可以使用.dtypes 属性来获取每一列的确切数据类型。.select_dtypes 方法可以在其 include 或 exclude 参数中接收一个列表或单个数据类型，并返回仅具有给定数据类型的列的 DataFrame（如果使用的是 exclude 参数，则返回一个不包含指定类型的列的 DataFrame）。列表值可以是数据类型的字符串名称，也可以是实际的 Python 对象。

.filter 方法仅通过检查列名而不是实际数据值来选择列。此外，.filter 方法具有 3 个互斥的参数，即 item、like 和 regex，互斥意味着一次只能使用其中一个。

like 参数接收一个字符串，并尝试查找名称中某处包含该确切字符串的所有列名称。为了获得更大的灵活性，也可以使用 regex 参数来通过正则表达式选择列名称。示例中的特殊正则表达式 r'\d' 表示 0～9 的所有数字，并且匹配其中至少包含一个数字的任何字符串。

.filter 方法还带有另一个参数，即 items，该参数接收一个确切的列名称的列表。这几乎是对索引操作的精确复制，只是如果其中一个字符串与列名称不匹配，则不会引发 KeyError。例如，movies.filter(items = ['actor_1_name', 'asdf'])运行时就不会出现错误，并且将返回单列的 DataFrame。

2.3.3 扩展知识

.select_dtypes 比较令人头疼的地方是它的灵活性，因为它可以同时接收字符串和 Python 对象。以下列表详细阐释了选择许多不同列数据类型的所有可能方法。在 Pandas 中并没有引用数据类型的标准或首选方法，因此你最好对这两种方式均有所了解。

- np.number、'number'：选择整数和浮点数，而不考虑大小。
- np.float64、np.float_、float、'float64'、'float_'、'float'：仅选择 64 位浮点值。
- np.float16、np.float32、np.float128、'float16'、'float32'、'float128'：分别选择精确的 16 位、32 位和 128 位浮点数。
- np.floating、'floating'：选择所有浮点值，无论大小。
- np.int0、np.int64、np.int_、int、'int0'、'int64'、'int_'、'int'：仅选择 64 位整数。
- np.int8、np.int16、np.int32、'int8'、'int16'、'int32'：分别选择 8 位、16 位和 32 位整数。
- np.integer、'integer'：选择所有整数，而不考虑大小。
- 'Int64'：选择可为空的整数；没有 NumPy 等效项。
- np.object、'object'、'O'：选择所有对象数据类型。
- np.datetime64、'datetime64'、'datetime'：所有日期时间均为 64 位。
- np.timedelta64、'timedelta64'、'timedelta'：所有时间增量均为 64 位。
- pd.Categorical、'category'：Pandas 特有分类；没有 NumPy 等效项。

由于所有整数和浮点数默认为 64 位，因此可以使用字符串'int'或'float'来选择它们。如果要选择所有整数和浮点数，而不管它们的具体大小如何，则可以使用字符串'number'。

2.4 排序列名称

在最初将数据集导入为 DataFrame 之后，首先要考虑的任务之一就是分析列的顺序。由于多数人的习惯是从左到右阅读文字资料，因此，列顺序会影响我们对数据的理解。如果有一个合理的列顺序，那么查找和解释信息要容易得多。

虽然 Pandas 没有标准的规则集来规定在数据集中应如何组织列，但是，最好的做法是制订一组你自己始终遵循的准则。如果你与一组分析师合作，并且需要共享大量的数据集，则尤其应该如此。

以下是对列进行排序的一般准则。
- 将每一列划分为分类列或连续列。
- 在分类列和连续列中，对公共列分组。
- 将最重要的列组放置于首位，然后放置分类列，最后放置连续列。

此秘笈将演示如何使用上述一般准则排序各列。需要说明的是，排序方法有很多种，只要方便你自己的应用即可。

2.4.1 实战操作

（1）读取 movie 数据集，然后扫描数据。

```
>>> movies = pd.read_csv("data/movie.csv")
>>> def shorten(col):
...     return col.replace("facebook_likes", "fb").replace(
...         "_for_reviews", ""
...     )
>>> movies = movies.rename(columns=shorten)
```

（2）输出所有列名称，并扫描相似的分类列和连续列。

```
>>> movies.columns
Index(['color', 'director_name', 'num_critic', 'duration',
'director_fb', 'actor_3_fb', 'actor_2_name', 'actor_1_fb', 'gross',
'genres', 'actor_1_name', 'movie_title', 'num_voted_users',
'cast_total_fb', 'actor_3_name', 'facenumber_in_poster',
'plot_keywords', 'movie_imdb_link', 'num_user', 'language', 'country',
'content_rating', 'budget', 'title_year', 'actor_2_fb', 'imdb_score',
'aspect_ratio', 'movie_fb'], dtype='object')
```

（3）这些列似乎没有什么逻辑顺序。我们可以将列名称以比较合理的方式组织到列表中，以遵循前面介绍的一般准则。

```
>>> cat_core = [
...     "movie_title",
...     "title_year",
...     "content_rating",
...     "genres",
... ]
>>> cat_people = [
...     "director_name",
...     "actor_1_name",
...     "actor_2_name",
...     "actor_3_name",
... ]
>>> cat_other = [
...     "color",
...     "country",
...     "language",
...     "plot_keywords",
```

```
...         "movie_imdb_link",
... ]
>>> cont_fb = [
...         "director_fb",
...         "actor_1_fb",
...         "actor_2_fb",
...         "actor_3_fb",
...         "cast_total_fb",
...         "movie_fb",
... ]
>>> cont_finance = ["budget", "gross"]
>>> cont_num_reviews = [
...         "num_voted_users",
...         "num_user",
...         "num_critic",
... ]
>>> cont_other = [
...         "imdb_score",
...         "duration",
...         "aspect_ratio",
...         "facenumber_in_poster",
... ]
```

(4) 将所有列表连接在一起以获得最终的列顺序。另外,请确保此列表包含原始数据集中的所有列。

```
>>> new_col_order = (
...     cat_core
...     + cat_people
...     + cat_other
...     + cont_fb
...     + cont_finance
...     + cont_num_reviews
...     + cont_other
... )
>>> set(movies.columns) == set(new_col_order)
True
```

(5) 将包含新列顺序的列表传递给 DataFrame 的索引操作符,以对列进行重新排序。

```
>>> movies[new_col_order].head()
   movie_title  title_year  ...  aspect_ratio  facenumber_in_poster
0       Avatar      2009.0  ...          1.78                   0.0
1  Pirates ...      2007.0  ...          2.35                   0.0
```

2	Spectre	2015.0	...	2.35	1.0
3	The Dark...	2012.0	...	2.35	0.0
4	Star War...	NaN	...	NaN	0.0

2.4.2 原理解释

分析人员可以使用包含特定列名称的列表从 DataFrame 中选择列的子集。例如，movies[['movie_title','director_name']]即可创建一个仅包含 movie_title 和 director_name 列的新 DataFrame。按名称选择列是 Pandas DataFrame 的索引操作符的默认行为。

步骤（3）根据类型（分类或连续列）以及它们的数据相似程度，将所有列名称整齐地组织到单独的列表中。最重要的列（如电影的片名）位于第一位。

步骤（4）连接了所有列名称列表，并验证此新列表包含与原始列名称相同的确切值。Python 数据集是无序的，而相等（==）语句则可以检查一个集合中的每个成员是否是另一个集合中的成员。手动排序此秘笈中的列容易受到人为错误的影响，因为我们很容易忘记新的列名称列表中的某一列。

步骤（5）通过将新的列顺序作为列表传递给索引操作符来完成重新排序。现在，这个新排序要比原来的排序合理得多。

2.4.3 扩展知识

除了前面提到的建议，还有其他排序列的准则。Hadley Wickham 在有关规整数据（tidy data）的开创性论文中建议将固定变量放在首位，然后放置测量变量。由于 movie 数据并非来自受控实验，因此在确定哪些变量是固定的和哪些变量是测量的方面都具有一定的灵活性。测量变量的良好候选者是我们希望预测的变量，如 gross（票房收入）、budget（预算）或 imdb_score（互联网电影资料库评分）等。以这种顺序，我们可以混合使用分类变量和连续变量。例如，在演员的姓名之后直接放置在 Facebook 平台上的喜欢/点赞数的列可能更有意义。当然，你可以提出自己的列排序准则，因为计算部分并不受列顺序的影响。

2.5 统计 DataFrame 摘要信息

在 1.7 节"调用 Series 方法"中，对单个列或 Series 数据使用了多种方法进行操作。其中许多是返回单个标量值的聚合（aggregation）或归约（reduce）方法。在使用 DataFrame 调用这些相同的方法时，它们会立即对每一列执行操作，并归约 DataFrame 中每一列的结果。它们将返回一个 Series，在索引中包含列名称，并汇总每一列作为值。

在本秘笈中，我们将对 movie 数据集探索各种最常见的 DataFrame 属性和方法。

2.5.1 实战操作

（1）读取 movie 数据集，检查其基本描述性属性，如.shape、.size 和.ndim，另外还可以运行 len 函数。

```
>>> movies = pd.read_csv("data/movie.csv")
>>> movies.shape
(4916, 28)
>>> movies.size
137648
>>> movies.ndim
2
>>> len(movies)
4916
```

（2）.count 方法可以显示每一列的非缺失值数。这是一种聚合方法，因为.count 方法可以将单个列汇总为一个值。其输出是一个以原始列名作为索引的 Series。

```
>>> movies.count()
color                        4897
director_name                4814
num_critic_for_reviews       4867
duration                     4901
director_facebook_likes      4814
                             ...
title_year                   4810
actor_2_facebook_likes       4903
imdb_score                   4916
aspect_ratio                 4590
movie_facebook_likes         4916
Length: 28, dtype: int64
```

（3）计算摘要统计信息的其他方法还包括.min（最小值）、.max（最大值）、.mean（平均值）、.median（中值）和.std（标准差），它们将返回一个 Series，其中包含列名称（作为索引），并使用聚合结果作为值。

```
>>> movies.min()
num_critic_for_reviews       1.00
duration                     7.00
```

```
director_facebook_likes       0.00
actor_3_facebook_likes        0.00
actor_1_facebook_likes        0.00
                               ...
title_year                 1916.00
actor_2_facebook_likes        0.00
imdb_score                    1.60
aspect_ratio                  1.18
movie_facebook_likes          0.00
Length: 16, dtype: float64
```

（4）.describe 方法非常强大，可以立即计算所有描述性统计数据和四分位数。最终结果是一个以描述性统计信息名称作为索引的 DataFrame。笔者喜欢使用.T 转置结果，因为这样做通常可以在屏幕上显示更多信息。

```
>>> movies.describe().T
                count         mean    ...         75%          max
num_criti...   4867.0   137.988905    ...      191.00        813.0
duration       4901.0   107.090798    ...      118.00        511.0
director_...   4814.0   691.014541    ...      189.75      23000.0
actor_3_f...   4893.0   631.276313    ...      633.00      23000.0
actor_1_f...   4909.0  6494.488491    ...    11000.00     640000.0
...                ...          ...    ...         ...          ...
title_year     4810.0  2002.447609    ...     2011.00       2016.0
actor_2_f...   4903.0  1621.923516    ...      912.00     137000.0
imdb_score     4916.0     6.437429    ...        7.20          9.5
aspect_ratio   4590.0     2.222349    ...        2.35         16.0
movie_fac...   4916.0  7348.294142    ...     2000.00     349000.0
```

（5）可以使用 percentiles 参数在.describe 方法中指定精确的分位数。

```
>>> movies.describe(percentiles=[0.01, 0.3, 0.99]).T
                count         mean    ...         99%          max
num_criti...   4867.0   137.988905    ...      546.68        813.0
duration       4901.0   107.090798    ...      189.00        511.0
director_...   4814.0   691.014541    ...    16000.00      23000.0
actor_3_f...   4893.0   631.276313    ...    11000.00      23000.0
actor_1_f...   4909.0  6494.488491    ...    44920.00     640000.0
...                ...          ...    ...         ...          ...
title_year     4810.0  2002.447609    ...     2016.00       2016.0
actor_2_f...   4903.0  1621.923516    ...    17000.00     137000.0
imdb_score     4916.0     6.437429    ...        8.50          9.5
aspect_ratio   4590.0     2.222349    ...        4.00         16.0
movie_fac...   4916.0  7348.294142    ...    93850.00     349000.0
```

2.5.2 原理解释

步骤（1）提供了有关数据集大小的基本信息。其中：.shape 属性可以返回包含行和列数的元组；.size 属性返回 DataFrame 中元素的总数，这其实就是行和列数的乘积；.ndim 属性返回维数，对于所有 DataFrame，维数均为 2。

将 DataFrame 传递给内置 len 函数时，该函数将返回行数。

步骤（2）和步骤（3）中的方法可以将每一列汇总为一个数字。现在，每个列名都是 Series 中的索引标签，其聚合结果为相应的值。

通过仔细观察可注意到，步骤（3）的输出缺少步骤（2）的所有对象列，这是因为此方法默认情况下忽略字符串列。

可以看到，虽然数字列中包含缺失值，但是.describe 方法也返回了结果。这是因为，默认情况下，Pandas 遇到数值列中的缺失值时会跳过。通过将 skipna 参数设置为 False，可以更改此行为。在将 skipna 参数改为 False 之后，只要存在一个缺失值，就会导致 Pandas 为所有这些聚合方法返回 NaN。

.describe 方法可以显示数字列的摘要统计信息。通过将 0～1 的数字列表传递给 percentiles 参数，可以扩展其摘要以包括更多分位数。有关.describe 方法的更多信息，请参见 4.2 节"开发数据分析例程"。

2.5.3 扩展知识

要查看.skipna 参数如何影响结果，我们可以将其值设置为 False，然后重新运行步骤（3）。在这种情况下，只有不包含缺失值的数字列才会返回计算的结果。

```
>>> movies.min(skipna=False)
num_critic_for_reviews        NaN
duration                      NaN
director_facebook_likes       NaN
actor_3_facebook_likes        NaN
actor_1_facebook_likes        NaN
                              ...
title_year                    NaN
actor_2_facebook_likes        NaN
imdb_score                    1.6
aspect_ratio                  NaN
movie_facebook_likes          0.0
Length: 16, dtype: float64
```

2.6 使用 DataFrame 方法链

在 1.9 节 "使用 Series 方法链"中展示了将若干个 Series 方法结合在一起形成方法链的示例。本章中的所有方法链都将从 DataFrame 开始。使用方法链的关键之一是知道在链条的每个步骤中返回的确切对象。在 Pandas 中，这总是某个 DataFrame、Series 或标量值。

在此秘笈中，我们将统计 movie 数据集每一列中的缺失值。

2.6.1 实战操作

（1）使用 .isnull 方法获取缺失值的计数。.isnull 方法会将每个值更改为布尔值，以指示是否为缺失值。

```
>>> movies = pd.read_csv("data/movie.csv")
>>> def shorten(col):
...     return col.replace("facebook_likes", "fb").replace(
...         "_for_reviews", "")
...
>>> movies = movies.rename(columns=shorten)
>>> movies.isnull().head()
   color  director_name  ...  aspect_ratio  movie_fb
0  False  False          ...  False         False
1  False  False          ...  False         False
2  False  False          ...  False         False
3  False  False          ...  False         False
4  True   False          ...  True          False
```

（2）在方法链上添加一个 .sum 方法，该方法将 True 和 False 分别解释为 1 和 0。由于这是一种归约方法，因此结果将被聚合为一个 Series。

```
>>> (movies.isnull().sum().head())
color              19
director_name     102
num_critic         49
duration           15
director_fb       102
dtype: int64
```

（3）这里还可以更进一步，取该 Series 的总和，这样就可以将整个 DataFrame 中缺失值总数的计数作为标量值返回。

```
>>> movies.isnull().sum().sum()
2654
```

（4）反过来，如果要确定整个 DataFrame 中是否包含缺失值，则可以在方法链中连续两次使用.any 方法。

```
>>> movies.isnull().any().any()
True
```

2.6.2 原理解释

.isnull 方法将返回一个与调用它的 DataFrame 相同大小的 DataFrame，只是其所有值都已转换为布尔值。可以通过查看以下数据类型的计数来验证这一点。

```
>>> movies.isnull().dtypes.value_counts()
bool    28
dtype: int64
```

在 Python 中，布尔值的取值为 0 和 1，这使得在步骤（2）中进行列求和成为可能。返回的 Series 本身也具有.sum 方法，因此可以通过这种方式统计在 DataFrame 中包含的全部缺失值的数量。

在步骤（4）中，DataFrame 中的.any 方法返回一个布尔值 Series，指示每列是否至少存在一个 True。再次将 .any 方法链接到生成的布尔值 Series 上，即可确定是否有任何列包含缺失值。如果步骤（4）的评估结果为 True，则整个 DataFrame 中至少有一个缺失值。

2.6.3 扩展知识

movie 数据集中包含 object 数据类型的大多数列都包含缺失值。默认情况下，聚合方法（.min、.max 和.sum）不会为 object 列返回任何内容。

在以下代码片段中，选择了 3 个 object 列并尝试查找每个 object 列的最大值。

```
>>> movies[["color", "movie_title", "color"]].max()
Series([], dtype: float64)
```

为了强制 Pandas 为每一列返回值，必须填充缺失值。在本示例中，我们选择给缺失值填充一个空字符串。

```
>>> movies.select_dtypes(["object"]).fillna("").max()
color                              Color
director_name                      Étienne Faure
actor_2_name                       Zubaida Sahar
```

```
genres                           Western
actor_1_name               Óscar Jaenada
                               ...
plot_keywords        zombie|zombie spoof
movie_imdb_link       http://www.imdb....
language                            Zulu
country                     West Germany
content_rating                         X
Length: 12, dtype: object
```

出于代码的可读性考虑，方法链通常被编写为每行一个方法调用，并通过括号将其括起来。在方法链每步返回的内容上，这将简化注释的阅读和插入；或者注释掉某些行进行调试。

```
>>> (movies.select_dtypes(["object"]).fillna("").max())
color                              Color
director_name              Étienne Faure
actor_2_name              Zubaida Sahar
genres                           Western
actor_1_name               Óscar Jaenada
                               ...
plot_keywords        zombie|zombie spoof
movie_imdb_link       http://www.imdb....
language                            Zulu
country                     West Germany
content_rating                         X
Length: 12, dtype: object
```

2.7　了解 DataFrame 的操作

在 1.8 节"了解 Series 的操作"中提供了有关运算符/操作符的入门知识，这将对你了解 DataFrame 的操作也有所帮助，因为 Python 的算术运算符和比较运算符可以与 DataFrame 协同工作，就像其与 Series 协同工作一样。

当算术运算符或比较运算符与 DataFrame 协同工作时，每列的每个值都会对其应用操作。通常而言，当运算符或比较运算符与 DataFrame 协同工作时，列全部为数字，或全部为对象（通常为字符串）。如果 DataFrame 确实不包含同类数据，则该操作很可能失败。

现在来看一个失败示例。在 college 数据集中，同时包含了数字和对象数据类型。尝试将 5 加到 DataFrame 的每个值上时，会引发 TypeError（类型错误），这是因为不能将整数和字符串加到一起。

```
>>> colleges = pd.read_csv("data/college.csv")
>>> colleges + 5
Traceback (most recent call last):
 ...
TypeError: can only concatenate str (not "int") to str
```

因此，要成功将运算符与 DataFrame 配合使用，首先需要选择同类数据。

在此秘笈中，我们将选择所有以 'UGDS_' 开头的列。这些列代表按种族划分的本科生（undergraduate student）的比例。我们将首先导入数据并使用机构名称作为索引的标签，然后使用 .filter 方法选择所需的列。

```
>>> colleges = pd.read_csv(
...     "data/college.csv", index_col="INSTNM"
... )
>>> college_ugds = colleges.filter(like="UGDS_")
>>> college_ugds.head()
               UGDS_WHITE   UGDS_BLACK   ...   UGDS_NRA   UGDS_UNKN
INSTNM                                   ...
Alabama A...       0.0333       0.9353   ...     0.0059      0.0138
Universit...       0.5922       0.2600   ...     0.0179      0.0100
Amridge U...       0.2990       0.4192   ...     0.0000      0.2715
Universit...       0.6988       0.1255   ...     0.0332      0.0350
Alabama S...       0.0158       0.9208   ...     0.0243      0.0137
```

此秘笈将使用多个运算符和一个 DataFrame 将本科生的列四舍五入到最接近的百分比值，其结果等效于 .round 方法。

2.7.1 实战操作

（1）Pandas 执行的是 Bankers 舍入规则，该规则简言之就是"四舍六入五成双"。"四舍六入"很好理解（例如，3.4 取整为 3，3.6 取整为 4），而"五成双"的意思则是保留位数后一位的数字为 5 时，根据前一位的奇偶性决定。为偶时向下取整，为奇数时向上取整。例如，3.5 取整为 4，而 4.5 同样取整为 4。也就是说，取整结果必定为偶数。

现在可以将该 Series 舍入到小数点后两位，看看 UGDS_BLACK 行会发生什么。

```
>>> name = "Northwest-Shoals Community College"
>>> college_ugds.loc[name]
UGDS_WHITE   0.7912
```

```
UGDS_BLACK    0.1250
UGDS_HISP     0.0339
UGDS_ASIAN    0.0036
UGDS_AIAN     0.0088
UGDS_NHPI     0.0006
UGDS_2MOR     0.0012
UGDS_NRA      0.0033
UGDS_UNKN     0.0324
Name: Northwest-Shoals Community College, dtype: float64

>>> college_ugds.loc[name].round(2)
UGDS_WHITE    0.79
UGDS_BLACK    0.12
UGDS_HISP     0.03
UGDS_ASIAN    0.00
UGDS_AIAN     0.01
UGDS_NHPI     0.00
UGDS_2MOR     0.00
UGDS_NRA      0.00
UGDS_UNKN     0.03
Name: Northwest-Shoals Community College, dtype: float64
```

如果在舍入之前先加.0001，则舍入结果会发生变化。

```
>>> (college_ugds.loc[name] + 0.0001).round(2)
UGDS_WHITE    0.79
UGDS_BLACK    0.13
UGDS_HISP     0.03
UGDS_ASIAN    0.00
UGDS_AIAN     0.01
UGDS_NHPI     0.00
UGDS_2MOR     0.00
UGDS_NRA      0.00
UGDS_UNKN     0.03
Name: Northwest-Shoals Community College, dtype: float64
```

（2）我们可以对 DataFrame 执行此操作。要开始使用运算符执行舍入计算，可以首先将.00501 添加到 college_ugds 的每个值中。

```
>>> college_ugds + 0.00501
               UGDS_WHITE   UGDS_BLACK  ...  UGDS_NRA   UGDS_UNKN
INSTNM                                  ...
Alabama A...      0.03831      0.94031  ...   0.01091     0.01881
Universit...      0.59721      0.26501  ...   0.02291     0.01501
```

```
Amridge U...     0.30401    0.42421    ...    0.00501    0.27651
Universit...     0.70381    0.13051    ...    0.03821    0.04001
Alabama S...     0.02081    0.92581    ...    0.02931    0.01871
       ...           ...        ...    ...        ...        ...
SAE Insti...         NaN        NaN    ...        NaN        NaN
Rasmussen...         NaN        NaN    ...        NaN        NaN
National ...         NaN        NaN    ...        NaN        NaN
Bay Area ...         NaN        NaN    ...        NaN        NaN
Excel Lea...         NaN        NaN    ...        NaN        NaN
```

（3）使用向下取整除法运算符（//）向下舍入到最接近的整数百分比。

```
>>> (college_ugds + 0.00501) // 0.01
              UGDS_WHITE   UGDS_BLACK   ...   UGDS_NRA   UGDS_UNKN
INSTNM                                  ...
Alabama A...         3.0         94.0   ...        1.0         1.0
Universit...        59.0         26.0   ...        2.0         1.0
Amridge U...        30.0         42.0   ...        0.0        27.0
Universit...        70.0         13.0   ...        3.0         4.0
Alabama S...         2.0         92.0   ...        2.0         1.0
       ...          ...          ...    ...        ...         ...
SAE Insti...         NaN          NaN   ...        NaN         NaN
Rasmussen...         NaN          NaN   ...        NaN         NaN
National ...         NaN          NaN   ...        NaN         NaN
Bay Area ...         NaN          NaN   ...        NaN         NaN
Excel Lea...         NaN          NaN   ...        NaN         NaN
```

（4）要完成舍入运算，需要除以 100。

```
>>> college_ugds_op_round = (
...     (college_ugds + 0.00501) // 0.01 / 100
... )
>>> college_ugds_op_round.head()
              UGDS_WHITE   UGDS_BLACK   ...   UGDS_NRA   UGDS_UNKN
INSTNM                                  ...
Alabama A...        0.03         0.94   ...       0.01        0.01
Universit...        0.59         0.26   ...       0.02        0.01
Amridge U...        0.30         0.42   ...       0.00        0.27
Universit...        0.70         0.13   ...       0.03        0.04
Alabama S...        0.02         0.92   ...       0.02        0.01
```

（5）使用 DataFrame 方法 round 自动进行舍入。由于 Pandas 采用的是 Bankers 舍入规则，因此在舍入前可以添加一个小数.00001。

```
>>> college_ugds_round = (college_ugds + 0.00001).round(2)
>>> college_ugds_round
                UGDS_WHITE    UGDS_BLACK  ...  UGDS_NRA   UGDS_UNKN
INSTNM                                    ...
Alabama A...       0.03          0.94     ...    0.01        0.01
Universit...       0.59          0.26     ...    0.02        0.01
Amridge U...       0.30          0.42     ...    0.00        0.27
Universit...       0.70          0.13     ...    0.03        0.04
Alabama S...       0.02          0.92     ...    0.02        0.01
    ...             ...           ...     ...     ...         ...
SAE Insti...       NaN           NaN      ...    NaN         NaN
Rasmussen...       NaN           NaN      ...    NaN         NaN
National ...       NaN           NaN      ...    NaN         NaN
Bay Area ...       NaN           NaN      ...    NaN         NaN
Excel Lea...       NaN           NaN      ...    NaN         NaN
```

（6）使用 DataFrame 方法 equals 测试两个 DataFrame 的相等性。

```
>>> college_ugds_op_round.equals(college_ugds_round)
True
```

2.7.2 原理解释

步骤（1）和步骤（2）均使用了加法运算符，该运算符尝试将标量值添加到 DataFrame 每列的每个值上。由于列都是数字，因此此操作可以按预期进行。每列中都有一些缺失值，但是在操作后这些缺失值将保持缺失。

从数学上讲，添加 .005 应该足够让下一步中的向下取整除法正确舍入到最接近的整数百分比。但是，这也可能由于浮点数的不精确性而出现问题。

```
>>> 0.045 + 0.005
0.04999999999999996
```

这就是为什么要在舍入之前对每个数字都加 .00001，以确保浮点表示的前四位数字与实际值相同。这样之所以可行，是因为数据集中所有点的最大精度是 4 个小数位。

步骤（3）将向下取整除法运算符（//）应用于 DataFrame 的所有值中。实际上，当我们使用小数除时，取整除法运算符会将每个值乘以 100 并舍去任何小数。在表达式的第一部分周围需要使用括号，这是因为向下取整除法的优先级高于加法。步骤（4）使用了除法运算符将小数返回正确的位置处。

在步骤（5）中，我们使用 round 方法重现了先前的步骤。在执行此操作之前，由于与步骤（2）不同的原因，我们必须再次向每个 DataFrame 值中添加一个额外的 .00001。

NumPy 和 Python 3 的舍入规则就是上面介绍过的 Bankers 舍入规则,该规则通常不是学校正式教授的规则(学校教授的规则一般是四舍五入)。

在本示例中,有必要舍入以使两个 DataFrame 值相等。.equals 方法将确定两个 DataFrame 之间的所有元素和索引是否完全相同,并返回一个布尔值。

2.7.3 扩展知识

与 Series 一样,DataFrame 也具有与运算符等效的方法。分析人员可以将运算符替换为其等效的方法。

```
>>> college2 = (
...     college_ugds.add(0.00501).floordiv(0.01).div(100)
... )
>>> college2.equals(college_ugds_op_round)
True
```

2.8 比较缺失值

Pandas 使用 NumPy NaN(np.nan)对象表示缺失值。这是一个不寻常的对象,并且具有非常有趣的数学特性。例如,np.nan 对象不等于自身。甚至 Python 的 None 对象在与自身进行比较时被评估为 True。

```
>>> np.nan == np.nan
False
>>> None == None
True
```

与 np.nan 对象进行的所有比较也返回 False,当然,不等于(!=)除外。

```
>>> np.nan > 5
False
>>> 5 > np.nan
False
>>> np.nan != 5
True
```

2.8.1 做好准备

Series 和 DataFrame 可以使用 equals 运算符(==)进行逐元素的比较。结果是一个包含相同维度的对象。此秘笈将展示如何使用 equals 运算符,该运算符与.equals 方法有很大的不同。

与前面的秘笈一样,我们将使用代表 college 数据集中各种族本科学生的百分比数的列。

```
>>> college = pd.read_csv(
...     "data/college.csv", index_col="INSTNM"
... )
>>> college_ugds = college.filter(like="UGDS_")
```

2.8.2 实战操作

(1)为了理解 equals 运算符的工作原理,可以将每个元素与一个标量值进行比较。

```
>>> college_ugds == 0.0019
              UGDS_WHITE  UGDS_BLACK  ...  UGDS_NRA  UGDS_UNKN
INSTNM                                ...
Alabama A...       False       False  ...     False      False
Universit...       False       False  ...     False      False
Amridge U...       False       False  ...     False      False
Universit...       False       False  ...     False      False
Alabama S...       False       False  ...     False      False
...                  ...         ...  ...       ...        ...
SAE Insti...       False       False  ...     False      False
Rasmussen...       False       False  ...     False      False
National ...       False       False  ...     False      False
Bay Area ...       False       False  ...     False      False
Excel Lea...       False       False  ...     False      False
```

(2)这可以按预期工作,但是当你尝试比较包含缺失值的 DataFrame 时,就会出现问题。对此,你可能很想使用 equals 运算符来逐个元素地比较两个 DataFrame。例如,使用 college_ugds 与自身进行比较。

```
>>> college_self_compare = college_ugds == college_ugds
>>> college_self_compare.head()
              UGDS_WHITE  UGDS_BLACK  ...  UGDS_NRA  UGDS_UNKN
INSTNM                                ...
Alabama A...        True        True  ...      True       True
Universit...        True        True  ...      True       True
Amridge U...        True        True  ...      True       True
Universit...        True        True  ...      True       True
Alabama S...        True        True  ...      True       True
```

(3) 乍一看, 所有值都与你期望的相同。但是, 使用 .all 方法确定每列是否仅包含 True 值, 则会产生意外的结果。

```
>>> college_self_compare.all()
UGDS_WHITE    False
UGDS_BLACK    False
UGDS_HISP     False
UGDS_ASIAN    False
UGDS_AIAN     False
UGDS_NHPI     False
UGDS_2MOR     False
UGDS_NRA      False
UGDS_UNKN     False
dtype: bool
```

(4) 发生这种情况是因为缺失值之间彼此进行比较时, 它们是不相等的。如果你尝试使用 equals 运算符对缺失值进行计数并对布尔值列求和, 则每列的计数总和都将得到 0。

```
>>> (college_ugds == np.nan).sum()
UGDS_WHITE    0
UGDS_BLACK    0
UGDS_HISP     0
UGDS_ASIAN    0
UGDS_AIAN     0
UGDS_NHPI     0
UGDS_2MOR     0
UGDS_NRA      0
UGDS_UNKN     0
dtype: int64
```

(5) 另外, 如果使用 .isna 方法代替 equals 运算符查找缺失值, 则结果将完全不同。

```
>>> college_ugds.isna().sum()
UGDS_WHITE    661
UGDS_BLACK    661
UGDS_HISP     661
UGDS_ASIAN    661
UGDS_AIAN     661
UGDS_NHPI     661
UGDS_2MOR     661
UGDS_NRA      661
UGDS_UNKN     661
dtype: int64
```

（6）比较两个完整的 DataFrame 的正确方法是使用 .equals 方法，而非使用 equals 运算符。.equals 方法可以将相同位置的 NaN 视为相等。

```
>>> college_ugds.equals(college_ugds)
True
```

值得一提的是，.eq 方法等效于 equals 运算符。

2.8.3 原理解释

步骤（1）将一个 DataFrame 与一个标量值进行了比较，而步骤（2）则将一个 DataFrame 与另一个 DataFrame 进行了比较。乍看之下，这两种操作都非常简单直观。步骤（2）的操作是检查 DataFrame 是否具有相同标签的索引，从而具有相同数量的元素；如果不是这种情况，则操作将失败。

步骤（3）验证了 DataFrame 中的列均不相等。
步骤（4）进一步显示了 np.nan 及其自身的不相等性。
步骤（5）验证了 DataFrame 中确实存在缺失值。
最后，步骤（6）显示了使用 .equals 方法比较 DataFrame 的正确方法，该方法将始终返回一个布尔标量值。

2.8.4 扩展知识

所有比较运算符都有对应的方法，可以使用更多功能。容易令人搞混的是，DataFrame 的 .eq 方法可以进行逐元素的比较，就像 equals 运算符一样。因此，.eq 方法与 .equals 方法是完全不同的。以下代码重复了步骤（1）。

```
>>> college_ugds.eq(0.0019) # same as college_ugds == .0019
              UGDS_WHITE  UGDS_BLACK  ...  UGDS_NRA  UGDS_UNKN
INSTNM                                ...
Alabama A...  False       False       ...  False     False
Universit...  False       False       ...  False     False
Amridge U...  False       False       ...  False     False
Universit...  False       False       ...  False     False
Alabama S...  False       False       ...  False     False
...           ...         ...         ...  ...       ...
SAE Insti...  False       False       ...  False     False
Rasmussen...  False       False       ...  False     False
National ...  False       False       ...  False     False
```

Bay Area ...	False		False	...	False	False
Excel Lea...	False		False	...	False	False

在 pandas.testing 子程序包中，在创建单元测试时，存在一个分析人员可以使用的函数。如果两个 DataFrame 不相等，则 assert_frame_equal 函数将引发 AssertionError；如果两个 DataFrame 相等，则返回 None。

```
>>> from pandas.testing import assert_frame_equal
>>> assert_frame_equal(college_ugds, college_ugds) is None
True
```

单元测试是软件开发中非常重要的部分，它将确保代码正确运行。Pandas 包含成千上万的单元测试，有助于确保其正常运行。要了解有关 Pandas 运行其单元测试的更多信息，可参考以下说明文档。

http://bit.ly/2vmCSU6

2.9 转置 DataFrame 操作的方向

许多 DataFrame 方法都有一个 axis 参数，此参数可控制操作发生的方向。此外，axis 参数可以是 'index'（或 0）或 'columns'（或 1）。笔者更喜欢使用字符串值，因为这些字符串值更明确，并且使代码更易于阅读。

几乎所有 DataFrame 方法都将 axis 参数默认为 0，该参数适用于沿着索引的操作。此秘笈将展示如何沿着两个轴调用相同的方法。

2.9.1 实战操作

（1）读取 college 数据集。以 UGDS 开头的列代表特定种族的本科生百分比。使用过滤器（filter）方法选择以下列。

```
>>> college = pd.read_csv(
...     "data/college.csv", index_col="INSTNM"
... )
>>> college_ugds = college.filter(like="UGDS_")
>>> college_ugds.head()
              UGDS_WHITE   UGDS_BLACK  ...  UGDS_NRA   UGDS_UNKN
INSTNM                                 ...
Alabama A...      0.0333       0.9353  ...    0.0059      0.0138
Universit...      0.5922       0.2600  ...    0.0179      0.0100
```

```
Amridge U...      0.2990    0.4192   ...   0.0000    0.2715
Universit...      0.6988    0.1255   ...   0.0332    0.0350
Alabama S...      0.0158    0.9208   ...   0.0243    0.0137
```

（2）现在，DataFrame 包含同类型的列数据，因此可以在垂直和水平方向上合理地进行操作。.count 方法可以返回非缺失值的数量。默认情况下，其 axis 参数被设置为 0。

```
>>> college_ugds.count()
UGDS_WHITE    6874
UGDS_BLACK    6874
UGDS_HISP     6874
UGDS_ASIAN    6874
UGDS_AIAN     6874
UGDS_NHPI     6874
UGDS_2MOR     6874
UGDS_NRA      6874
UGDS_UNKN     6874
dtype: int64
```

axis 参数总是被设置为 0。因此，步骤（2）等效于 college_ugds.count(axis=0) 和 college_ugds.count(axis='index')。

（3）将 axis 参数更改为 'columns' 会改变操作的方向，这样就可以获取每行中非缺失值的计数。

```
>>> college_ugds.count(axis="columns").head()
INSTNM
Alabama A & M University                 9
University of Alabama at Birmingham      9
Amridge University                       9
University of Alabama in Huntsville      9
Alabama State University                 9
dtype: int64
```

（4）我们可以对每行中的所有值求和，而不是计算非缺失值。每行的百分比数加起来应该为 1。.sum 方法可用于验证这一点。

```
>>> college_ugds.sum(axis="columns").head()
INSTNM
Alabama A & M University                 1.0000
University of Alabama at Birmingham      0.9999
Amridge University                       1.0000
University of Alabama in Huntsville      1.0000
```

```
Alabama State University              1.0000
dtype: float64
```

（5）要了解每一列的分布，可以使用.median 方法。

```
>>> college_ugds.median(axis="index")
UGDS_WHITE   0.55570
UGDS_BLACK   0.10005
UGDS_HISP    0.07140
UGDS_ASIAN   0.01290
UGDS_AIAN    0.00260
UGDS_NHPI    0.00000
UGDS_2MOR    0.01750
UGDS_NRA     0.00000
UGDS_UNKN    0.01430
dtype: float64
```

2.9.2 原理解释

轴上的操作方向是 Pandas 比较混乱的方面之一。许多 Pandas 用户很难记住 axis 参数的含义。这里，笔者通过提醒自己，一个 Series 只有一个轴，这个轴就是索引（或 0），以此记住 axis 参数的含义。DataFrame 除了有索引（轴 0）外，还有列（轴 1）。

2.9.3 扩展知识

在使用.cumsum 方法时，设置 axis=1 参数可以跨越每行累计（accumulate）本科生的种族百分比。.cumsum 方法给出的数据视图略有不同。例如，.cumsum 方法很容易看到每所学校的白人和黑人学生的确切百分比。

```
>>> college_ugds_cumsum = college_ugds.cumsum(axis=1)
>>> college_ugds_cumsum.head()
              UGDS_WHITE   UGDS_BLACK   ...   UGDS_NRA   UGDS_UNKN
INSTNM                                  ...
Alabama A...     0.0333       0.9686    ...    0.9862     1.0000
Universit...     0.5922       0.8522    ...    0.9899     0.9999
Amridge U...     0.2990       0.7182    ...    0.7285     1.0000
Universit...     0.6988       0.8243    ...    0.9650     1.0000
Alabama S...     0.0158       0.9366    ...    0.9863     1.0000
```

2.10 确定大学校园的多样性

每年都会有很多文章,讨论多样性对大学校园的不同方面的影响。各种组织已经开发了不同的度量标准,试图度量这种多样性。

US news(美国新闻)为许多不同类别的大学提供了排名,其中一个指标就是多样性(diversity)。美国新闻的多样性指数排名前 10 位的大学如下。

```
>>> pd.read_csv(
...     "data/college_diversity.csv", index_col="School"
... )
```

	Diversity Index
School	
Rutgers University--Newark Newark, NJ	0.76
Andrews University Berrien Springs, MI	0.74
Stanford University Stanford, CA	0.74
University of Houston Houston, TX	0.74
University of Nevada--Las Vegas Las Vegas, NV	0.74
University of San Francisco San Francisco, CA	0.74
San Francisco State University San Francisco, CA	0.73
University of Illinois--Chicago Chicago, IL	0.73
New Jersey Institute of Technology Newark, NJ	0.72
Texas Woman's University Denton, TX	0.72

college 数据集将种族分为 9 个不同的类别。当尝试对没有明显定义的事物(如多样性)进行量化时,进行一些简化会更好。在此秘笈中,多样性指标将等于学生人数超过 15%的种族数。

2.10.1 实战操作

(1)读取 college 数据集,并仅针对本科生的种族列进行过滤(种族列是包含 UGDS_前缀的列,所以可指定 like="UGDS_")。

```
>>> college = pd.read_csv(
...     "data/college.csv", index_col="INSTNM"
... )
>>> college_ugds = college.filter(like="UGDS_")
```

（2）这些大学中有许多都在种族列中包含缺失值。可以针对每一行计算种族的所有缺失值，并对所得的 Series 从最高到最低进行排序。这将揭示出那些包含缺失值的大学。

```
>>> (
...     college_ugds.isnull()
...     .sum(axis="columns")
...     .sort_values(ascending=False)
...     .head()
... )
INSTNM
Excel Learning Center-San Antonio South          9
Philadelphia College of Osteopathic Medicine     9
Assemblies of God Theological Seminary           9
Episcopal Divinity School                        9
Phillips Graduate Institute                      9
dtype: int64
```

（3）种族列（包含 UGDS_前缀的列）总共才 9 列，因此可以看到有些大学在其所有种族列中都包含缺失值，可以使用.dropna 方法删除在所有 9 个种族列中都包含缺失值的所有行，然后来看缺失值的情况。

```
>>> college_ugds = college_ugds.dropna(how="all")
>>> college_ugds.isnull().sum()
UGDS_WHITE    0
UGDS_BLACK    0
UGDS_HISP     0
UGDS_ASIAN    0
UGDS_AIAN     0
UGDS_NHPI     0
UGDS_2MOR     0
UGDS_NRA      0
UGDS_UNKN     0
dtype: int64
```

（4）在 college 数据集中已经没有缺失值了。现在，我们可以计算多样性指标。

首先，我们将使用 DataFrame 的大于或等于方法（.ge）为每个单元格返回一个带有布尔值的 DataFrame。

```
>>> college_ugds.ge(0.15)
        UGDS_WHITE   UGDS_BLACK  ...  UGDS_NRA   UGDS_UNKN
INSTNM                           ...
```

```
Alabama A...       False      True    ...    False      False
Universit...       True       True    ...    False      False
Amridge U...       True       True    ...    False      True
Universit...       True       False   ...    False      False
Alabama S...       False      True    ...    False      False
      ...           ...       ...     ...     ...        ...
Hollywood...       True       True    ...    False      False
Hollywood...       False      True    ...    False      False
Coachella...       True       False   ...    False      False
Dewey Uni...       False      False   ...    False      False
Coastal P...       True       True    ...    False      False
```

（5）在这里，我们可以使用 .sum 方法为每所大学计算 True 值。请注意，这里返回的是一个 Series。

```
>>> diversity_metric = college_ugds.ge(0.15).sum(
...      axis="columns"
... )
>>> diversity_metric.head()
INSTNM
Alabama A & M University                   1
University of Alabama at Birmingham        2
Amridge University                         3
University of Alabama in Huntsville        1
Alabama State University                   1
dtype: int64
```

（6）要了解分布情况，可以在此 Series 上使用 .value_counts 方法。

```
>>> diversity_metric.value_counts()
1    3042
2    2884
3     876
4      63
0       7
5       2
dtype: int64
```

（7）令人惊讶的是，有两所学校在 5 个不同种族类别中的比例超过 15%。因此，可以对 diversity_metric 这个 Series 进行排序，以找出它们是哪些学校。

```
>>> diversity_metric.sort_values(ascending=False).head()
```

```
INSTNM
Regency Beauty Institute-Austin         5
Central Texas Beauty College-Temple     5
Sullivan and Cogliano Training Center   4
Ambria College of Nursing               4
Berkeley College-New York               4
dtype: int64
```

（8）学校可以如此多样化似乎有点可疑。下面查看这两所排名最高学校的原始百分比。这里将使用.loc 属性根据索引标签选择行。

```
>>> college_ugds.loc[
...     [
...         "Regency Beauty Institute-Austin",
...         "Central Texas Beauty College-Temple",
...     ]
... ]
              UGDS_WHITE  UGDS_BLACK  ...  UGDS_NRA  UGDS_UNKN
INSTNM                                ...
Regency  B...     0.1867      0.2133  ...       0.0     0.2667
Central  T...     0.1616      0.2323  ...       0.0     0.1515
```

（9）似乎有若干个种族聚合到 UNKN（未知）和 2MOR（两个或多个种族）列中。无论如何，它们看起来确实富于多样性。我们可以看看美国新闻排名前 5 的学校在多样性指标上的表现如何。

```
>>> us_news_top = [
...     "Rutgers University-Newark",
...     "Andrews University",
...     "Stanford University",
...     "University of Houston",
...     "University of Nevada-Las Vegas",
... ]
>>> diversity_metric.loc[us_news_top]
INSTNM
Rutgers University-Newark         4
Andrews University                3
Stanford University               3
University of Houston             3
University of Nevada-Las Vegas    3
dtype: int64
```

2.10.2 原理解释

步骤（2）对缺失值最多的学校进行了统计和显示。由于 DataFrame 中有 9 列，因此每所学校的缺失值最大数目为 9。可以看到，许多学校在每列中都包含缺失值。

步骤（3）删除了所有值均缺失的行。该步骤中的.dropna 方法具有 how 参数，该参数默认为字符串'any'，但也可以将其更改为'all'。当将 how 参数设置为'any'时，.dropna 方法将删除包含一个或多个缺失值的行；当将 how 参数设置为'all'时，.dropna 方法仅删除缺少所有值的行。

在本示例中，我们保守地删除了所有值都缺失的行。这是因为某些缺失值可能代表0%。在这里并没有这种情况，因为在执行.dropna 方法之后没有丢失任何值。如果仍然有缺失值，则可以运行.fillna(0)方法以用 0 填充所有余下的值。

步骤（5）使用了大于或等于方法（.ge）执行多样性指标的计算。这将产生全部包含布尔值的 DataFrame，可以通过设置 axis='columns' 进行横向求和。

在步骤（6）中，.value_counts 方法用于生成多样性指标的分布。对于学校来说，很少有 3 个种族的本科生人数占总人数的比例均超过 15%或更多。

步骤（7）和步骤（8）根据指标找到了两所最多样化的学校。尽管这两所学校的多样性指数排名前 2，但这似乎只是由于许多种族并没有得到明确的申报，因为其中两个超过 15%的种族是 UNKN（未知）和 2MOR（两个或多个种族）列。

步骤（9）从美国新闻的文章中选择了排名前 5 的学校，然后从新创建的 Series 中选择了其多样性指标。事实证明，这些学校在我们的简单排名系统中得分也很高。

2.10.3 扩展知识

作为一种替代方法，我们还可以通过按最大种族百分比对学校进行排序来找到多样性最差的学校。

```
>>> (
...     college_ugds.max(axis=1)
...     .sort_values(ascending=False)
...     .head(10)
... )
INSTNM
Dewey University-Manati                         1.0
Yeshiva and Kollel Harbotzas Torah              1.0
```

```
Mr Leon's School of Hair Design-Lewiston              1.0
Dewey University-Bayamon                              1.0
Shepherds Theological Seminary                        1.0
Yeshiva Gedolah Kesser Torah                          1.0
Monteclaro Escuela de Hoteleria y Artes Culinarias    1.0
Yeshiva Shaar Hatorah                                 1.0
Bais Medrash Elyon                                    1.0
Yeshiva of Nitra Rabbinical College                   1.0
dtype: float64
```

我们还可以确定是否有任何一所学校的 9 个种族类别都超过 1%。

```
>>> (college_ugds > 0.01).all(axis=1).any()
True
```

第 3 章　创建和保留 DataFrame

本章包含以下秘笈。
- 从头开始创建 DataFrame。
- 编写 CSV。
- 读取大型 CSV 文件。
- 使用 Excel 文件。
- 使用 ZIP 文件。
- 与数据库协同工作。
- 读取 JSON。
- 读取 HTML 表格。

3.1　介　　绍

在实际操作中，有多种创建 DataFrame 的方法。本章将讨论一些最常见的操作。另外，本章还将演示如何保留 DataFrame。

3.2　从头开始创建 DataFrame

一般来说，分析人员可以从现有文件或数据库中创建一个 DataFrame，当然，也可以从头开始创建一个 DataFrame。此秘笈将介绍如何从并行数据列表中创建一个 DataFrame。

3.2.1　实战操作

（1）创建包含数据的并行列表。这些列表中的每个都将是 DataFrame 中的一列，因此它们应具有相同的类型。

```
>>> import pandas as pd
>>> import numpy as np
>>> fname = ["Paul", "John", "Richard", "George"]
>>> lname = ["McCartney", "Lennon", "Starkey", "Harrison"]
```

```
>>> birth = [1942, 1940, 1940, 1943]
```

（2）从列表中创建字典，将列名称映射到列表中。

```
>>> people = {"first": fname, "last": lname, "birth": birth}
```

（3）从字典中创建一个 DataFrame。

```
>>> beatles = pd.DataFrame(people)
>>> beatles
    first     last  birth
0    Paul  McCartney   1942
1    John     Lennon   1940
2  Richard    Starkey   1940
3   George   Harrison   1943
```

3.2.2 原理解释

默认情况下，当调用构造函数时，Pandas 将为 DataFrame 创建一个 RangeIndex。

```
>>> beatles.index
RangeIndex(start=0, stop=4, step=1)
```

当然，如果需要，也可以为 DataFrame 指定其他索引。

```
>>> pd.DataFrame(people, index=["a", "b", "c", "d"])
    first     last  birth
a    Paul  McCartney   1942
b    John     Lennon   1940
c  Richard    Starkey   1940
d   George   Harrison   1943
```

3.2.3 扩展知识

还可以从词典列表中创建一个 DataFrame。

```
>>> pd.DataFrame(
...     [
...         {
...             "first": "Paul",
...             "last": "McCartney",
...             "birth": 1942,
...         },
...         {
```

```
...             "first": "John",
...             "last": "Lennon",
...             "birth": 1940,
...         },
...         {
...             "first": "Richard",
...             "last": "Starkey",
...             "birth": 1940,
...         },
...         {
...             "first": "George",
...             "last": "Harrison",
...             "birth": 1943,
...         },
...     ]
... )
   birth    first      last
0   1942     Paul  McCartney
1   1940     John     Lennon
2   1940  Richard    Starkey
3   1943   George   Harrison
```

请注意，当使用字典作为行时，列是按字典键的字母顺序排序的。例如，在上面的示例中，birth 在字典中是第 3 个键，而在 DataFrame 中则被排在前面。如果这个顺序很重要，则可以使用 columns 参数指定列顺序。

```
>>> pd.DataFrame(
...     [
...         {
...             "first": "Paul",
...             "last": "McCartney",
...             "birth": 1942,
...         },
...         {
...             "first": "John",
...             "last": "Lennon",
...             "birth": 1940,
...         },
...         {
...             "first": "Richard",
...             "last": "Starkey",
...             "birth": 1940,
...         },
```

```
...            {
...                "first": "George",
...                "last": "Harrison",
...                "birth": 1943,
...            },
...        ]
...        columns=["last", "first", "birth"],
... )
        last      first   birth
0   McCartney      Paul    1942
1      Lennon      John    1940
2     Starkey   Richard    1940
3    Harrison    George    1943
```

3.3 编写 CSV

逗号分隔值（comma-separated value，CSV）文件以纯文本形式存储表格数据（数字和文本）。无论你对它的评价如何，这个世界上都有很多的 CSV 文件。像大多数技术一样，CSV 文件也有其优缺点。优点是，CSV 文件是人类可读的，可以在任何文本编辑器中将其打开，并且大多数电子表格软件都可以对其进行加载；缺点是，CSV 文件没有标准，虽然其名称是逗号分隔值，而实际上其分隔字符也可以非逗号，因此其编码可能很奇怪，无法强制执行类型，并且 CSV 文件可能很大，因为此类文件是基于文本的（尽管可以压缩）。

在本秘笈中，我们将展示如何从 Pandas DataFrame 中创建 CSV 文件。

DataFrame 有若干个以 to_ 开头的方法。这些是导出 DataFrame 的方法。我们将使用 .to_csv 方法。请注意，在本秘笈的示例中，我们会将写入信息输出到一个字符串缓冲区中，但是在实际操作中，通常用文件名替代。

3.3.1 实战操作

（1）将 DataFrame 写入 CSV 文件中。

```
>>> beatles
     first        last  birth
0     Paul   McCartney   1942
1     John      Lennon   1940
2  Richard     Starkey   1940
3   George    Harrison   1943
```

```
>>> from io import StringIO
>>> fout = StringIO()
>>> beatles.to_csv(fout)        # 实际操作中一般将 fout 替换为文件名
```

（2）查看输出内容。

```
>>> print(fout.getvalue())
,first,last,birth
0,Paul,McCartney,1942
1,John,Lennon,1940
2,Richard,Starkey,1940
3,George,Harrison,1943
```

3.3.2 扩展知识

.to_csv 方法包含一些选项。可以注意到，.to_csv 方法在输出中包括了索引，但没有给索引指定列名称。如果使用 read_csv 函数将此 CSV 文件读取到 DataFrame 中，则默认情况下 read_csv 函数不会将其用作索引。相反，你将获得一个名为 Unnamed:0 的列作为索引，因此出现了多余的列。

```
>>> _ = fout.seek(0)
>>> pd.read_csv(fout)
   Unnamed: 0    first       last  birth
0           0     Paul  McCartney   1942
1           1     John     Lennon   1940
2           2  Richard    Starkey   1940
3           3   George   Harrison   1943
```

read_csv 函数具有一个 index_col 参数，可用于指定索引的位置。

```
>>> _ = fout.seek(0)
>>> pd.read_csv(fout, index_col=0)
     first       last  birth
0     Paul  McCartney   1942
1     John     Lennon   1940
2  Richard    Starkey   1940
3   George   Harrison   1943
```

另外，如果在编写 CSV 文件时不想包含索引，则可以将 index 参数设置为 False。

```
>>> fout = StringIO()
>>> beatles.to_csv(fout, index=False)
>>> print(fout.getvalue())
```

```
first,last,birth
Paul,McCartney,1942
John,Lennon,1940
Richard,Starkey,1940
George,Harrison,1943
```

3.4 读取大型 CSV 文件

Pandas 库是一个内存中（in-memory）工具。你需要能够将数据容纳在内存中，才能与 Pandas 一起使用。如果遇到要处理的大型 CSV 文件，则有几种选择。如果一次可以处理其中的一部分，则可以将其读取成块（chunk）并处理每个块。另外，如果你知道应该有足够的内存来加载文件，则也可以通过一些技巧来帮助减少文件大小。

请注意，一般来说，内存量应为要处理的 DataFrame 大小的 3～10 倍。额外的内存应提供足够的额外空间来执行许多常见操作。

3.4.1 实战操作

在本节中，我们将查看一个 diamonds（钻石）数据集。将该数据集可以轻松放入我们的 MacBook 笔记本电脑的内存中，但是我们会假装该文件比实际大小大很多，或者我们的计算机内存有限，这样当 Pandas 尝试使用 read_csv 函数加载该文件时，我们就会得到内存溢出的错误提示。

（1）确定整个文件将占用多少内存。我们将使用 read_csv 函数的 nrows 参数来限制向一个小样本中加载的数据量。

```
>>> diamonds = pd.read_csv("data/diamonds.csv", nrows=1000)
>>> diamonds
     carat      cut color clarity  ...  price     x     y     z
0     0.23    Ideal     E     SI2  ...    326  3.95  3.98  2.43
1     0.21  Premium     E     SI1  ...    326  3.89  3.84  2.31
2     0.23     Good     E     VS1  ...    327  4.05  4.07  2.31
3     0.29  Premium     I     VS2  ...    334  4.20  4.23  2.63
4     0.31     Good     J     SI2  ...    335  4.34  4.35  2.75
..     ...      ...   ...     ...  ...    ...   ...   ...   ...
995   0.54    Ideal     D    VVS2  ...   2897  5.30  5.34  3.26
996   0.72    Ideal     E     SI1  ...   2897  5.69  5.74  3.57
997   0.72     Good     F     VS1  ...   2897  5.82  5.89  3.48
998   0.74  Premium     D     VS2  ...   2897  5.81  5.77  3.58
999   1.12  Premium     J     SI2  ...   2898  6.68  6.61  4.03
```

（2）使用.info 方法查看数据样本使用了多少内存。

```
>>> diamonds.info()
<class 'pandas.core.frame.DataFrame'>
RangeIndex: 1000 entries, 0 to 999
Data columns (total 10 columns):
carat      1000 non-null float64
cut        1000 non-null object
color      1000 non-null object
clarity    1000 non-null object
depth      1000 non-null float64
table      1000 non-null float64
price      1000 non-null int64
x          1000 non-null float64
y          1000 non-null float64
z          1000 non-null float64
dtypes: float64(6), int64(1), object(3)
memory usage: 78.2+ KB
```

可以看到，这里的 1000 行数据使用了大约 78.2KB 的内存。如果有 10 亿行的话，那将需要大约 78 GB 的内存。目前高端计算机的主流内存配置大约为 64GB，所以只能考虑在云中租用具有如此大内存的计算机。当然，我们也可以看看是否能够将其内存需求减少一些。

（3）使用 dtype 参数通知 read_csv 方法使用正确的（或更小的）数字类型。

```
>>> diamonds2 = pd.read_csv(
...     "data/diamonds.csv",
...     nrows=1000,
...     dtype={
...         "carat": np.float32,
...         "depth": np.float32,
...         "table": np.float32,
...         "x": np.float32,
...         "y": np.float32,
...         "z": np.float32,
...         "price": np.int16,
...     },
... )

>>> diamonds2.info()
<class 'pandas.core.frame.DataFrame'>
RangeIndex: 1000 entries, 0 to 999
```

```
Data columns (total 10 columns):
carat      1000 non-null float32
cut        1000 non-null object
color      1000 non-null object
clarity    1000 non-null object
depth      1000 non-null float32
table      1000 non-null float32
price      1000 non-null int16
x          1000 non-null float32
y          1000 non-null float32
z          1000 non-null float32
dtypes: float32(6), int16(1), object(3)
memory usage: 49.0+ KB
```

确保新数据集与原始数据集的摘要统计信息相似。

```
>>> diamonds.describe()
            carat          depth    ...            y            z
count  1000.000000    1000.000000   ...  1000.000000  1000.000000
mean      0.689280      61.722800   ...     5.599180     3.457530
std       0.195291       1.758879   ...     0.611974     0.389819
min       0.200000      53.000000   ...     3.750000     2.270000
25%       0.700000      60.900000   ...     5.630000     3.450000
50%       0.710000      61.800000   ...     5.760000     3.550000
75%       0.790000      62.600000   ...     5.910000     3.640000
max       1.270000      69.500000   ...     7.050000     4.330000


>>> diamonds2.describe()
            carat          depth    ...            y            z
count  1000.000000    1000.000000   ...  1000.000000  1000.000000
mean      0.689453      61.718750   ...     5.601562     3.457031
std       0.195312       1.759766   ...     0.611816     0.389648
min       0.199951      53.000000   ...     3.750000     2.269531
25%       0.700195      60.906250   ...     5.628906     3.449219
50%       0.709961      61.812500   ...     5.761719     3.550781
75%       0.790039      62.593750   ...     5.910156     3.640625
max       1.269531      69.500000   ...     7.050781     4.328125
```

通过更改数字类型，新数据集使用了 49KB 的内存，大约仅为原始内存使用量的（49/78.2≈）62%。值得一提的是，新的数据集有一些精度上的损失，这可能会被接收，也可能不被接收。

（4）使用 dtype 参数将对象类型更改为分类对象。首先，检查对象列的 .value_counts 方法。如果这些列的基数（cardinality）较低，则可以将其转换为分类列以节省更多内存。

```
>>> diamonds2.cut.value_counts()
Ideal        333
Premium      290
Very Good    226
Good          89
Fair          62
Name: cut, dtype: int64

>>> diamonds2.color.value_counts()
E    240
F    226
G    139
D    129
H    125
I     95
J     46
Name: color, dtype: int64

>>> diamonds2.clarity.value_counts()
SI1     306
VS2     218
VS1     159
SI2     154
VVS2     62
VVS1     58
I1       29
IF       14
Name: clarity, dtype: int64
```

由于这些列的基数较低，因此可以将其转换为 category（分类）对象类型。

```
>>> diamonds3 = pd.read_csv(
...     "data/diamonds.csv",
...     nrows=1000,
...     dtype={
...         "carat": np.float32,
...         "depth": np.float32,
...         "table": np.float32,
...         "x": np.float32,
...         "y": np.float32,
```

```
...             "z": np.float32,
...             "price": np.int16,
...             "cut": "category",
...             "color": "category",
...             "clarity": "category",
...         },
... )

>>> diamonds3.info()
<class 'pandas.core.frame.DataFrame'>
RangeIndex: 1000 entries, 0 to 999
Data columns (total 10 columns):
carat       1000 non-null float32
cut         1000 non-null category
color       1000 non-null category
clarity     1000 non-null category
depth       1000 non-null float32
table       1000 non-null float32
price       1000 non-null int16
x           1000 non-null float32
y           1000 non-null float32
z           1000 non-null float32
dtypes: category(3), float32(6), int16(1)
memory usage: 29.4 KB
```

可以看到，现在的内存使用量只有29.4KB，大约为原始内存使用量的（29.4/78.2≈）37%。

（5）如果存在可以忽略的列，则可以使用 usecols 参数指定要加载的列。在本示例中，可以忽略 x、y 和 z 列。

```
>>> cols = [
...     "carat",
...     "cut",
...     "color",
...     "clarity",
...     "depth",
...     "table",
...     "price",
... ]
>>> diamonds4 = pd.read_csv(
...     "data/diamonds.csv",
...     nrows=1000,
...     dtype={
...         "carat": np.float32,
```

```
...         "depth": np.float32,
...         "table": np.float32,
...         "price": np.int16,
...         "cut": "category",
...         "color": "category",
...         "clarity": "category",
...     },
...     usecols=cols,
... )

>>> diamonds4.info()
<class 'pandas.core.frame.DataFrame'>
RangeIndex: 1000 entries, 0 to 999
Data columns (total 7 columns):
carat      1000 non-null float32
cut        1000 non-null category
color      1000 non-null category
clarity    1000 non-null category
depth      1000 non-null float32
table      1000 non-null float32
price      1000 non-null int16
dtypes: category(3), float32(3), int16(1)
memory usage: 17.7 KB
```

可以看到，现在内存的使用量进一步降低，只有17.7KB，大约只是原始内存使用量的（17.7/78.2≈）22%。

（6）如果上述步骤仍不足以创建足够小的DataFrame，那么也不用慌，因为我们还有手段，那就是一次处理一些数据块，并且不需要在内存中存储所有数据。该功能可以使用chunksize参数来设置。

```
>>> cols = [
...     "carat",
...     "cut",
...     "color",
...     "clarity",
...     "depth",
...     "table",
...     "price",
... ]
>>> diamonds_iter = pd.read_csv(
...     "data/diamonds.csv",
```

```
...         nrows=1000,
...         dtype={
...             "carat": np.float32,
...             "depth": np.float32,
...             "table": np.float32,
...             "price": np.int16,
...             "cut": "category",
...             "color": "category",
...             "clarity": "category",
...         },
...         usecols=cols,
...         chunksize=200,
... )

>>> def process(df):
...     return f"processed {df.size} items"

>>> for chunk in diamonds_iter:
...     process(chunk)
```

3.4.2 原理解释

由于 CSV 文件不包含有关类型的信息，因此 Pandas 会尝试推断列的类型。如果一列的所有值都是整数，并且都不缺失，则它将使用 int64 类型；如果该列包含的都是数字但不是整数，或者有缺失值，则使用 float64。这些数据类型可能存储你需要的更多信息。例如，如果数字都小于 200，则可以使用较小的类型，如 np.int16（如果数字均为正整数，则还可以使用 np.int8）。

从 Pandas 0.24 开始，有一个新类型'Int64'（请注意区分大小写），它支持包含缺失值的整数类型。如果要使用此类型，则需要使用 dtype 参数指定它，因为 Pandas 默认会将包含缺失值的整数列转换为 float64 类型。

如果该列证明是非数字列，则 Pandas 会将其转换为 object 列，并将值视为字符串。Pandas 中的字符串值会占用大量内存，因为每个值都将被存储为 Python 字符串。因此，如果将它们转换为分类对象，则 Pandas 将只存储一次字符串，而不是为每行创建新的字符串（即使它们是重复的），这样也可以使用更少的内存。

Pandas 库还可以读取在 Internet 上找到的 CSV 文件。你可以将 read_csv 函数直接指向某个 URL。

3.4.3 扩展知识

如果对于 price（价格）列使用 int8 类型，则很可能会丢失信息，因此可以使用 NumPy iinfo 函数列出 NumPy 整数类型的限制。

```
>>> np.iinfo(np.int8)
iinfo(min=-128, max=127, dtype=int8)
```

也可以使用 finfo 函数获取有关浮点数的信息。

```
>>> np.finfo(np.float16)
finfo(resolution=0.001, min=-6.55040e+04,
    max=6.55040e+04, dtype=float16)
```

还可以通过.memory_usage 方法询问 DataFrame 或 Series 使用了多少字节。请注意，这还包括索引的内存要求。另外，需要传递 deep=True 来获取包含 object 数据类型的 Series 的内存使用量。

```
>>> diamonds.price.memory_usage()
8080

>>> diamonds.price.memory_usage(index=False)
8000

>>> diamonds.cut.memory_usage()
8080

>>> diamonds.cut.memory_usage(deep=True)
63413
```

在获得所需格式的数据后，可以将其保存为可跟踪类型的二进制格式，如 Feather 格式（Pandas 可利用 pyarrow 库执行此操作）。此格式旨在实现语言之间的结构化数据在内存中的传输并进行了优化，以便无须内部转换即可直接使用数据。一旦定义了类型，从这种格式中读取就会变得更加快捷和容易。

```
>>> diamonds4.to_feather("d.arr")
>>> diamonds5 = pd.read_feather("d.arr")
```

另一个二进制选项是 Parquet 格式。Feather 针对内存中结构优化了二进制数据，而 Parquet 则是针对磁盘上的格式进行了优化。很多大数据产品都使用 Parquet。Pandas 库也支持 Parquet。

```
>>> diamonds4.to_parquet("/tmp/d.pqt")
```

现在，Pandas 需要进行一些转换才能从 Parquet 和 Feather 中加载数据。但是二者都比 CSV 更快，并且可以持久化类型。

3.5 使用 Excel 文件

尽管 CSV 文件很常见，但是电子表格最常用的工具仍然是 Excel。有很多家公司都将 Excel 用作数据分析和决策的关键工具。

在本秘笈中，我们将展示如何创建和读取 Excel 文件。你可能需要安装 xlwt 或 openpyxl 才能分别写入 XLS 或 XLSX 文件。

3.5.1 实战操作

（1）使用.to_excel 方法创建一个 Excel 文件。你可以写入 xls 或 xlsx 两种格式的文件。xls 是 Excel 早期版本支持的格式，xlsx 则是 Excel 2007 版本之后支持的格式。

```
>>> beatles.to_excel("beat.xls")
>>> beatles.to_excel("beat.xlsx")
```

写入的 beat.xls 文件如图 3-1 所示。

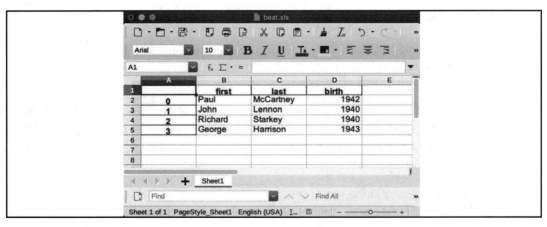

图 3-1 Excel 文件

（2）使用 read_excel 函数读取 Excel 文件。

```
>>> beat2 = pd.read_excel("/tmp/beat.xls")
```

```
>>> beat2
   Unnamed: 0    first       last  birth
0          0     Paul  McCartney   1942
1          1     John     Lennon   1940
2          2  Richard    Starkey   1940
3          3   George   Harrison   1943
```

（3）因为 beat.xls 文件本身已经包含索引列，所以可使用 index_col 参数指定该列。

```
>>> beat2 = pd.read_excel("/tmp/beat.xls", index_col=0)
>>> beat2
     first       last  birth
0     Paul  McCartney   1942
1     John     Lennon   1940
2  Richard    Starkey   1940
3   George   Harrison   1943
```

（4）查看文件的数据类型以检查 Excel 是否保留了这些类型。

```
>>> beat2.dtypes
first    object
last     object
birth     int64
dtype: object
```

3.5.2 原理解释

Python 生态系统具有许多软件包，其中包括读取和写入 Excel 中的功能。此功能已被集成到 Pandas 中，你只需要确保具有适当的库用于读写 Excel 即可。

3.5.3 扩展知识

分析人员可以使用 Pandas 编写电子表格。例如，可以将 sheet_name 参数传递给 .to_excel 方法，以告诉该方法要创建的工作表（sheet）的名称。

```
>>> xl_writer = pd.ExcelWriter("beat2.xlsx")
>>> beatles.to_excel(xl_writer, sheet_name="All")
>>> beatles[beatles.birth < 1941].to_excel(
...     xl_writer, sheet_name="1940"
... )
>>> xl_writer.save()
```

上述文件具有两个工作表：一个工作表的名称为 All，该工作表包含整个 DataFrame；另一个工作表的名称为 1940，该工作表提取 birth 列小于 1941 的数据。

3.6 使用 ZIP 文件

如前文所述，CSV 文件在共享数据中很常见。因为 CSV 文件是纯文本文件，所以该 CSV 文件可能很大。解决 CSV 文件大小问题的一种方法是对其进行压缩。在本秘笈中，我们将研究如何从 ZIP 压缩包中加载文件。

我们将加载一个已经被压缩为 ZIP 格式的 CSV 文件，该 CSV 文件是压缩包中的唯一内容。如果你是在 Mac 系统的 Finder（访达）程序中，则可右击 beatles.csv 文件，然后单击 Compress beatles.csv（压缩 beatles.csv），获得的就是这样的压缩包。本节秘笈还将研究从包含多个文件的 ZIP 包中读取 CSV 文件。

第一个文件来自 fueleconomy.gov 网站。该文件已包含 1984—2018 年在美国市场上提供的所有汽车制造商的列表。

第二个文件是对 Kaggle 网站用户的调查。该文件旨在获取有关用户的信息、背景以及喜欢的工具等。

3.6.1 实战操作

（1）如果 CSV 文件是 ZIP 包中唯一的文件，则可以直接通过该文件调用 read_csv 函数。

```
>>> autos = pd.read_csv("data/vehicles.csv.zip")
>>> autos
       barrels08  barrelsA08  ...  phevHwy  phevComb
0      15.695714         0.0  ...        0         0
1      29.964545         0.0  ...        0         0
2      12.207778         0.0  ...        0         0
3      29.964545         0.0  ...        0         0
4      17.347895         0.0  ...        0         0
...          ...         ...  ...      ...       ...
41139  14.982273         0.0  ...        0         0
41140  14.330870         0.0  ...        0         0
41141  15.695714         0.0  ...        0         0
41142  15.695714         0.0  ...        0         0
41143  18.311667         0.0  ...        0         0

>>> autos.modifiedOn.dtype
dtype('O')
```

（2）要注意的一件事是，如果 CSV 文件中有日期列，那么这些列将作为字符串被保留。可通过两种方式对其进行转换。一种方式是可以使用 read_csv 函数中的 parse_dates 参数并在加载文件时对其进行转换，另一种方式是可以在文件被加载后使用功能更强大的 to_datetime 函数。

```
>>> autos.modifiedOn
0            Tue Jan 01 00:00:00    EST 2013
1            Tue Jan 01 00:00:00    EST 2013
2            Tue Jan 01 00:00:00    EST 2013
3            Tue Jan 01 00:00:00    EST 2013
4            Tue Jan 01 00:00:00    EST 2013
                    ...
39096        Tue Jan 01 00:00:00    EST 2013
39097        Tue Jan 01 00:00:00    EST 2013
39098        Tue Jan 01 00:00:00    EST 2013
39099        Tue Jan 01 00:00:00    EST 2013
39100        Tue Jan 01 00:00:00    EST 2013
Name: modifiedOn, Length: 39101, dtype: object

>>> pd.to_datetime(autos.modifiedOn)
0            2013-01-01
1            2013-01-01
2            2013-01-01
3            2013-01-01
4            2013-01-01
                ...
39096        2013-01-01
39097        2013-01-01
39098        2013-01-01
39099        2013-01-01
39100        2013-01-01
Name: modifiedOn, Length: 39101, dtype: datetime64[ns]
```

以下是在加载文件期间进行转换的代码。

```
>>> autos = pd.read_csv(
...       "data/vehicles.csv.zip", parse_dates=["modifiedOn"]
... )
>>> autos.modifiedOn 0    2013-01-0...
1         2013-01-0...
2         2013-01-0...
3         2013-01-0...
```

```
4          2013-01-0...
              ...
41139      2013-01-0...
41140      2013-01-0...
41141      2013-01-0...
41142      2013-01-0...
41143      2013-01-0...
Name: modifiedOn, Length: 41144, dtype: datetime64[ns, tzlocal()]
```

（3）如果 ZIP 包中包含许多文件，那么从其中读取 CSV 文件会更复杂一些。read_csv 函数无法指定 ZIP 包中的文件。因此，可以考虑使用 Python 标准库中的 zipfile 模块。

建议输出 ZIP 包中的文件名，这样可以轻松查看要选择的文件名。请注意，在本示例中，multipleChoiceResponses.csv 文件的第二行是一个比较长的问题（第一行是一个问题的标识符，可以将其保留为列名），因此需要将第二行提取出来作为 kag_questions。相应的回答则被存储在 survey 变量中。

```
>>> import zipfile

>>> with zipfile.ZipFile(
...     "data/kaggle-survey-2018.zip"
... ) as z:
...     print("\n".join(z.namelist()))
...     kag = pd.read_csv(
...         z.open("multipleChoiceResponses.csv")
...     )
...     kag_questions = kag.iloc[0]
...     survey = kag.iloc[1:]
multipleChoiceResponses.csv
freeFormResponses.csv
SurveySchema.csv

>>> survey.head(2).T
                    1              2
Time from...      710            434
Q1              Female           Male
Q1_OTHER_...        -1             -1
Q2               45-49          30-34
Q3          United S...       Indonesia
...                ...            ...
Q50_Part_5         NaN            NaN
Q50_Part_6         NaN            NaN
```

Q50_Part_7	NaN	NaN
Q50_Part_8	NaN	NaN
Q50_OTHER...	-1	-1

3.6.2 原理解释

可以使用 read_csv 函数直接读取只有一个文件的 ZIP 文件。如果 ZIP 文件中包含多个文件，则需要使用另一种机制来读取数据。Python 标准库中包括 zipfile 模块，该模块可以从 ZIP 文件中提取文件。

遗憾的是，zipfile 模块将无法使用 URL（这与 read_csv 函数不同）。因此，如果你的 ZIP 文件位于 URL 中，则需要先下载它。

3.6.3 扩展知识

read_csv 函数也可以处理其他一些压缩类型。例如，如果你具有 GZIP、BZ2 或 XZ 文件，则 Pandas 可以处理这些文件，只要它们仅压缩 CSV 文件而不是目录即可。

3.7 与数据库协同工作

如前文所述，Pandas 对于表格或结构化数据很有用。许多组织都使用了数据库来存储表格数据。在本秘笈中，我们将使用数据库来插入和读取数据。

请注意，此示例使用了 Python 随附的 SQLite 数据库。但是，Python 也能够连接大多数 SQL 数据库，因此，Pandas 可以反过来利用这一点。

3.7.1 实战操作

（1）创建一个 SQLite 数据库来存储 Beatles（披头士乐队）的信息。

```
>>> import sqlite3
>>> con = sqlite3.connect("data/beat.db")
>>> with con:
...     cur = con.cursor()
...     cur.execute("""DROP TABLE Band""")
...     cur.execute(
...         """CREATE TABLE Band(id INTEGER PRIMARY KEY,
```

```
...         fname TEXT, lname TEXT, birthyear INT)"""
...     )
...     cur.execute(
...         """INSERT INTO Band VALUES(
...         0, 'Paul', 'McCartney', 1942)"""
...     )
...     cur.execute(
...         """INSERT INTO Band VALUES(
...         1, 'John', 'Lennon', 1940)"""
...     )
...     _ = con.commit()
```

(2) 将表从数据库读取到 DataFrame 中。

请注意，如果要读取表，则需要使用 SQLAlchemy 连接。SQLAlchemy 库为我们抽象了数据库。

```
>>> import sqlalchemy as sa
>>> engine = sa.create_engine(
...     "sqlite:///data/beat.db", echo=True
... )
>>> sa_connection = engine.connect()

>>> beat = pd.read_sql(
...     "Band", sa_connection, index_col="id"
... )
>>> beat
     fname     lname    birthyear
id
0    Paul      McCartney    1942
1    John      Lennon       1940
```

(3) 使用 SQL 查询从表中读取数据。这可以使用 SQLite 连接或 SQLAlchemy 连接来完成。示例如下。

```
>>> sql = """SELECT fname, birthyear from Band"""
>>> fnames = pd.read_sql(sql, con)
>>> fnames
   fname   birthyear
0  Paul         1942
1  John         1940
```

3.7.2 原理解释

Pandas 库利用了可以与大多数 SQL 数据库通信的 SQLAlchemy 库。这使分析人员可以从数据库表中创建 DataFrame，或者可以运行 SQL select 查询并从查询创建 DataFrame。

3.8 读取 JSON

JavaScript 对象表示法（JavaScript object notation，JSON）是一种通用格式，适用于通过 Internet 传输数据。不要被 JSON 的名称所迷惑，其实 JSON 不需要 JavaScript 即可读取或创建。Python 标准库随附 json 库，该 json 库可以执行 JSON 编码和解码操作。

```
>>> import json
>>> encoded = json.dumps(people)
>>> encoded
'{"first": ["Paul", "John", "Richard", "George"], "last": ["McCartney",
"Lennon", "Starkey", "Harrison"], "birth": [1942, 1940, 1940, 1943]}'

>>> json.loads(encoded)
{'first': ['Paul', 'John', 'Richard', 'George'], 'last': ['McCartney',
'Lennon', 'Starkey', 'Harrison'], 'birth': [1942, 1940, 1940, 1943]}
```

3.8.1 实战操作

（1）使用 read_json 函数读取数据。如果该 JSON 已经是将字典映射到列的列表中的形式，则无须转换即可接收它。此方向在 Pandas 中称为 columns（列）。

```
>>> beatles = pd.read_json(encoded)
>>> beatles
     first       last  birth
0     Paul  McCartney   1942
1     John     Lennon   1940
2  Richard    Starkey   1940
3   George   Harrison   1943
```

（2）读取 JSON 时要注意的一件事是，它必须是 Pandas 能够加载的特定格式。Pandas 可以支持的数据方向如下。

❑ columns（列）：这是默认方向，将列名称映射到列的值列表中。

- records（记录）：行列表。每行都是将一列映射到一个值中的字典。
- split（拆分）：将 columns 映射到列名称中，index 映射到索引值中，data 映射到每行数据的列表（每一行也是一个列表）中。
- index（索引）：索引值映射到行中。行是将列映射到值中的字典。
- values（值）：每行数据的列表（每行也是一个列表）。这不包括列或索引值。
- table（表）：schema 映射到 DataFrame 模式中，data 映射到字典列表中。

以下分别是这些样式的示例。至于 columns 样式，可参考之前显示的示例。

```
>>> records = beatles.to_json(orient="records")
>>> records
'[{"first":"Paul","last":"McCartney","birth":1942},{"first":"John",
"last":"Lennon","birth":1940},{"first":"Richard","last":"Starkey",
"birth":1940},{"first":"George","last":"Harrison","birth":1943}]'

>>> pd.read_json(records, orient="records")
   birth   first       last
0   1942    Paul    McCartney
1   1940    John       Lennon
2   1940 Richard      Starkey
3   1943  George     Harrison


>>> split = beatles.to_json(orient="split")
>>> split
'{"columns":["first","last","birth"],"index":[0,1,2,3],"data":[["Paul",
"McCartney",1942],["John","Lennon",1940],["Richard","Starkey",1940],
["George","Harrison",1943]]}'

>>> pd.read_json(split, orient="split")
     first       last  birth
0     Paul  McCartney   1942
1     John     Lennon   1940
2  Richard    Starkey   1940
3   George   Harrison   1943

>>> index = beatles.to_json(orient="index")
>>> index
'{"0":{"first":"Paul","last":"McCartney","birth":1942},"1":{"first":
"John","last":"Lennon","birth":1940},"2":{"first":"Richard","last":
"Starkey","birth":1940},"3":{"first":"George","last":"Harrison",
"birth":1943}}'
```

```
>>> pd.read_json(index, orient="index")
   birth    first       last
0   1942     Paul  McCartney
1   1940     John     Lennon
2   1940  Richard    Starkey
3   1943   George   Harrison

>>> values = beatles.to_json(orient="values")
>>> values
'[["Paul","McCartney",1942],["John","Lennon",1940],["Richard",
"Starkey",1940],["George","Harrison",1943]]'
>>> pd.read_json(values, orient="values")
         0          1     2
0     Paul  McCartney  1942
1     John     Lennon  1940
2  Richard    Starkey  1940
3   George   Harrison  1943

>>> (
...     pd.read_json(values, orient="values").rename(
...         columns=dict(
...             enumerate(["first", "last", "birth"])
...         )
...     )
... )
     first       last  birth
0     Paul  McCartney   1942
1     John     Lennon   1940
2  Richard    Starkey   1940
3   George   Harrison   1943

>>> table = beatles.to_json(orient="table")
>>> table
'{"schema": {"fields":[{"name":"index","type":"integer"},{"name":
"first","type":"string"},{"name":"last","type":"string"},{"name":
"birth","type":"integer"}],"primaryKey":["index"],"pandas_version":
"0.20.0"}, "data": [{"index":0,"first":"Paul","last":"McCartney",
"birth":1942},{"index":1,"first":"John","last":"Lennon",
"birth":1940},{"index":2,"first":"Richard","last":"Starkey",
"birth":1940},{"index":3,"first":"George","last":"Harrison",
"birth":1943}]}'
>>> pd.read_json(table, orient="table")
```

```
        first       last  birth
0        Paul  McCartney   1942
1        John     Lennon   1940
2     Richard    Starkey   1940
3      George   Harrison   1943
```

3.8.2 原理解释

JSON 可以通过多种方式格式化。因此，分析人员最好让需要使用的 JSON 具有受支持的方向。如果没有，则使用标准 Python 在字典中创建数据，字典会将列名称映射到值并将其传递到 DataFrame 构造函数中。

如果你需要生成 JSON（例如，假设你正在创建 Web 服务），则建议使用 columns 或 records 方向。

3.8.3 扩展知识

如果你正在使用 Web 服务，并且需要向 JSON 中添加其他数据，则只需使用.to_dict 方法即可生成字典。你可以将新数据添加到字典中，然后将该字典转换为 JSON。

```
>>> output = beat.to_dict()
>>> output
{'fname': {0: 'Paul', 1: 'John'}, 'lname': {0: 'McCartney', 1: 'Lennon'},
'birthyear': {0: 1942, 1: 1940}}

>>> output["version"] = "0.4.1"
>>> json.dumps(output)
'{"fname": {"0": "Paul", "1": "John"}, "lname": {"0": "McCartney", "1":
"Lennon"}, "birthyear": {"0": 1942, "1": 1940}, "version": "0.4.1"}'
```

3.9 读取 HTML 表格

可以使用 Pandas 从 Web 站点上读取 HTML 表格。这样可以很容易地提取诸如在 Wikipedia 或其他网站上找到的表格。

在此秘笈中，我们将从 The Beatles Discography（披头士唱片目录）这一 Wikipedia 条目中抓取表格。特别是，我们要提取图 3-2 中的表格。

第 3 章 创建和保留 DataFrame

List of studio albums,[A] with selected chart positions and certifications		Peak chart positions							Certifications
Title	Release	UK [1][2]	AUS [3]	CAN [4]	FRA [5]	GER [6]	NOR [7]	US [8][9]	
Please Please Me ‡	• Released: 22 March 1963 • Label: Parlophone (UK)	1	—	—	5	5	—	—	• BPI: Gold[10] • ARIA: Gold[11] • MC: Gold[12] • RIAA: Platinum[13]
With the Beatles[B] ‡	• Released: 22 November 1963 • Label: Parlophone (UK), Capitol (CAN), Odeon (FRA)	1	—	—	5	1	—	—	• BPI: Gold[10] • ARIA: Gold[11] • BVMI: Gold[15] • MC: Gold[12] • RIAA: Gold[13]

图 3-2 Wikipedia 上的专辑表

3.9.1 实战操作

（1）使用 read_html 函数从 https://en.wikipedia.org/wiki/The_Beatles_discography 网址中加载所有表。

```
>>> url = https://en.wikipedia.org/wiki/The_Beatles_discography
>>> dfs = pd.read_html(url)
>>> len(dfs)
51
```

（2）检查第一个 DataFrame。

```
>>> dfs[0]
    The Beatles discography The Beatles discography.1
0       The Beat...           The Beat...
1       Studio a...                    23
2       Live albums                     5
3       Compilat...                    53
4       Video al...                    15
5       Music vi...                    64
6       EPs                            21
7       Singles                        63
8       Mash-ups                        2
9       Box sets                       15
```

（3）上表汇总了工作室专辑、现场专辑、合辑专辑等的数量。这不是我们想要的表。我们可以遍历 read_html 创建的每个表，或者可以提示它查找特定的表。

read_html 函数具有 match 参数，该参数可以是字符串或正则表达式。另外，read_html 函数还具有 attrs 参数，该参数允许传入 HTML 标记属性键和值（在字典中），并将使用该键标识表。

我们使用了 Chrome 浏览器检查 HTML 代码，以查看 table 元素上是否存在属性或表中是否存在要使用的唯一字符串。

以下是 HTML 代码的一部分。

```
<table class="wikitable plainrowheaders" style="text-
align:center;">
    <caption>List of studio albums,<sup id="cite_ref-1"
class="reference"><a href="#cite_note-1">[A]</a></sup> with
selected chart positions and certifications
    </caption>
    <tbody>
        <tr>
            <th scope="col" rowspan="2" style="width:20em;">Title
            </th>
            <th scope="col" rowspan="2" style="width:20em;">Release
            ...
```

可以看到该网页表格上没有属性，但是可以使用字符串 List of studio albums（工作室专辑列表）来匹配表格。另外还可以保留从 Wikipedia 页面中复制的 na_values 的值。

```
>>> url = https://en.wikipedia.org/wiki/The_Beatles_discography
>>> dfs = pd.read_html(
...     url, match="List of studio albums", na_values="—"
... )
>>> len(dfs)
1

>>> dfs[0].columns
Int64Index([0, 1, 2, 3, 4, 5, 6, 7, 8, 9], dtype='int64')
```

（4）列有点乱了。我们可以尝试使用前两行作为列，但是它们仍然很混乱。

```
>>> url = https://en.wikipedia.org/wiki/The_Beatles_discography
>>> dfs = pd.read_html(
...     url,
...     match="List of studio albums",
...     na_values="—",
...     header=[0, 1],
... )
```

```
>>> len(dfs)
1

>>> dfs[0]
        Title      Release   ...  Peak chart positions  Certifications
        Title      Release   ...                US[8][9]  Certifications
0   Please P... Released... ...                    NaN     BPI: Gol...
1   With the... Released... ...                    NaN     BPI: Gol...
2   Introduc... Released... ...                      2     RIAA: Pl...
3   Meet the... Released... ...                      1     MC: Plat...
4   Twist an... Released... ...                    NaN     MC: 3×P...
..        ...         ...   ...                    ...            ...
22  The Beat... Released... ...                      1     BPI: 2×...
23  Yellow S... Released... ...                      2     BPI: Gol...
24  Abbey Road  Released... ...                      1     BPI: 2×...
25   Let It Be  Released... ...                      1     BPI: Gol...
26   "—" deno...  "—" deno... ...              "—" deno...   "—" deno...

>>> dfs[0].columns
MultiIndex(levels=[['Certifications', 'Peak chart positions',
'Release', 'Title'], ['AUS[3]', 'CAN[4]', 'Certifications',
'FRA[5]', 'GER[6]', 'NOR[7]', 'Release', 'Title', 'UK[1][2]',
'US[8][9]']],
  codes=[[3, 2, 1, 1, 1, 1, 1, 1, 1, 0], [7, 6, 8, 0, 1, 3, 4, 5,
9, 2]])
```

这不是通过编程方式就很容易解决的问题。在这种情况下，最简单的解决方案是手动更新列。

```
>>> df = dfs[0]
>>> df.columns = [
...     "Title",
...     "Release",
...     "UK",
...     "AUS",
...     "CAN",
...     "FRA",
...     "GER",
...     "NOR",
...     "US",
...     "Certifications",
... ]
```

```
>>> df
       Title        Release    ...    US    Certifications
0   Please P...    Released...  ...   NaN    BPI: Gol...
1   With the...    Released...  ...   NaN    BPI: Gol...
2   Introduc...    Released...  ...    2     RIAA: Pl...
3   Meet the...    Released...  ...    1     MC: Plat...
4   Twist an...    Released...  ...   NaN    MC: 3× P...
..      ...           ...       ...   ...       ...
22  The Beat...    Released...  ...    1     BPI: 2× ...
23  Yellow S...    Released...  ...    2     BPI: Gol...
24  Abbey Road     Released...  ...    1     BPI: 2× ...
25  Let It Be      Released...  ...    1     BPI: Gol...
26  "—" deno...    "—" deno...  ...   "—" deno...   "—" deno...
```

（5）我们还应该对数据进行更多清理。标题以 Released 开头的任何行都是上一行的另一个版本。Pandas 无法解析 rowspan 大于 1 的行（Release 行正是这种情况）。在 Wikipedia 页面中，这些行的代码如下。

```
<th scope="row" rowspan="2">
    <i><a href="/wiki/A_Hard_Day%27s_Night_(album)" title="A Hard
Day's Night (album)">A Hard Day's Night</a></i>
    <img alt="double-dagger" src="//upload.wikimedia.org/wikipedia/
commons/f/f9/Double-dagger-14-plain.png" decoding="async"
width="9" height="14" data-file-width="9" data-file-height="14">
</th>
```

我们将跳过这些行，因为相关内容将 Pandas 陷入混乱之中，并且 Pandas 在这些行中放入的数据不正确。我们将 release 列拆分为两列，即 release_date（发布日期）和 label（标签）。

```
>>> res = (
...    df.pipe(
...        lambda df_: df_[
...            ~df_.Title.str.startswith("Released")
...        ]
...    )
...    .assign(
...        release_date=lambda df_: pd.to_datetime(
...            df_.Release.str.extract(
...                r"Released: (.*) Label"
...            )[0].str.replace(r"\[E\]", "")
...        ),
...        label=lambda df_: df_.Release.str.extract(
```

```
...                 r"Label: (.*)"
...             ),
...         )
...         .loc[
...             :,
...             [
...                 "Title",
...                 "UK",
...                 "AUS",
...                 "CAN",
...                 "FRA",
...                 "GER",
...                 "NOR",
...                 "US",
...                 "release_date",
...                 "label",
...             ],
...         ]
...     )
>>> res
           Title    UK  ...  release_date         label
0     Please P...    1  ...    1963-03-22   Parlopho...
1     With the...    1  ...    1963-11-22   Parlopho...
2     Introduc...  NaN  ...    1964-01-10    Vee-Jay...
3     Meet the...  NaN  ...    1964-01-20    Capitol...
4     Twist an...  NaN  ...    1964-02-03    Capitol...
..            ...  ...  ...           ...           ...
21      Magical...   31  ...    1967-11-27   Parlopho...
22     The Beat...    1  ...    1968-11-22         Apple
23      Yellow S...   3  ...    1969-01-13   Apple (U...
24      Abbey Road    1  ...    1969-09-26         Apple
25       Let It Be    1  ...    1970-05-08         Apple
```

3.9.2 原理解释

read_html 函数可以遍历 HTML 代码查找 table 标签，然后将内容解析为 DataFrame，这可以简化网站的抓取。遗憾的是，如本示例所示，有时 HTML 表中的数据可能难以解析。rowspan 和多行标题可能会将 Pandas 陷入混乱之中。分析人员需要对结果执行健全性检查。

当然，对于非常简单的 HTML 表格（即没有进行过单元格行的拆分和合并操作的表格），Pandas 可轻松提取其数据。至于像本示例中这样的表格，则需要对输出结果执行一些操作以进行清理。

3.9.3 扩展知识

还可以使用 attrs 参数从页面中选择一个表格。接下来，我们将从 GitHub 的 CSV 文件视图中选择读取数据。请注意，我们不是从原始 CSV 数据中读取数据，而是从 GitHub 的在线文件查看器中读取数据。我们检查了该表，并注意到该表有一个 class 属性，其值是 csv-data。使用该值可以限制选择的表格。

```
>>> url = https://github.com/mattharrison/datasets/blob/master/data/
anscombes.csv
>>> dfs = pd.read_html(url, attrs={"class": "csv-data"})
>>> len(dfs)
1
>>> dfs[0]
    Unnamed: 0  quadrant     x     y
0          NaN         I  10.0  8.04
1          NaN         I  14.0  9.96
2          NaN         I   6.0  7.24
3          NaN         I   9.0  8.81
4          NaN         I   4.0  4.26
..         ...       ...   ...   ...
39         NaN        IV   8.0  6.58
40         NaN        IV   8.0  7.91
41         NaN        IV   8.0  8.47
42         NaN        IV   8.0  5.25
43         NaN        IV   8.0  6.89
```

请注意，GitHub 劫持了一个 td 元素以显示行号，因此显示了 Unnamed:0 列。它似乎正在使用 JavaScript 以动态方式向网页中添加行号，因此，当网页显示行号时，源代码具有空单元格，也因此该列中包含的是 NaN 值。你可能需要删除该列，因为它没有用。

还需要注意的一件事是：网站是会改变的。不要指望下周还可以从同一个地方提取到相同的数据。因此，在检索数据后可以将其保存到本地。

有时你可能还需要使用其他工具。如果 read_html 函数无法从网站获取数据，则可能需要求助于屏幕抓取功能。幸运的是，Python 也有相应的工具。可以使用请求库进行简单的抓取操作。另外，Beautiful Soup 库也是一个使 HTML 内容更容易浏览的工具。

第 4 章 开始数据分析

本章包含以下秘笈。
- 开发数据分析例程。
- 数据字典。
- 通过更改数据类型减少内存使用量。
- 从最大中选择最小。
- 通过排序选择每组中的最大值。
- 使用 sort_values 复制 nlargest。
- 计算追踪止损单价格。

4.1 介 绍

作为一名数据分析人员,在将数据集作为 DataFrame 导入工作区后,首先要考虑的是遇到数据集时应采取的步骤。是否有一组通常负责检查数据的任务?你是否了解所有可能的数据类型?对此,本章将为你介绍首次接触新数据集时可能要执行的任务。此外,本章还将解决一些与数据分析有关的常见问题,这些问题在 Pandas 中解决起来并不那么容易。

4.2 开发数据分析例程

尽管开始进行数据分析时并没有一个标准方法,但是一般来说,最好还是在首次检查数据集时为自己开发一个例程。这有点类似于我们日常的起床、洗嗽、上班、就餐等例程,数据分析例程可以帮助你快速熟悉新数据集。该例程也可以表现为动态任务清单,将随着你对 Pandas 的熟悉和数据分析的扩展而不断发展。

探索性数据分析(exploratory data analysis,EDA)是一个术语,用于描述分析数据集的过程。一般来说,探索性数据分析不涉及模型创建,而是汇总数据特征并对其进行可视化。这并不是什么新鲜事物,John Tukey 早在 1977 年的 *Exploratory Data Analysis*(《探索性数据分析》)一书中就对此进行了推广。

这些过程有许多在今天仍然适用，并且对于理解数据集非常有帮助。实际上，它们还可以在以后帮助创建机器学习（machine learning，ML）模型。

本秘笈讨论探索性数据分析的一小部分，也是最基础的一部分：以常规和系统性的方式收集元数据（metadata）和描述性统计信息（descriptive statistics）。我们简要介绍了第一次将任何数据集作为 Pandas DataFrame 导入时可以执行的一组标准任务。此秘笈可能有助于形成你在首次检查数据集时实现的例程的基础。

元数据可以描述数据集，或者更恰当地说，它是关于数据的数据。元数据的示例包括列数/行数、列名称、每列的数据类型、数据集的来源、收集日期、各个列的可接收值等。

单变量描述性统计信息（univariate descriptive statistics）是有关数据集变量（列）的摘要统计信息，独立于所有其他变量。

4.2.1 实战操作

首先，我们将收集 college 数据集中的一些元数据，然后是每列的基本摘要统计信息。

（1）读取数据集，并使用 .sample 方法查看行的样本。

```
>>> import pandas as pd
>>> import numpy as np
>>> college = pd.read_csv("data/college.csv")
>>> college.sample(random_state=42)
           INSTNM         CITY ...  MD_EARN_WNE_P10  GRAD_DEBT_MDN_SUPP
3649   Career P...  San Antonio ...            20700                14977
```

（2）使用 .shape 属性获取 DataFrame 的维度。

```
>>> college.shape
(7535, 27)
```

（3）使用 .info 方法列出每列的数据类型、非缺失值的数量以及内存使用情况。

```
>>> college.info()
<class 'pandas.core.frame.DataFrame'>
RangeIndex: 7535 entries, 0 to 7534
Data columns (total 27 columns):
 #   Column                  Non-Null Count  Dtype
---  ------                  --------------  -----
 0   INSTNM                  7535 non-null   object
 1   CITY                    7535 non-null   object
 2   STABBR                  7535 non-null   object
```

```
3   HBCU               7164 non-null   float64
4   MENONLY            7164 non-null   float64
5   WOMENONLY          7164 non-null   float64
6   RELAFFIL           7535 non-null   int64
7   SATVRMID           1185 non-null   float64
8   SATMTMID           1196 non-null   float64
9   DISTANCEONLY       7164 non-null   float64
10  UGDS               6874 non-null   float64
11  UGDS_WHITE         6874 non-null   float64
12  UGDS_BLACK         6874 non-null   float64
13  UGDS_HISP          6874 non-null   float64
14  UGDS_ASIAN         6874 non-null   float64
15  UGDS_AIAN          6874 non-null   float64
16  UGDS_NHPI          6874 non-null   float64
17  UGDS_2MOR          6874 non-null   float64
18  UGDS_NRA           6874 non-null   float64
19  UGDS_UNKN          6874 non-null   float64
20  PPTUG_EF           6853 non-null   float64
21  CURROPER           7535 non-null   int64
22  PCTPELL            6849 non-null   float64
23  PCTFLOAN           6849 non-null   float64
24  UG25ABV            6718 non-null   float64
25  MD_EARN_WNE_P10    6413 non-null   object
26  GRAD_DEBT_MDN_SUPP 7503 non-null   object
dtypes: float64(20), int64(2), object(5)
memory usage: 1.6+ MB
```

（4）获取数字列的摘要统计信息，并转置 DataFrame 以获得更容易阅读的输出。

```
>>> college.describe(include=[np.number]).T
            count     mean    ...       75%      max
HBCU       7164.0   0.014238 ...    0.000000   1.0
MENONLY    7164.0   0.009213 ...    0.000000   1.0
WOMENONLY  7164.0   0.005304 ...    0.000000   1.0
RELAFFIL   7535.0   0.190975 ...    0.000000   1.0
SATVRMID   1185.0 522.819409 ...  555.000000 765.0
    ...      ...      ...    ...      ...      ...
PPTUG_EF   6853.0   0.226639 ...    0.376900   1.0
CURROPER   7535.0   0.923291 ...    1.000000   1.0
PCTPELL    6849.0   0.530643 ...    0.712900   1.0
PCTFLOAN   6849.0   0.522211 ...    0.745000   1.0
UG25ABV    6718.0   0.410021 ...    0.572275   1.0
```

（5）获取对象（字符串）列的摘要统计信息。

```
>>> college.describe(include=[np.object]).T
              count  unique         top  freq
INSTNM         7535    7535   Academy ...     1
CITY           7535    2514   New York    87
STABBR         7535      59          CA   773
MD_EARN_W...   6413     598   PrivacyS...  822
GRAD_DEBT...   7503    2038   PrivacyS... 1510
```

4.2.2 原理解释

在导入数据集后，一个常见的任务是输出 DataFrame 的行样本，以使用.sample 方法进行手动检查。.shape 属性可以返回一些元数据；另外，.shape 属性也是一个包含行数和列数的元组。

.info 方法是一次获取更多元数据的方法。该方法可以提供每列的名称、非缺失值的数量、每列的数据类型以及 DataFrame 大致的内存使用情况。一般来说，Pandas 中的列只有一个类型（但是，也可能会有包含混合类型的列，并将其报告为 object）。总的而言，DataFrame 可能由具有不同数据类型的列组成。

上述步骤（4）和步骤（5）生成了有关不同类型列的描述性统计信息。默认情况下，.describe 方法可以输出所有数字列的摘要，并静默删除所有非数字列。你也可以将其他选项传递给 include 参数，以使其包含具有非数字数据类型的列的计数和频率。从技术上讲，数据类型是分层结构的一部分，在该分层结构中，np.number 位于整数和浮点数的上方。

如前文所述，我们可以将数据分类为连续数据或分类数据。连续数据始终是数字，通常可以具有无限数量的可能性，如身高、体重和薪水等；分类数据则表示具有有限可能性的离散值，如种族、就业状况和汽车颜色等。分类数据可以用数字或字符表示。

分类列通常是 np.object 类型或 pd.Categorical 类型。步骤（5）中的项目同时代表了这两种类型。

在步骤（4）和步骤（5）中，输出 DataFrame 都通过.T 属性进行转置。这可以使包含许多列的 DataFrame 更易于阅读，因为.T 属性通常允许更多数据在屏幕上显示而无须滚动。

4.2.3 扩展知识

与数字列一起使用时，可以指定从.describe 方法返回的精确分位数。

第 4 章 开始数据分析

```
>>> college.describe(
...     include=[np.number],
...     percentiles=[
...         0.01,
...         0.05,
...         0.10,
...         0.25,
...         0.5,
...         0.75,
...         0.9,
...         0.95,
...         0.99,
...     ],
... ).T
            count        mean   ...         99%    max
HBCU       7164.0    0.014238   ...    1.000000    1.0
MENONLY    7164.0    0.009213   ...    0.000000    1.0
WOMENONLY  7164.0    0.005304   ...    0.000000    1.0
RELAFFIL   7535.0    0.190975   ...    1.000000    1.0
SATVRMID   1185.0  522.819409   ...  730.000000  765.0
...           ...         ...   ...         ...    ...
PPTUG_EF   6853.0    0.226639   ...    0.946724    1.0
CURROPER   7535.0    0.923291   ...    1.000000    1.0
PCTPELL    6849.0    0.530643   ...    0.993908    1.0
PCTFLOAN   6849.0    0.522211   ...    0.986368    1.0
UG25ABV    6718.0    0.410021   ...    0.917383    1.0
```

4.3 数据字典

数据分析的关键部分涉及创建和维护数据字典（data dictionary）。数据字典是元数据表和每列数据上的注释。数据字典的主要目的之一是解释列名称的含义。例如，college 数据集使用了许多缩写词，这对于首次检查该数据集的分析人员而言可能是陌生的。

以下 college_data_dictionary.csv 文件中提供了 college 数据集的数据字典。

```
>>> pd.read_csv("data/college_data_dictionary.csv")
    column_name  description
0        INSTNM  Institut...
1          CITY  City Loc...
2        STABBR  State Ab...
3          HBCU  Historic...
```

```
4           MENONLY    0/1 Men ...
..              ...            ...
22           PCTPELL    Percent ...
23          PCTFLOAN    Percent ...
24           UG25ABV    Percent ...
25         MD_EARN_...  Median E...
26         GRAD_DEB...  Median d...
```

可以看到，数据字典在解读缩写的列名称方面非常有用。当然，DataFrame 并不是存储数据字典的最佳位置。诸如 Excel 或 Google 表格之类的平台具有轻松编辑值和附加列的能力，因此最好选择此类平台存储数据字典。

或者，可以在 Jupyter 的 Markdown 单元格中描述数据字典。数据字典是你可以与协作者共享的第一批内容之一。

一般来说，你正在使用的数据集源自数据库，你必须联系该数据库的管理员才能获取更多信息。数据库具有其数据的表示形式，称为模式（schema）。如果可能的话，你还可以尝试与行业专家（subject matter expert，SME）一起调查你的数据集。

4.4 通过更改数据类型减少内存使用量

Pandas 对许多数据类型都有精确的技术定义。但是，当你从无类型的格式（如 CSV）中加载数据时，Pandas 必须推断其类型。

此秘笈将 college 数据集中的 object 列之一的数据类型更改为特殊的 Pandas 分类数据类型，以大大减少其内存使用量。

4.4.1 实战操作

（1）读取 college 数据集后，可以选择一些不同数据类型的列，这些列将清楚地显示可以节省多少内存。

```
>>> college = pd.read_csv("data/college.csv")
>>> different_cols = [
...     "RELAFFIL",
...     "SATMTMID",
...     "CURROPER",
...     "INSTNM",
...     "STABBR",
... ]
```

```
>>> col2 = college.loc[:, different_cols]
>>> col2.head()
   RELAFFIL  SATMTMID  ...    INSTNM   STABBR
0         0     420.0  ...   Alabama ...   AL
1         0     565.0  ...   Universi...   AL
2         1       NaN  ...   Amridge ...   AL
3         0     590.0  ...   Universi...   AL
4         0     430.0  ...   Alabama ...   AL
```

（2）检查每列的数据类型。

```
>>> col2.dtypes
RELAFFIL       int64
SATMTMID     float64
CURROPER       int64
INSTNM        object
STABBR        object
dtype: object
```

（3）使用.memory_usage方法查找每列的内存使用情况。

```
>>> original_mem = col2.memory_usage(deep=True)
>>> original_mem
Index            128
RELAFFIL       60280
SATMTMID       60280
CURROPER       60280
INSTNM        660240
STABBR        444565
dtype: int64
```

（4）RELAFFIL列不需要使用64位，因为该列仅包含0或1。可以使用.astype方法将此列转换为8位（1字节）整。

```
>>> col2["RELAFFIL"] = col2["RELAFFIL"].astype(np.int8)
```

（5）使用.dtypes属性查看对数据类型进行的更改。

```
>>> col2.dtypes
RELAFFIL        int8
SATMTMID     float64
CURROPER       int64
INSTNM        object
STABBR        object
dtype: object
```

（6）再次查找每列的内存使用情况，并注意减少的地方。

```
>>> col2.memory_usage(deep=True)
Index            128
RELAFFIL        7535
SATMTMID       60280
CURROPER       60280
INSTNM        660240
STABBR        444565
dtype: int64
```

（7）为了节省更多的内存，如果对象数据类型的基数（唯一值数量）相当低，则可以考虑将其更改为分类对象。首先让我们检查两个对象列的唯一值数量。

```
>>> col2.select_dtypes(include=["object"]).nunique()
INSTNM  7535
STABBR    59
dtype: int64
```

（8）可以看到，STABBR 列是转换为分类对象列的很好的候选者，因为其值的唯一性不足百分之一。

```
>>> col2["STABBR"] = col2["STABBR"].astype("category")
>>> col2.dtypes
RELAFFIL          int8
SATMTMID       float64
CURROPER         int64
INSTNM          object
STABBR        category
dtype: object
```

（9）再次计算内存的使用情况。

```
>>> new_mem = col2.memory_usage(deep=True)
>>> new_mem
Index            128
RELAFFIL        7535
SATMTMID       60280
CURROPER       60280
INSTNM        660699
STABBR         13576
dtype: int64
```

（10）比较原始内存使用情况和更新后的内存使用情况。RELAFFIL 列是原始大小的八分之一，而 STABBR 列则已缩小到原始大小的百分之三。

```
>>> new_mem / original_mem
Index         1.000000
RELAFFIL      0.125000
SATMTMID      1.000000
CURROPER      1.000000
INSTNM        1.000695
STABBR        0.030538
dtype: float64
```

4.4.2 原理解释

无论特定 DataFrame 所需的最大内存量如何，Pandas 都会默认将整数和浮点数据类型设置为 64 位。可以使用 .astype 方法将整数、浮点数甚至布尔值强制转换为不同的数据类型，然后将此数据类型作为字符串或特定对象的确切类型传递给它，步骤（4）就是这样操作的。

RELAFFIL 列是转换为较小整数类型的一个不错的选择，因为数据字典说明其值必须为 0 或 1。RELAFFIL 列的内存当前为 CURROPER 列的八分之一，后者仍然保持着其原来的类型。

具有 object 数据类型（如 INSTNM）的列与其他 Pandas 数据类型不同。对于所有其他 Pandas 数据类型，该列中的每个值都是相同的数据类型，例如当列具有 int64 类型时，该列中的每个值也都是 int64；对于具有 object 数据类型的列，情况并非如此，该列中的每个值可以是任何类型，这些值可以混合使用字符串、数字、日期时间，甚至其他 Python 对象（如列表或元组）。因此，有时将 object 数据类型称为与所有其他数据类型都不匹配的兜底（catch all）类型。当然，绝大多数时候，object 数据类型列都是字符串。

因此，object 数据类型列中每个值的存储都不一致。像其他数据类型一样，每个值都没有预定义的内存量。为了让 Pandas 提取 object 数据类型列的确切内存量，必须在 .memory_usage 方法中将 deep 参数设置为 True。

object 列是节省内存最大的目标。Pandas 具有 NumPy 中不可用的 category 数据类型。当将 object 数据类型转换为 category 数据类型时，Pandas 将在内部创建从整数到每个唯一字符串值的映射。因此，每个字符串仅需要在内存中保留一次。在前面的示例中可见，这种数据类型的更改将内存使用量减少了 97%。当然，这个效率也跟列中数据类型的基数有关。

你可能已经注意到，索引使用的内存量极低。如果在创建 DataFrame 的过程中未指定索引（本秘笈就是这样），则 Pandas 会将索引默认为 RangeIndex。RangeIndex 与内置 range 函数非常相似。RangeIndex 按需生成值，并且仅存储创建索引所需的最少信息量。

4.4.3 扩展知识

为了更好地了解对象数据类型的列与整数和浮点数列之间的区别，可以修改这些列中每一列的单个值，并显示结果的内存使用情况。

CURROPER 和 INSTNM 列分别为 int64 和 object 类型。

```
>>> college.loc[0, "CURROPER"] = 10000000
>>> college.loc[0, "INSTNM"] = (
...     college.loc[0, "INSTNM"] + "a"
... )
>>> college[["CURROPER", "INSTNM"]].memory_usage(deep=True)
Index             80
CURROPER       60280
INSTNM        660804
dtype: int64
```

CURROPER 的内存使用保持不变，因为 64 位整数足以容纳更大的数字。另外，仅向一个值中添加一个字母，INSTNM 的内存使用量就会增加 105 个字节。

Python 3 使用 Unicode，这是一种标准的字符表示形式，旨在对世界上所有的书写系统进行编码。机器上的 Unicode 字符串占用多少内存取决于 Python 的构建方式。在本台机器上，每个字符最多使用 4 个字节。当对字符值进行第一次修改时，Pandas 有一些开销（100 字节）。之后，每个字符增加 5 个字节。

并非所有的列都可以被强制转换为所需的类型。来看 MENONLY 列，该列在数据字典中似乎只包含 0 或 1，但导入时此列的实际数据类型意外地是 float64，出现这种情况的原因是该列包含了缺失值，缺失值用 np.nan 表示。对于 int64 类型来说，并没有整数表示缺失值（请注意，在 Pandas 0.24 及其以上版本中的 int64 类型确实支持缺失值，但默认情况下是不使用的）。对于数字列来说，即使只有一个缺失值也将会被转换为浮点数列。此外，如果其中有一个值缺失，则整数数据类型的任何列都将自动强制转换为浮点类型。

```
>>> college["MENONLY"].dtype
dtype('float64')
>>> college["MENONLY"].astype(np.int8)
Traceback (most recent call last):
  ...
ValueError: Cannot convert non-finite values (NA or inf) to integer
```

第 4 章 开始数据分析

此外，在引用数据类型时，可以用字符串名称代替 Python 对象。例如，当在 .describe DataFrame 方法中使用 include 参数时，可以传递 NumPy 或 Pandas 对象或其等效字符串表示形式的列表。例如，以下每个命令都将产生相同的结果。

```
college.describe(include=['int64', 'float64']).T

college.describe(include=[np.int64, np.float64]).T

college.describe(include=['int', 'float']).T

college.describe(include=['number']).T
```

类型字符串也可以与 .astype 方法结合使用。

```
>>> college.assign(
...     MENONLY=college["MENONLY"].astype("float16"),
...     RELAFFIL=college["RELAFFIL"].astype("int8"),
... )
            INSTNM         CITY   ...  MD_EARN_WNE_P10  GRAD_DEBT_MDN_SUPP
0        Alabama ...      Normal  ...            30300               33888
1        Universi...  Birmingham  ...            39700             21941.5
2        Amridge ...  Montgomery  ...            40100               23370
3        Universi...  Huntsville  ...            45500               24097
4        Alabama ...  Montgomery  ...            26600             33118.5
...          ...          ...     ...              ...                 ...
7530    SAE Inst...   Emeryville  ...              NaN                9500
7531    Rasmusse...   Overland... ...              NaN               21163
7532    National...   Highland... ...              NaN                6333
7533    Bay Area...     San Jose  ...              NaN           PrivacyS...
7534    Excel Le...  San Antonio  ...              NaN               12125
```

最后，你可能会看到最小的 RangeIndex 和 Int64Index 之间存在巨大的内存使用量差异，后者将每个行索引存储在内存中。

```
>>> college.index = pd.Int64Index(college.index)
>>> college.index.memory_usage()     # 前者内存使用量仅需 80
60280
```

4.5 从最大中选择最小

此秘笈可用于创建一些很吸引眼球的新闻头条，例如"排名前 100 的大学中，这 5 所大学的学费最低"或"最宜居的前 50 强城市中，这 10 座城市的生活成本最低"。

在分析过程中，可能需要先找到一个数据组，该数据组在单个列中包含前 n 个值，然后从该子集中，根据不同的列查找前 m 个值。

在此秘笈中，我们将利用便捷方法从评分最高的 100 部电影中找到 5 部预算最低的电影。这两个便捷方法分别为.nlargest 和.nsmallest。

4.5.1 实战操作

（1）读取 movie 数据集，然后分别选择 movie_title、imdb_score 和 budget 列。

```
>>> movie = pd.read_csv("data/movie.csv")
>>> movie2 = movie[["movie_title", "imdb_score", "budget"]]
>>> movie2.head()
    movie_title    imdb_score      budget
0        Avatar           7.9  237000000.0
1      Pirates ...        7.1  300000000.0
2        Spectre         6.8  245000000.0
3   The Dark...            8.5  250000000.0
4   Star War...            7.1          NaN
```

（2）使用.nlargest 方法通过 imdb_score 选择前 100 部电影。

```
>>> movie2.nlargest(100, "imdb_score").head()
            movie_title    imdb_score      budget
2725       Towering Inferno        9.5         NaN
1920  The Shawshank Redemption    9.3  25000000.0
3402           The Godfather      9.2   6000000.0
2779                 Dekalog      9.1         NaN
4312      Kickboxer: Vengeance    9.1  17000000.0
```

（3）链接.nsmallest 方法，以从评分最高的 100 部电影中返回预算最低的 5 部。

```
>>> (
...     movie2.nlargest(100, "imdb_score").nsmallest(
...         5, "budget")
... )
            movie_title    imdb_score    budget
4804       Butterfly Girl        8.7  180000.0
4801   Children of Heaven       8.5  180000.0
4706          12 Angry Men      8.9  350000.0
4550          A Separation      8.4  500000.0
4636    The Other Dream Team   8.4  500000.0
```

4.5.2 原理解释

.nlargest 方法的第一个参数 n 必须为整数，并选择要返回的行数。第二个参数 columns 可以采用列名称作为字符串。步骤（2）返回了评分最高的 100 部电影。我们可以以将该中间结果另存为自己的变量，然后在步骤（3）中将.nsmallest 方法链接到该变量，该方法恰好返回 5 行，按 budget（预算）排序。

4.5.3 扩展知识

可以将列名称列表分别传递给.nlargest 和.nsmallest 方法的 columns 参数，当出现并列排名的情况时，这有助于区分共享排名的项。

4.6 通过排序选择每组中的最大值

在数据分析期间执行的最基本、最常见的操作之一是选择包含组中某个列的最大值的行。例如，查找每年评分最高的电影或按内容分级的票房收入最高的电影。要完成此任务，我们需要对组以及用于对组中每个成员排名的列进行排序，然后提取每个组中的最高成员。

在此秘笈中，我们将找到评分最高的电影。

4.6.1 实战操作

（1）读取 movie 数据集，并将该数据集精简为我们关心的 3 列。

```
>>> movie = pd.read_csv("data/movie.csv")
>>> movie[["movie_title", "title_year", "imdb_score"]]
                                    movie_title  ...
0                                        Avatar  ...
1      Pirates of the Caribbean: At World's End  ...
2                                       Spectre  ...
3                         The Dark Knight Rises  ...
4     Star Wars: Episode VII - The Force Awakens ...
...                                          ...  ...
4911                      Signed Sealed Delivered ...
4912                                The Following ...
```

```
4913                    A Plague So Pleasant  ...
4914                       Shanghai Calling   ...
4915                       My Date with Drew  ...
```

（2）使用.sort_values 方法按 title_year 对 DataFrame 进行排序。默认行为从最小到最大。使用 ascending = True 参数可以反转此行为。

```
>>> (
...     movie[
...         ["movie_title", "title_year", "imdb_score"]
...     ].sort_values("title_year", ascending=True)
... )
                                        movie_title  ...
4695   Intolerance: Love's Struggle Throughout the Ages ...
4833                    Over the Hill to the Poorhouse ...
4767                              The Big Parade  ...
2694                                   Metropolis  ...
4697                          The Broadway Melody  ...
...                                           ...  ...
4683                                       Heroes  ...
4688                                  Home Movies  ...
4704                                   Revolution  ...
4752                                 Happy Valley  ...
4912                                The Following  ...
```

（3）注意如何只对年份进行排序。要一次对多列进行排序，可以使用一个列表。让我们看看如何同时对年份和电影评分进行排序。

```
>>> (
...     movie[
...         ["movie_title", "title_year", "imdb_score"]
...     ].sort_values(
...         ["title_year", "imdb_score"], ascending=False
...     )
... )
                         movie_title  title_year  imdb_score
4312            Kickboxer: Vengeance      2016.0         9.1
4277       A Beginner's Guide to Snuff  2016.0         8.7
3798                         Airlift      2016.0         8.5
27          Captain America: Civil War  2016.0         8.2
98               Godzilla Resurgence    2016.0         8.2
...                              ...         ...         ...
1391                       Rush Hour         NaN         5.8
```

```
4031                  Creature            NaN         5.0
2165              Meet the Browns         NaN         3.5
3246       The Bold and the Beautiful    NaN         3.5
2119                The Bachelor          NaN         2.9
```

（4）现在可以使用.drop_duplicates 方法仅保留每年的第一行。

```
>>> (
...     movie[["movie_title", "title_year", "imdb_score"]]
...     .sort_values(
...         ["title_year", "imdb_score"], ascending=False
...     )
...     .drop_duplicates(subset="title_year")
... )
       movie_title   title_year   imdb_score
4312   Kickboxe...     2016.0        9.1
3745   Running ...     2015.0        8.6
4369   Queen of...     2014.0        8.7
3935   Batman: ...     2013.0        8.4
3      The Dark...     2012.0        8.5
...       ...            ...         ...
2694   Metropolis      1927.0        8.3
4767   The Big ...     1925.0        8.3
4833   Over the...     1920.0        4.8
4695   Intolera...     1916.0        8.0
2725   Towering...      NaN          9.5
```

4.6.2 原理解释

本示例展示了如何使用链接构建和测试一系列的 Pandas 操作。

在步骤（1）中，我们将数据集精简为仅需要关注的重要列。当然，此秘笈使用的就是整个 DataFrame。

步骤（2）显示了如何按单个列对 DataFrame 进行排序，这并不是我们想要的结果。

步骤（3）同时对多个列进行排序。首先，对所有 title_year 进行排序，然后在 title_year 的每个值内按 imdb_score 进行排序。

.drop_duplicates 方法的默认行为是保留每个唯一行的第一次出现，因为每个行都是唯一的，所以这不会删除任何行。但是，subset 参数会将其更改为仅考虑为其分配的列（或列的列表）。在此示例中，每年仅返回一行。当我们在最后一步中按年份和评分排序时，获得的就是每年评分最高的电影。

4.6.3 扩展知识

像大多数 Pandas 操作一样,有多种方法可以进行排序选择。如果你发现自己更喜欢分组操作,则也可以使用.groupby 方法执行此操作。

```
>>> (
...     movie[["movie_title", "title_year", "imdb_score"]]
...     .groupby("title_year", as_index=False)
...     .apply(
...         lambda df: df.sort_values(
...             "imdb_score", ascending=False
...         ).head(1)
...     )
...     .droplevel(0)
...     .sort_values("title_year", ascending=False)
... )
          movie_title    title_year   imdb_score
90  4312  Kickboxe...      2016.0         9.1
89  3745  Running ...      2015.0         8.6
88  4369  Queen of...      2014.0         8.7
87  3935  Batman: ...      2013.0         8.4
86  3     The Dark...      2012.0         8.5
..  ...   ...              ...            ...
4   4555  Pandora'...      1929.0         8.0
3   2694   Metropolis      1927.0         8.3
2   4767  The Big ...      1925.0         8.3
1   4833  Over the...      1920.0         4.8
0   4695  Intolera...      1916.0         8.0
```

可以按升序对一列进行排序,而同时按降序对另一列进行排序。要完成此操作,可以将一个布尔值列表传递给 ascending 参数(该参数与你希望的各列排序方式相对应)。在下面的示例中,即按降序分别对 title_year 和 content_rating 进行排序,并按升序对 budget 进行排序,这样就可以在年度和内容分级组中找到预算最低的电影。

```
>>> (
...     movie[
...         [
...             "movie_title",
...             "title_year",
...             "content_rating",
...             "budget",
...         ]
```

```
...         ]
...     .sort_values(
...         ["title_year", "content_rating", "budget"],
...         ascending=[False, False, True],
...     )
...     .drop_duplicates(
...         subset=["title_year", "content_rating"]
...     )
... )
       movie_title   title_year   content_rating        budget
4026     Compadres       2016.0                R     3000000.0
4658    Fight to...     2016.0            PG-13      150000.0
4661    Rodeo Girl     2016.0               PG       500000.0
3252   The Wailing     2016.0        Not Rated            NaN
4659    Alleluia...    2016.0              NaN       500000.0
 ...          ...         ...              ...            ...
2558    Lilyhammer        NaN           TV-MA     34000000.0
807     Sabrina,...       NaN            TV-G      3000000.0
848     Stargate...       NaN           TV-14      1400000.0
2436        Carlos        NaN        Not Rated           NaN
2119    The Bach...       NaN             NaN      3000000.0
```

默认情况下，.drop_duplicates 方法会保留值的第一个外观，但这也是可以修改的。例如，可以通过传递 keep ='last'选择每个组的最后一行，或者通过传递 keep = False 完全删除所有重复项。

4.7 使用 sort_values 复制 nlargest

前两个秘笈的工作方式相似，它们以略有不同的方式对值进行排序。查找数据列的前 n 个值实际上等效于对整个列进行降序排序并获取前 n 个值。Pandas 有许多操作都是可以按多种方式执行的。

在此秘笈中，我们将使用.sort_values 方法复制 4.5 节"从最大中选择最小"秘笈的操作，并探索二者的区别。

4.7.1 实战操作

（1）让我们从 4.5 节"从最大中选择最小"秘笈的步骤（3）中来重新创建结果。

```
>>> movie = pd.read_csv("data/movie.csv")
```

```
>>> (
...     movie[["movie_title", "imdb_score", "budget"]]
...     .nlargest(100, "imdb_score")
...     .nsmallest(5, "budget")
... )
            movie_title  imdb_score    budget
4804      Butterfly Girl         8.7  180000.0
4801  Children of Heaven         8.5  180000.0
4706        12 Angry Men         8.9  350000.0
4550        A Separation         8.4  500000.0
4636  The Other Dream Team        8.4  500000.0
```

（2）使用.sort_values方法复制表达式的第一部分，并使用.head方法获取前100行。

```
>>> (
...     movie[["movie_title", "imdb_score", "budget"]]
...     .sort_values("imdb_score", ascending=False)
...     .head(100)
... )
         movie_title  imdb_score      budget
2725     Towering...         9.5         NaN
1920     The Shaw...         9.3  25000000.0
3402     The Godf...         9.2   6000000.0
2779         Dekalog         9.1         NaN
4312     Kickboxe...         9.1  17000000.0
...              ...         ...         ...
3799     Anne of ...         8.4         NaN
3777     Requiem ...         8.4   4500000.0
3935     Batman: ...         8.4   3500000.0
4636     The Othe...         8.4    500000.0
2455          Aliens         8.4  18500000.0
```

（3）现在我们拥有评分最高的100部电影，可以再次将.sort_values方法与.head方法结合使用，以按budget排序，获得预算最低的5部电影。

```
>>> (
...     movie[["movie_title", "imdb_score", "budget"]]
...     .sort_values("imdb_score", ascending=False)
...     .head(100)
...     .sort_values("budget")
...     .head(5)
... )
                    movie_title  imdb_score    budget
4815  A Charlie Brown Christmas         8.4  150000.0
```

```
4801    Children of Heaven         8.5    180000.0
4804         Butterfly Girl        8.7    180000.0
4706           12 Angry Men        8.9    350000.0
4636    The Other Dream Team       8.4    500000.0
```

4.7.2 原理解释

如步骤（2）所示，通过在操作后链接.head 方法，.sort_values 方法几乎可以复制.nlargest。步骤（3）则通过链接另一个.sort_values方法复制了.nsmallest，并通过使用.head 方法仅获取前 5 行完成了查询。

现在可以仔细查看步骤（1）中第一个 DataFrame 的输出，并将其与步骤（3）中的输出进行比较。什么？它们竟然是不一样的？为什么？当理解为什么两个结果不一致时，可以查看每个操作秘笈的中间步骤的尾部。

```
>>> (
...     movie[["movie_title", "imdb_score", "budget"]]
...     .nlargest(100, "imdb_score")
...     .tail()
... )
           movie_title        imdb_score    budget
4023              Oldboy           8.4    3000000.0
4163  To Kill a Mockingbird        8.4    2000000.0
4395        Reservoir Dogs         8.4    1200000.0
4550           A Separation        8.4     500000.0
4636   The Other Dream Team        8.4     500000.0

>>> (
...     movie[["movie_title", "imdb_score", "budget"]]
...     .sort_values("imdb_score", ascending=False)
...     .head(100)
...     .tail()
... )
        movie_title    imdb_score     budget
3799     Anne of ...      8.4          NaN
3777     Requiem ...      8.4      4500000.0
3935      Batman: ...     8.4      3500000.0
4636     The Othe...     8.4       500000.0
2455          Aliens     8.4     18500000.0
```

可以看到，这是由于存在超过 100 部电影的评分至少为 8.4 而引起的问题。.nlargest 和.sort_values 方法打破这种并列排名时处理方式不同，这导致 100 行 DataFrame 略有不同。

如果将 kind ='mergsort' 传递给 .sort_values 方法，则将获得与.nlargest 方法相同的结果。

4.8 计算追踪止损单价格

　　交易股票有许多策略。许多投资者采用的一种基本交易类型是止损单（stop order）。止损单是投资者下达的买入或卖出股票的命令，每当市场价格达到特定点时，该命令就会执行。止损单对于防止巨大损失和保护收益都是有用的。

　　在此秘笈中，我们将仅检查用于出售当前拥有的股票的止损单。在典型的止损单中，价格在挂单的整个生命周期内都不会改变。例如，如果你以每股 100 元的价格购买了股票，则可能希望将止损单设置为每股 90 元，以将下行空间限制为 10%。

　　更高级的策略是不断修改止损单的售出价格，以跟踪股票价格的增加，这称为追踪止损单（trailing stop order）。

　　具体来说，如果相同的股票已经由 100 元增加到 120 元，那么低于当前市价 10%的追踪止损单将使售出价格上涨到 108 元。

　　需要重要的是，自购买之日起，追踪止损单永远不会向下移动，并始终与最大值挂钩。如果股票价格从 120 元跌至 110 元，那么止损单仍将保持在 108 元。只有价格上涨至 120 元以上时，止损单才会增加。

　　此秘笈需要使用第三方软件包 pandas-datareader，该软件包可以在线获取股票市场价格。pandas-datareader 软件包没有随 Pandas 一起预装。要安装此软件包，请使用命令行并运行以下命令。

```
conda install pandas-datareader
```

或

```
pip install pandas-datareader
```

你可能还需要安装 request_cache 库。

此秘笈将在给定任何股票的初始购买价格时确定其追踪止损单价格。

4.8.1 实战操作

　　（1）我们将使用特斯拉汽车公司的股票（TSLA）作为示例，并假设在 2017 年的第一个交易日进行购买。

```
>>> import datetime
```

```
>>> import pandas_datareader.data as web
>>> import requests_cache
>>> session = requests_cache.CachedSession(
...     cache_name="cache",
...     backend="sqlite",
...     expire_after=datetime.timedelta(days=90),
... )
>>> tsla = web.DataReader(
...     "tsla",
...     data_source="yahoo",
...     start="2017-1-1",
...     session=session,
... )
>>> tsla.head(8)
                  High         Low  ...     Volume    Adj Close
Date                                ...
2017-01-03  220.330002  210.960007  ...    5923300   216.990005
2017-01-04  228.000000  214.309998  ...   11213500   226.990005
2017-01-05  227.479996  221.949997  ...    5911700   226.750000
2017-01-06  230.309998  225.449997  ...    5527900   229.009995
2017-01-09  231.919998  228.000000  ...    3979500   231.279999
2017-01-10  232.000000  226.889999  ...    3660000   229.869995
2017-01-11  229.979996  226.679993  ...    3650800   229.729996
2017-01-12  230.699997  225.580002  ...    3790200   229.589996
```

（2）为简单起见，我们将使用每个交易日的收盘价。

```
>>> tsla_close = tsla["Close"]
```

（3）使用 .cummax 方法跟踪直到当前日期的最高收盘价。

```
>>> tsla_cummax = tsla_close.cummax()
>>> tsla_cummax.head()
Date
2017-01-03    216.990005
2017-01-04    226.990005
2017-01-05    226.990005
2017-01-06    229.009995
2017-01-09    231.279999
Name: Close, dtype: float64
```

（4）为了将下行限制为 10%，可以将结果乘以 0.9。这将创建追踪止损单。现在可以将所有步骤链接在一起。

```
>>> (tsla["Close"].cummax().mul(0.9).head())
Date
2017-01-03    195.291005
2017-01-04    204.291005
2017-01-05    204.291005
2017-01-06    206.108995
2017-01-09    208.151999
Name: Close,dtype: float64
```

4.8.2 原理解释

.cummax 方法的工作原理是，保留直至当前遇到的最大值（包括当前值）。将该 Series 乘以 0.9 或你要使用的任何停损比例，以创建跟踪止损单。在此特定示例中，TSLA 股票的价格增加了，因此其跟踪止损价格也增加了。

4.8.3 扩展知识

此秘笈简单介绍了如何使用 Pandas 交易证券，在触发止损单时应立即卖出。

在减肥计划中可以使用非常相似的策略。当你偏离最小体重太远时，可以设置警告。Pandas 为你提供了 .cummin 方法来追踪最小值。如果你连续跟踪每天的体重，则以下代码可提供比迄今为止最低记录体重高出 5% 的追踪减肥指标。

```
weight.cummin() * 1.05
```

第 5 章 探索性数据分析

本章包含以下秘笈。
- 摘要统计。
- 查看列类型。
- 分类数据。
- 连续数据。
- 跨越分类比较连续值。
- 比较两个连续列。
- 使用分类值比较分类值。
- 使用 Pandas 分析库。

5.1 介 绍

本章将深入讨论探索性数据分析。这是筛查数据试图弄清各个列的意义及其之间关系的过程。

这项工作可能很耗时，但也会带来丰厚的回报。你越了解数据，就越能很好地利用它。如果你打算建立机器学习模型，那么对数据的深入了解有助于建立性能更高的模型并准确理解预测结果。

本章将使用来自 www.fueleconomy.gov 的数据集，该数据集提供了 1984—2018 年汽车的制造商和型号信息。通过探索性数据分析，分析人员将可以了解到该数据中许多列的意义及其之间的关系。

5.2 摘要统计

摘要统计信息包括平均值、分位数和标准差等。.describe 方法将在 DataFrame 中的所有数字列上计算这些度量。

5.2.1 实战操作

(1) 加载数据集。

```
>>> import pandas as pd
>>> import numpy as np
>>> fueleco = pd.read_csv("data/vehicles.csv.zip")
>>> fueleco
       barrels08  barrelsA08  ...  phevHwy  phevComb
0      15.695714         0.0  ...        0         0
1      29.964545         0.0  ...        0         0
2      12.207778         0.0  ...        0         0
3      29.964545         0.0  ...        0         0
4      17.347895         0.0  ...        0         0
...          ...         ...  ...      ...       ...
39096  14.982273         0.0  ...        0         0
39097  14.330870         0.0  ...        0         0
39098  15.695714         0.0  ...        0         0
39099  15.695714         0.0  ...        0         0
39100  18.311667         0.0  ...        0         0
```

(2) 调用单个摘要统计方法,如.mean(平均值)、.std(标准差)和 .quantile(分位数)等。

```
>>> fueleco.mean()
barrels08          17.442712
barrelsA08          0.219276
charge120           0.000000
charge240           0.029630
city08             18.077799
                     ...
youSaveSpend    -3459.572645
charge240b          0.005869
phevCity            0.094703
phevHwy             0.094269
phevComb            0.094141
Length: 60, dtype: float64

>>> fueleco.std()
barrels08           4.580230
barrelsA08          1.143837
```

```
charge120              0.000000
charge240              0.487408
city08                 6.970672
                         ...
youSaveSpend        3010.284617
charge240b             0.165399
phevCity               2.279478
phevHwy                2.191115
phevComb               2.226500
Length: 60, dtype: float64

>>> fueleco.quantile(
...     [0, 0.25, 0.5, 0.75, 1]
... )
       barrels08   barrelsA08  ...  phevHwy   phevComb
0.00    0.060000    0.000000  ...      0.0       0.0
0.25   14.330870    0.000000  ...      0.0       0.0
0.50   17.347895    0.000000  ...      0.0       0.0
0.75   20.115000    0.000000  ...      0.0       0.0
1.00   47.087143   18.311667  ...     81.0      88.0
```

（3）调用.describe 方法。

```
>>> fueleco.describe()
         barrels08    barrelsA08  ...    phevHwy      phevComb
count  39101.00...   39101.00...  ...  39101.00...   39101.00...
mean      17.442712     0.219276  ...     0.094269     0.094141
std        4.580230     1.143837  ...     2.191115     2.226500
min        0.060000     0.000000  ...     0.000000     0.000000
25%       14.330870     0.000000  ...     0.000000     0.000000
50%       17.347895     0.000000  ...     0.000000     0.000000
75%       20.115000     0.000000  ...     0.000000     0.000000
max       47.087143    18.311667  ...    81.000000    88.000000
```

（4）要获取有关 object 列的摘要统计信息，可使用.include 参数。

```
>>> fueleco.describe(include=object)
            drive     eng_dscr  ...  modifiedOn    startStop
count       37912        23431  ...       39101         7405
unique          7          545  ...          68            2
top     Front-Wh...       (FFS) ...    Tue Jan ...        N
freq        13653         8827  ...       29438         5176
```

5.2.2 原理解释

笔者曾经举办过数据分析方面的培训，在介绍完.describe 方法的应用之后，有些学生狠狠地拍了拍自己的脑袋，一副恍然大悟而又后悔不迭的模样，当笔者问他们原委时，他们回答说，他们刚刚花了两周的时间来为数据库实现该功能。

默认情况下，.describe 方法将计算数字列上的摘要统计信息。你可以传递 include 参数来告诉该方法包括非数字数据类型。请注意，这将显示唯一值的计数、最频繁出现的值（顶部）以及对象列的频率计数。

5.2.3 扩展知识

通常会用到的一个在屏幕上显示更多数据的技巧是转置 DataFrame。这对于.describe 方法的输出很有用。

```
>>> fueleco.describe().T
                count        mean      ...        75%          max
barrels08      39101.0    17.442712   ...     20.115       47.087143
barrelsA08     39101.0     0.219276   ...      0.000       18.311667
charge120      39101.0     0.000000   ...      0.000        0.000000
charge240      39101.0     0.029630   ...      0.000       12.000000
city08         39101.0    18.077799   ...     20.000      150.000000
...                ...          ...   ...        ...             ...
youSaveSpend   39101.0 -3459.572645   ...  -1500.000     5250.000000
charge240b     39101.0     0.005869   ...      0.000        7.000000
phevCity       39101.0     0.094703   ...      0.000       97.000000
phevHwy        39101.0     0.094269   ...      0.000       81.000000
phevComb       39101.0     0.094141   ...      0.000       88.000000
```

5.3 查看列类型

通过查看列的类型也可以收集到有关 Pandas 数据的信息。在本秘笈中，我们将深入探讨列的类型。

5.3.1 实战操作

（1）检查.dtypes 属性。

```
>>> fueleco.dtypes
barrels08          float64
barrelsA08         float64
charge120          float64
charge240          float64
city08               int64
                    ...
modifiedOn          object
startStop           object
phevCity             int64
phevHwy              int64
phevComb             int64
Length: 83, dtype: object
```

(2)汇总列的类型。

```
>>> fueleco.dtypes.value_counts()
float64    32
int64      27
object     23
bool        1
dtype: int64
```

5.3.2 原理解释

在 Pandas 中读取 CSV 文件时,该文件必须推断出列的类型。该过程如下。

- ❑ 如果列中的所有值看起来都像是整数值,则将它们转换为整数,并将列的类型设置为 int64。
- ❑ 如果这些值看起来像是浮点数,则将其类型设置为 float64。
- ❑ 如果值是数字,看起来像是浮点或整数值,但是包含了缺失值,则将它们分配为 float64 类型,因为通常用于缺失值的值 np.nan 是浮点类型的。
- ❑ 如果值中包含 False 或 True,则将它们分配为布尔类型。
- ❑ 其他情况下,将该列视为字符串值,并为其指定 object 类型(对于缺失值可以使用 float64 类型)。

请注意,如果使用 parse_dates 参数,则某些列可能会被转换为日期时间。本书第 12 章"时间序列分析"和第 13 章"使用 Matplotlib、Pandas 和 Seaborn 进行可视化"均显示了解析日期的示例。

仅通过查看 .dtypes 的输出,即可了解到有关数据的更多信息,而不只是数据类型那么简单。例如，我们可以看到某些东西是字符串还是缺失值。object 类型可以是字符串或分类数据，但它们也可以是类似数字的值，只需要稍微调整，它们就可能变成数字。我们通常会让整数列保持原样不动，或者将它们视为连续值。如果某一列中的值是浮点值，则表明该列可能有以下 3 种情况。

- 所有值都是浮点值，并且没有缺失值。
- 有浮点值，但是也包含缺失值。
- 多数都是整数值，只是因为包含缺失值而被转换为浮点类型。

5.3.3 扩展知识

当 Pandas 将列转换为浮点数或整数类型时，Pandas 将使用这些类型的 64 位版本。如果你知道整数的取值限制在某个范围内（或者你愿意牺牲浮点数的精度），则可以通过将这些列的数值类型转换为使用较少内存的类型来节省内存使用量。

```
>>> fueleco.select_dtypes("int64").describe().T
                count        mean   ...      75%      max
city08        39101.0   18.077799   ...     20.0    150.0
cityA08       39101.0    0.569883   ...      0.0    145.0
co2           39101.0   72.538989   ...     -1.0    847.0
co2A          39101.0    5.543950   ...     -1.0    713.0
comb08        39101.0   20.323828   ...     23.0    136.0
...               ...         ...   ...      ...      ...
year          39101.0 2000.635406   ...   2010.0   2018.0
youSaveSpend  39101.0 -3459.572645  ...  -1500.0   5250.0
phevCity      39101.0    0.094703   ...      0.0     97.0
phevHwy       39101.0    0.094269   ...      0.0     81.0
phevComb      39101.0    0.094141   ...      0.0     88.0
```

在上面的示例中可以看到，city08 和 comb08 列的值都没有超过 150。NumPy 中的 iinfo 函数可以向我们显示整数类型的限制。在了解了这一限制之后，我们就会明白，在本示例中，虽然不能为这些列使用 int8，但是可以使用 int16。通过从 int64 转换为 int16 类型，这些列的内存使用量将只有原来的 25%。

```
>>> np.iinfo(np.int8)
iinfo(min=-128, max=127, dtype=int8)
>>> np.iinfo(np.int16)
iinfo(min=-32768, max=32767, dtype=int16)

>>> fueleco[["city08", "comb08"]].info(memory_usage="deep")
```

```
<class 'pandas.core.frame.DataFrame'>
RangeIndex: 39101 entries, 0 to 39100
Data columns (total 2 columns):
 #   Column  Non-Null Count  Dtype
---  ------  --------------  -----
 0   city08  39101 non-null  int64
 1   comb08  39101 non-null  int64
dtypes: int64(2)
memory usage: 611.1 KB

>>> (
...     fueleco[["city08", "comb08"]]
...     .assign(
...         city08=fueleco.city08.astype(np.int16),
...         comb08=fueleco.comb08.astype(np.int16),
...     )
...     .info(memory_usage="deep")
... )
<class 'pandas.core.frame.DataFrame'>
RangeIndex: 39101 entries, 0 to 39100
Data columns (total 2 columns):
 #   Column  Non-Null Count  Dtype
---  ------  --------------  -----
 0   city08  39101 non-null  int16
 1   comb08  39101 non-null  int16
dtypes: int16(2)
memory usage: 152.9 KB
```

请注意，NumPy 中有一个类似的 finfo 函数，可用于检索浮点信息。

为字符串列节省内存的方法之一是将此列转换为分类。如果字符串列中的每个值都是唯一的，那么这将降低 Pandas 的速度并且使用更多的内存；但是，如果基数较低，则转换为分类可以节省很多内存。例如，在下面的示例中，make 列的基数较低（134），而 model 列的基数较高（3816），则该列转换为分类时可节省的内存较少。

以下我们将仅提取出这两列。但是，我们将获得仅包含该列名称的列表，而不是获得 Series，这将使我们返回一个具有单个列的 DataFrame。我们将列类型更新为分类，并查看其内存使用情况。

请记住传递 memory_usage ='deep' 来获取 object 列的内存使用情况。

```
>>> fueleco.make.nunique()
134
>>> fueleco.model.nunique()
```

3816

```
>>> fueleco[["make"]].info(memory_usage="deep")
<class 'pandas.core.frame.DataFrame'>
RangeIndex: 39101 entries, 0 to 39100
Data columns (total 1 columns):
 #   Column  Non-Null Count  Dtype
---  ------  --------------  -----
 0   make    39101 non-null  object
dtypes: object(1)
memory usage: 2.4 MB

>>> (
...     fueleco[["make"]]
...     .assign(make=fueleco.make.astype("category"))
...     .info(memory_usage="deep")
... )
<class 'pandas.core.frame.DataFrame'>
RangeIndex: 39101 entries, 0 to 39100
Data columns (total 1 columns):
 #   Column  Non-Null Count  Dtype
---  ------  --------------  -----
 0   make    39101 non-null  category
dtypes: category(1)
memory usage: 90.4 KB

>>> fueleco[["model"]].info(memory_usage="deep")
<class 'pandas.core.frame.DataFrame'>
RangeIndex: 39101 entries, 0 to 39100
Data columns (total 1 columns):
 #   Column  Non-Null Count  Dtype
---  ------  --------------  -----
 0   model   39101 non-null  object
dtypes: object(1)
memory usage: 2.5 MB

>>> (
...     fueleco[["model"]]
...     .assign(model=fueleco.model.astype("category"))
...     .info(memory_usage="deep")
... )
<class 'pandas.core.frame.DataFrame'>
```

```
RangeIndex: 39101 entries, 0 to 39100
Data columns (total 1 columns):
 #   Column  Non-Null Count  Dtype
---  ------  --------------  -----
 0   model   39101 non-null  category
dtypes: category(1)
memory usage: 496.7 KB
```

5.4 分 类 数 据

我们可以将数据大致分为日期值、连续值和分类值。在本节中，我们将探索量化和可视化分类数据。

5.4.1 实战操作

（1）选择具有 object 数据类型的列。

```
>>> fueleco.select_dtypes(object).columns
Index(['drive', 'eng_dscr', 'fuelType', 'fuelType1', 'make', 'model',
    'mpgData', 'trany', 'VClass', 'guzzler', 'trans_dscr', 'tCharger',
    'sCharger', 'atvType', 'fuelType2', 'rangeA', 'evMotor', 'mfrCode',
    'c240Dscr', 'c240bDscr', 'createdOn', 'modifiedOn', 'startStop'],
    dtype='object')
```

（2）使用.nunique 方法确定基数。

```
>>> fueleco.drive.nunique()
7
```

（3）使用.sample 方法查看一些值。

```
>>> fueleco.drive.sample(5, random_state=42)
4217     4-Wheel ...
1736     4-Wheel ...
36029    Rear-Whe...
37631    Front-Wh...
1668     Rear-Whe...
Name: drive, dtype: object
```

（4）确定缺失值的数量和百分比。

```
>>> fueleco.drive.isna().sum()
```

```
1189
```

```
>>> fueleco.drive.isna().mean() * 100
3.0408429451932175
```

（5）使用 .value_counts 方法汇总一列。

```
>>> fueleco.drive.value_counts()
Front-Wheel Drive              13653
Rear-Wheel Drive               13284
4-Wheel or All-Wheel Drive      6648
All-Wheel Drive                 2401
4-Wheel Drive                   1221
2-Wheel Drive                    507
Part-time 4-Wheel Drive          198
Name: drive, dtype: int64
```

（6）如果摘要中的值太多，则可能要查看前 6 个值并折叠其余值。

```
>>> top_n = fueleco.make.value_counts().index[:6]
>>> (
...     fueleco.assign(
...         make=fueleco.make.where(
...             fueleco.make.isin(top_n), "Other"
...         )
...     ).make.value_counts()
... )
Other        23211
Chevrolet     3900
Ford          3208
Dodge         2557
GMC           2442
Toyota        1976
BMW           1807
Name: make, dtype: int64
```

（7）使用 Pandas 绘制计数并将其可视化。

```
>>> import matplotlib.pyplot as plt
>>> fig, ax = plt.subplots(figsize=(10, 8))
>>> top_n = fueleco.make.value_counts().index[:6]
>>> (
...     fueleco.assign(
...         make=fueleco.make.where(
...             fueleco.make.isin(top_n), "Other"
```

```
...             )
...         )
...         .make.value_counts()
...         .plot.bar(ax=ax)
... )
>>> fig.savefig("c5-catpan.png", dpi=300)
```

其输出结果如图 5-1 所示。

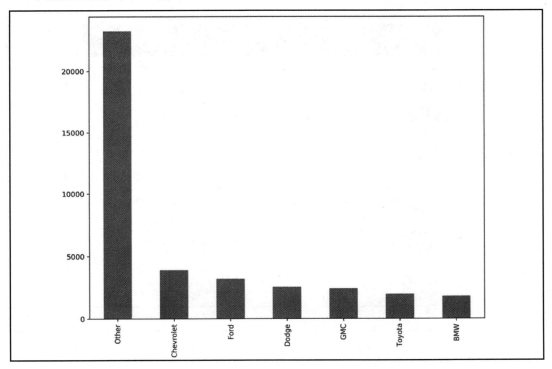

图 5-1　使用 Pandas 可视化分类

（8）使用 seaborn 绘制计数并将其可视化。

```
>>> import seaborn as sns
>>> fig, ax = plt.subplots(figsize=(10, 8))
>>> top_n = fueleco.make.value_counts().index[:6]
>>> sns.countplot(
...     y="make",
...     data=(
...         fueleco.assign(
...             make=fueleco.make.where(
```

```
...                     fueleco.make.isin(top_n), "Other"
...                )
...            )
...        ),
...    )
>>> fig.savefig("c5-catsns.png", dpi=300)
```

其输出结果如图 5-2 所示。

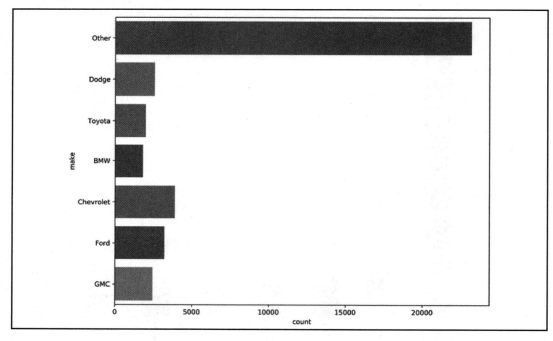

图 5-2　使用 Seaborn 可视化分类

5.4.2　原理解释

当检查分类变量时，我们想知道有多少个唯一值。如果这是一个很大的值，则该列可能不是分类的，而是自由文本或数字列（这可能是因为 Pandas 遇到无效的数字，Pandas 不知道如何将其存储为数字）。

.sample 方法可以让我们对某些值做一些研究。对于大多数列来说，确定有多少缺失值很重要。看起来好像有超过 1000 行，或大约 3% 的缺失值。一般来说，我们需要与行业专家（SME）进行交流，以确定为什么会有这些缺失值，以及是否需要评估或丢弃它们。

下面就来研究 drive 缺失的行。

```
>>> fueleco[fueleco.drive.isna()]
       barrels08  barrelsA08  ...  phevHwy  phevComb
7138    0.240000         0.0  ...        0         0
8144    0.312000         0.0  ...        0         0
8147    0.270000         0.0  ...        0         0
18215  15.695714         0.0  ...        0         0
18216  14.982273         0.0  ...        0         0
...          ...         ...  ...      ...       ...
23023   0.240000         0.0  ...        0         0
23024   0.546000         0.0  ...        0         0
23026   0.426000         0.0  ...        0         0
23031   0.426000         0.0  ...        0         0
23034   0.204000         0.0  ...        0         0
```

我们最喜欢使用的检查分类列的方法是 .value_counts 方法，这是一个很好的起点，因为可以从该方法的输出中找到许多其他问题的答案。默认情况下，.value_counts 方法不显示缺失值，但是可以使用 dropna 参数来解决此问题。

```
>>> fueleco.drive.value_counts(dropna=False)
Front-Wheel Drive             13653
Rear-Wheel Drive              13284
4-Wheel or All-Wheel Drive     6648
All-Wheel Drive                2401
4-Wheel Drive                  1221
NaN                            1189
2-Wheel Drive                   507
Part-time 4-Wheel Drive         198
Name: drive, dtype: int64
```

最后，可以使用 Pandas 或 Seaborn 可视化此输出。条形图是执行此操作的合适图。但是，如果这是基数较高的列，则对于有效图而言，可能有太多条形。你可以按照步骤（6）中的操作来限制列数。在使用 Seaborn 的情况下，则可以使用 countplot 的 order 参数。

使用 Pandas 可以进行快速绘图，因为这通常是一个方法调用。当然，Seaborn 库也有自己的特色技巧并且是在 Pandas 中不容易做到的，在后面的秘笈中将会看到这一点。

5.4.3　扩展知识

某些列报告为 object 数据类型，但此类列并不是真正的分类。在此数据集中，rangeA 列具有 object 数据类型。但是，如果使用分类方法 .value_counts 进行检查，则会发现

rangeA 并不是真正的分类，而是一个貌似分类的数字列。

这是因为，正如我们从.value_counts 方法的输出中看到的那样，某些条目中有斜杠（/）和短横（-），而 Pandas 不知道如何将这些值转换为数字，因此它将整列保留为字符串列。

```
>>> fueleco.rangeA.value_counts()
290       74
270       56
280       53
310       41
277       38
          ..
328        1
250/370    1
362/537    1
310/370    1
340-350    1
Name: rangeA, Length: 216, dtype: int64
```

查找特殊字符的另一种方法是使用.str.extract 方法（该方法可以使用正则表达式）。

```
>>> (
...     fueleco.rangeA.str.extract(r"([^0-9.])")
...     .dropna()
...     .apply(lambda row: "".join(row), axis=1)
...     .value_counts()
... )
/    280
-     71
Name: rangeA, dtype: int64
```

这意味着 rangeA 列实际上具有两种类型，即 float 和 string。其数据类型被报告为 object，因为该类型可以包含异构类型的列。缺失值被存储为 NaN，而非缺失值则是字符串。

```
>>> set(fueleco.rangeA.apply(type))
{<class 'str'>, <class 'float'>}
```

以下是缺失值的计数。

```
>>> fueleco.rangeA.isna().sum()
37616
```

根据 fueleconomy.gov 网站的数据，rangeA 列值表示混合动力车辆的第二种燃料类型的范围（包括 E85、电力、天然气和液化石油气等）。使用 Pandas 时，可以将缺失值替换为 0，将短横（-）替换为斜杠（/），然后拆分并取每一行的平均值（在出现了短横/斜杠的情况下）。

```
>>> (
...     fueleco.rangeA.fillna("0")
...     .str.replace("-", "/")
...     .str.split("/", expand=True)
...     .astype(float)
...     .mean(axis=1)
... )
0        0.0
1        0.0
2        0.0
3        0.0
4        0.0
         ...
39096    0.0
39097    0.0
39098    0.0
39099    0.0
39100    0.0
Length: 39101, dtype: float64
```

还可以通过对数字列进行分箱（binning）处理来将它们视为分类。Pandas 有两个强大的函数来辅助划分，即 cut 和 qcut。可以使用 cut 函数来切成等宽的箱子或由我们指定箱子的宽度。对于 rangeA 列来说，大多数值都是空的，所以可将该列值替换为 0，并据此划分为 10 个等宽的箱子。具体划分如下。

```
>>> (
...     fueleco.rangeA.fillna("0")
...     .str.replace("-", "/")
...     .str.split("/", expand=True)
...     .astype(float)
...     .mean(axis=1)
...     .pipe(lambda ser_: pd.cut(ser_, 10))
...     .value_counts()
... )
(-0.45, 44.95]      37688
(269.7, 314.65]       559
(314.65, 359.6]       352
(359.6, 404.55]       205
(224.75, 269.7]       181
```

```
(404.55, 449.5]    82
(89.9, 134.85]     12
(179.8, 224.75]     9
(44.95, 89.9]       8
(134.85, 179.8]     5
dtype: int64
```

另外，qcut 函数——表示分位数剪切（quantile cut）——会将条目切成相同大小的箱子。但是，由于 rangeA 列严重偏斜，并且大多数条目为 0，而我们无法将 0 量化到多个箱子中，因此这种方法是失败的。但是它确实（某种程度上）适用于 city08。之所以出现这种情况是因为 city08 的值是整数，因此它们不会平均划分到 10 个存储桶中，但是大小很接近。

```
>>> (
...     fueleco.rangeA.fillna("0")
...     .str.replace("-", "/")
...     .str.split("/", expand=True)
...     .astype(float)
...     .mean(axis=1)
...     .pipe(lambda ser_: pd.qcut(ser_, 10))
...     .value_counts()
... )
Traceback (most recent call last):
  ...
ValueError: Bin edges must be unique: array([   0. ,    0. ,    0. ,    0.,
    0. ,    0. ,    0. ,    0. ,    0. ,
    0. ,  449.5]).

>>> (
...     fueleco.city08.pipe(
...         lambda ser: pd.qcut(ser, q=10)
...     ).value_counts()
... )
(5.999, 13.0]    5939
(19.0, 21.0]     4477
(14.0, 15.0]     4381
(17.0, 18.0]     3912
(16.0, 17.0]     3881
(15.0, 16.0]     3855
(21.0, 24.0]     3676
(24.0, 150.0]    3235
(13.0, 14.0]     2898
```

```
(18.0, 19.0]    2847
Name: city08, dtype: int64
```

5.5 连 续 数 据

笔者对连续数据的广义定义是存储为数字（整数或浮点数）的数据。在分类数据和连续数据之间存在一些灰色区域。例如，年级就可以表示为数字（忽略幼儿园，或使用 0 表示）。在这种情况下，年级列既可以是分类的，也可以是连续的，因此，本节和 5.4 节"分类数据"中的技术都可以应用。

本节将检查燃油经济性数据集中的连续列。在 city08 列中，列出了城市道路条件下以较低速度驾驶汽车时预期的每加仑英里数。

5.5.1 实战操作

（1）选择数字列（通常是 int64 或 float64）。

```
>>> fueleco.select_dtypes("number")
       barrels08  barrelsA08  ...  phevHwy  phevComb
0      15.695714         0.0  ...        0         0
1      29.964545         0.0  ...        0         0
2      12.207778         0.0  ...        0         0
3      29.964545         0.0  ...        0         0
4      17.347895         0.0  ...        0         0
...          ...         ...  ...      ...       ...
39096  14.982273         0.0  ...        0         0
39097  14.330870         0.0  ...        0         0
39098  15.695714         0.0  ...        0         0
39099  15.695714         0.0  ...        0         0
39100  18.311667         0.0  ...        0         0
```

（2）使用 .sample 方法查看部分值。

```
>>> fueleco.city08.sample(5, random_state=42)
4217     11
1736     21
36029    16
37631    16
1668     17
Name: city08, dtype: int64
```

（3）确定缺失值的数量和百分比。

```
>>> fueleco.city08.isna().sum()
0

>>> fueleco.city08.isna().mean() * 100
0.0
```

（4）获取摘要统计信息。

```
>>> fueleco.city08.describe()
count    39101.000000
mean        18.077799
std          6.970672
min          6.000000
25%         15.000000
50%         17.000000
75%         20.000000
max        150.000000
Name: city08, dtype: float64
```

（5）使用 Pandas 绘制直方图。

```
>>> import matplotlib.pyplot as plt
>>> fig, ax = plt.subplots(figsize=(10, 8))
>>> fueleco.city08.hist(ax=ax)
>>> fig.savefig(
...     "c5-conthistpan.png", dpi=300
... )
```

其输出结果如图 5-3 所示。

（6）图 5-3 看起来很偏斜，因此可以增加直方图中箱子（bin）的数量，以查看偏斜是否隐藏了行为（因为偏斜会使箱子更宽）。

```
>>> import matplotlib.pyplot as plt
>>> fig, ax = plt.subplots(figsize=(10, 8))
>>> fueleco.city08.hist(ax=ax, bins=30)
>>> fig.savefig(
...     "c5-conthistpanbins.png", dpi=300
... )
```

其输出结果如图 5-4 所示。

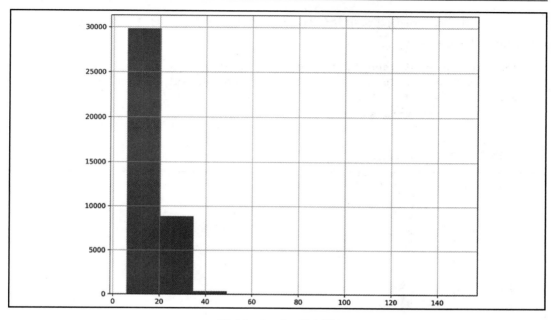

图 5-3　Pandas 直方图

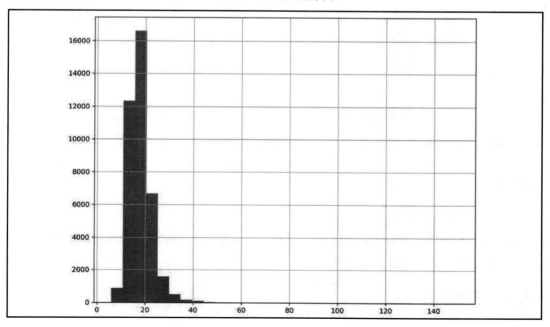

图 5-4　修改之后的 Pandas 直方图

（7）使用 Seaborn 创建一个分布图，其中包括直方图、核密度估计（kernel density estimation，KDE）和 rug 图。

```
>>> fig, ax = plt.subplots(figsize=(10, 8))
>>> sns.distplot(fueleco.city08, rug=True, ax=ax)
>>> fig.savefig(
...     "c5-conthistsns.png", dpi=300
... )
```

其输出结果如图 5-5 所示。

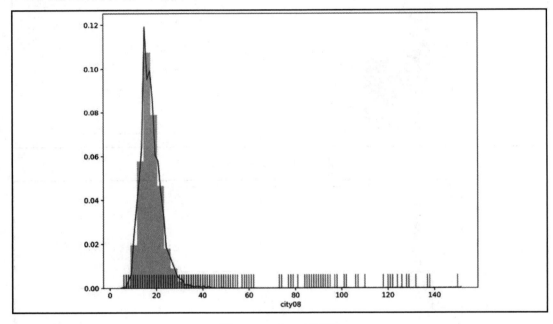

图 5-5　Seaborn 直方图

💡 提示：
- ❑ 直方图显示的是箱子。
- ❑ KDE 显示的是密度曲线。
- ❑ rug 图可以标示数据分布情况。元素出现一次，就会绘制一个小竖杠。因此，从 rug 的疏密可以看出元素的疏密。

5.5.2　原理解释

分析人员应该对数字代表的意义有良好的感觉，而研究数据的样本将使你找到这种感觉，知道某些值代表的意义。另外，我们还需要知道缺失值的情况。回想一下，当我们对列执行操作时，Pandas 将忽略缺失值。

.describe 方法提供的摘要统计信息非常有用，这可能是我们最喜欢的检查连续值的方法。一定要检查最小值和最大值，这对于理解数字的意义很有帮助。例如，如果在 miles per gallon（每加仑英里数）这一列中的最小值出现了负值，那就很奇怪。

分位数向我们表明了数据的偏斜度。由于分位数是数据趋势的可靠指标，因此它们不受异常值的影响。

还需要注意的另一件事是无穷值，无论是正无穷还是负无穷。在我们的示例中没有出现无穷值，如果有这些值，那么它可能会导致某些数学运算或绘图失败。如果数据中包含无穷值，则需要考虑如何对这些值进行处理。Pandas 的常见做法是对无穷值进行修剪和删除。

我们是绘图功能的忠实拥护者，Pandas 和 Seaborn 都使可视化连续数据的分布变得非常容易。俗话说"一图胜千言"，这样的俗话用在数据分析领域真是再恰当不过了。

5.5.3 扩展知识

Seaborn 库有许多汇总连续数据的选项。除 distplot 函数外，还有一些用于创建箱形图（box plot）、增强箱形图（boxen plot）和小提琴图（violin plot）的函数。

R 语言是用于统计分析、绘图的语言和操作环境。R 语言创建了一个称为字母值图（letter value plot）的图，Seaborn 的作者复制了该图，但是将名称更改为增强箱形图。在增强箱形图中，它的中值（median value）是黑色线。中值以 50 表示，从 50~0 和 100 各划分一半，从而划分出四分位数线。箱子从上到下展示的是上四分位数（75 分位数，这是箱子的上边线）、中位数（median，即箱子中间的线）和下四分位数（25 分位数，这是箱子的下边线），上四分位数和下四分位数之间的距离称为四分位距（inter quartile range，IQR），也就是 25~75 分位数的范围。最高的块显示的就是 25~75 分位数的范围。在下边缘的下一个箱子是从 25 到该值的一半（也就是 12.5），所以显示的就是 12.5~25 分位数。重复此模式，下一个箱子就是 6.25~12.5 分位数，以此类推。

小提琴图基本上是一个直方图，其另一侧翻转了一个副本。如果你有双模型直方图，则该直方图看起来像小提琴，这也是它名称的由来。

```
>>> fig, axs = plt.subplots(nrows=3, figsize=(10, 8))
>>> sns.boxplot(fueleco.city08, ax=axs[0])
>>> sns.violinplot(fueleco.city08, ax=axs[1])
>>> sns.boxenplot(fueleco.city08, ax=axs[2])
>>> fig.savefig("c5-contothersns.png", dpi=300)
```

其输出结果如图 5-6 所示。

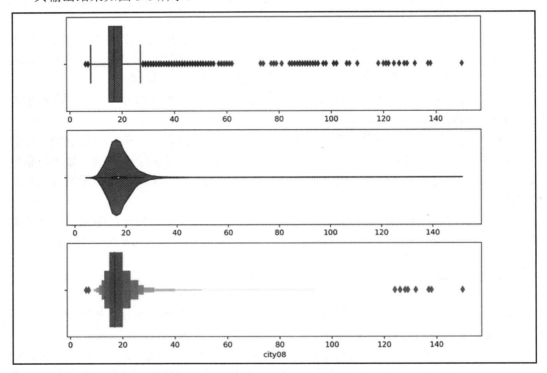

图 5-6 用 Seaborn 创建的箱形图、小提琴图和增强箱形图

如果你担心数据是否正常，则可以使用 SciPy 库通过数字和可视化对其进行量化。

K-S 检验（kolmogorov-smirnov test）可以评估数据是否为正态分布。K-S 检验为我们提供了 p 值。如果此值是显著的（<0.05），则数据就不是正态分布的。

```
>>> from scipy import stats
>>> stats.kstest(fueleco.city08, cdf="norm")
KstestResult(statistic=0.9999999990134123, pvalue=0.0)
```

我们可以绘制概率图（probability plot）以查看值是否为正态分布。正态概率图将是一条直线，如果样本跟踪这条线，则数据是正态分布的。

```
>>> from scipy import stats
>>> fig, ax = plt.subplots(figsize=(10, 8))
>>> stats.probplot(fueleco.city08, plot=ax)
>>> fig.savefig("c5-conprob.png", dpi=300)
```

其输出结果如图 5-7 所示。

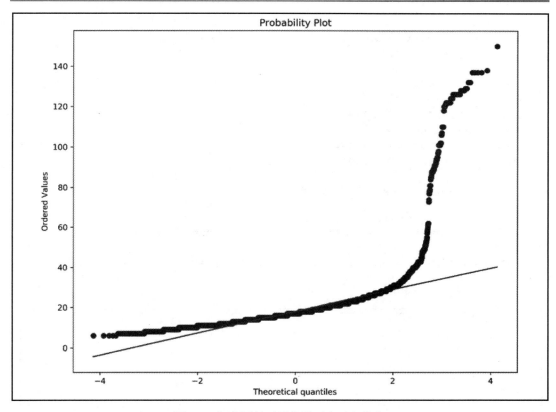

图 5-7　概率图显示了值是否为正态分布

5.6　跨越分类比较连续值

前面的小节讨论了单列数据的研究。本节将显示如何比较不同分类中的连续变量。我们将研究不同汽车品牌（福特、本田、特斯拉和宝马）的里程油耗（mileage）指标。

5.6.1　实战操作

（1）为想要比较的汽车品牌制作一个 mask，然后按汽车品牌分组以查看每组汽车的 city08 列的均值（Mean）和标准差（Std）。

```
>>> mask = fueleco.make.isin(
...     ["Ford", "Honda", "Tesla", "BMW"]
```

```
...   )
>>> fueleco[mask].groupby("make").city08.agg(
...       ["mean", "std"]
...   )
          mean       std
make
BMW    17.817377  7.372907
Ford   16.853803  6.701029
Honda  24.372973  9.154064
Tesla  92.826087  5.538970
```

（2）使用 Seaborn 可视化每个品牌的 city08 值。

```
>>> g = sns.catplot(
...       x="make", y="city08", data=fueleco[mask], kind="box"
...   )
>>> g.ax.figure.savefig("c5-catbox.png", dpi=300)
```

其输出结果如图 5-8 所示。

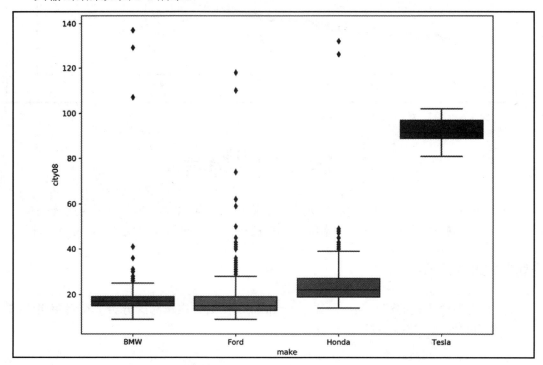

图 5-8　每个品牌的箱形图

5.6.2 原理解释

如果汇总统计信息针对不同的品牌出现了变化，则表明该品牌具有不同的特征。中心趋势（均值或中位数）和方差（或标准偏差）是做比较的良好指标。我们可以看到，本田汽车在城市道路条件下的里程油耗指标要比宝马和福特都好，但方差更大，而特斯拉则要比所有其他品牌都更好，并且方差最小。

使用像 Seaborn 这样的可视化库，可以快速查看到分类中的差异。如图 5-8 所示，这 4 种汽车品牌之间的差异是巨大的，但你也可以看到，非特斯拉品牌汽车还有一些离群值（以菱形表示），其里程油耗指标甚至比特斯拉还要更好。

5.6.3 扩展知识

箱形图的一个缺点是，尽管它指示了数据的传播，但并没有显示出每个品牌中有多少个样本。你可能很自然地认为每个箱形图都具有相同数量的样本，但是，通过使用 Pandas 进行量化，我们可以得出"并非如此"的结论。

```
>>> mask = fueleco.make.isin(
...     ["Ford", "Honda", "Tesla", "BMW"]
... )
>>> (fueleco[mask].groupby("make").city08.count())
make
BMW      1807
Ford     3208
Honda     925
Tesla      46
Name: city08, dtype: int64
```

另一种选择是在箱形图的上方绘制群图（swarm plot，也称为分簇散点图）。

```
>>> g = sns.catplot(
...     x="make", y="city08", data=fueleco[mask], kind="box"
... )
>>> sns.swarmplot(
...     x="make",
...     y="city08",
...     data=fueleco[mask],
...     color="k",
...     size=1,
...     ax=g.ax,
```

```
...    )
>>> g.ax.figure.savefig(
...        "c5-catbox2.png", dpi=300
...    )
```

其输出结果如图 5-9 所示。

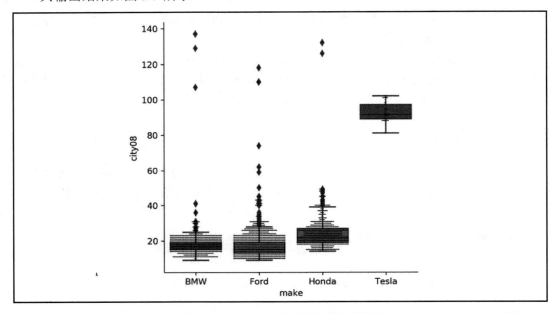

图 5-9 在 Seaborn 箱形图上叠加的群图

此外，catplot 函数还有很多技巧。例如，当前显示了只有两个维度，即城市里程油耗和品牌。事实上，我们还可以在图中添加更多维度。

我们可以通过其他功能对网格进行分面。例如，可以使用 col 参数将新图划分到自己的网格中。

```
>>> g = sns.catplot(
...        x="make",
...        y="city08",
...        data=fueleco[mask],
...        kind="box",
...        col="year",
...        col_order=[2012, 2014, 2016, 2018],
...        col_wrap=2,
...    )
```

```
>>> g.axes[0].figure.savefig(
...     "c5-catboxcol.png", dpi=300
... )
```

其输出结果如图 5-10 所示。

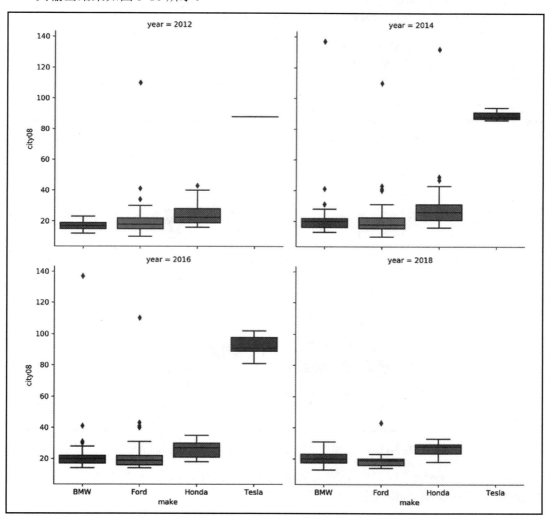

图 5-10　按年份划分网格的 Seaborn 箱形图

另外，还可以使用 hue 参数将新维度嵌入同一绘图中。

```
>>> g = sns.catplot(
```

```
...        x="make",
...        y="city08",
...        data=fueleco[mask],
...        kind="box",
...        hue="year",
...        hue_order=[2012, 2014, 2016, 2018],
... )
>>> g.ax.figure.savefig(
...        "c5-catboxhue.png", dpi=300
... )
```

其输出结果如图 5-11 所示。

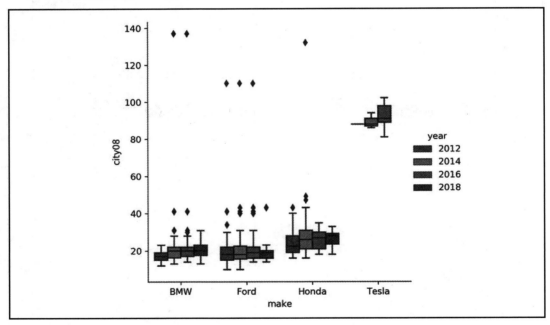

图 5-11　一个按年份给每个品牌上色的 Seaborn 箱形图

如果你在 Jupyter 中，则可以设置 groupby 调用的输出样式，以突出显示极端值。使用 .style.background_gradient 方法即可执行此操作。

```
>>> mask = fueleco.make.isin(
...        ["Ford", "Honda", "Tesla", "BMW"]
... )
>>> (
...        fueleco[mask]
```

```
...     .groupby("make")
...     .city08.agg(["mean", "std"])
...     .style.background_gradient(cmap="RdBu", axis=0)
... )
```

其输出结果如图 5-12 所示。

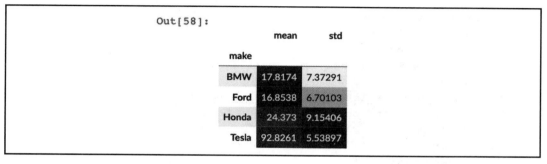

图 5-12　使用 Pandas 样式功能突出显示均值和标准差的最小值和最大值

5.7　比较两个连续列

评估两个连续列如何相互关联是回归分析（regression analysis）的本质，当然我们要做的还不止于此。如果你经常有两列相互之间具有高度相关性，则可以将其中一列作为冗余列删除。本节将讨论连续列对的探索性数据分析。

5.7.1　实战操作

（1）如果两个数字的比例相同，则查看这两个数字的协方差（covariance）。

```
>>> fueleco.city08.cov(fueleco.highway08)
46.33326023673625

>>> fueleco.city08.cov(fueleco.comb08)
47.41994667819079

>>> fueleco.city08.cov(fueleco.cylinders)
-5.931560263764761
```

（2）查看两个数字之间的皮尔逊相关系数（pearson correlation coefficient）。

```
>>> fueleco.city08.corr(fueleco.highway08)
```

```
0.932494506228495

>>> fueleco.city08.corr(fueleco.cylinders)
-0.701654842382788
```

(3) 使用 Seaborn 绘制热图（heatmap）以可视化相关性。

```
>>> import seaborn as sns
>>> fig, ax = plt.subplots(figsize=(8, 8))
>>> corr = fueleco[
...     ["city08", "highway08", "cylinders"]
... ].corr()
>>> mask = np.zeros_like(corr, dtype=np.bool)
>>> mask[np.triu_indices_from(mask)] = True
>>> sns.heatmap(
...     corr,
...     mask=mask,
...     fmt=".2f",
...     annot=True,
...     ax=ax,
...     cmap="RdBu",
...     vmin=-1,
...     vmax=1,
...     square=True,
... )
>>> fig.savefig(
...     "c5-heatmap.png", dpi=300, bbox_inches="tight"
... )
```

其输出结果如图 5-13 所示。

(4) 使用 Pandas 绘制散点图（scatter plot）来表现关系。

先绘制城市道路里程油耗和高速公路（highway）里程油耗之间关系的散点图。

```
>>> fig, ax = plt.subplots(figsize=(8, 8))
>>> fueleco.plot.scatter(
...     x="city08", y="highway08", alpha=0.1, ax=ax
... )
>>> fig.savefig(
...     "c5-scatpan.png", dpi=300, bbox_inches="tight"
... )
```

其输出结果如图 5-14 所示。

第 5 章　探索性数据分析

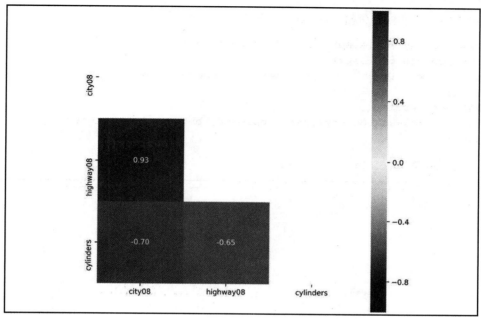

图 5-13　Seaborn 热图

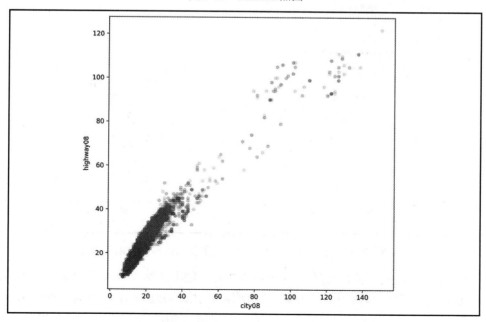

图 5-14　Pandas 散点图，查看城市道路和高速公路里程油耗之间的关系

再绘制里程油耗和气缸（cylinder）之间关系的散点图。

```
>>> fig, ax = plt.subplots(figsize=(8, 8))
>>> fueleco.plot.scatter(
...     x="city08", y="cylinders", alpha=0.1, ax=ax
... )
>>> fig.savefig(
...     "c5-scatpan-cyl.png", dpi=300, bbox_inches="tight"
... )
```

其输出结果如图 5-15 所示。

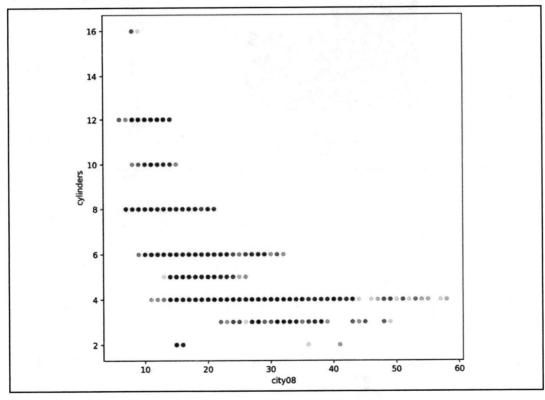

图 5-15　Pandas 散点图，查看城市道路里程油耗和气缸之间的关系

（5）填充一些缺失值。从图 5-15 中可以看出，城市道路里程油耗的某些高端值缺失了，这是因为这些汽车往往是电动的，没有汽缸，因此可通过用 0 填充这些值来解决此问题。

```
>>> fueleco.cylinders.isna().sum()
145

>>> fig, ax = plt.subplots(figsize=(8, 8))
>>> (
...     fueleco.assign(
...         cylinders=fueleco.cylinders.fillna(0)
...     ).plot.scatter(
...         x="city08", y="cylinders", alpha=0.1, ax=ax
...     )
... )
>>> fig.savefig(
...     "c5-scatpan-cyl0.png", dpi=300, bbox_inches="tight"
... )
```

其输出结果如图 5-16 所示。

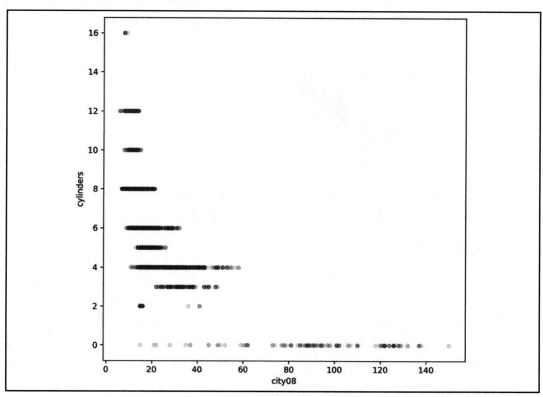

图 5-16　将气缸缺失值填充为 0 后，使用 Pandas 绘制的里程油耗与气缸之间关系的散点图

（6）使用 Seaborn 为城市道路里程油耗和高速公路里程油耗之间关系添加回归线。

```
>>> res = sns.lmplot(
...     x="city08", y="highway08", data=fueleco
... )
>>> res.fig.savefig(
...     "c5-lmplot.png", dpi=300, bbox_inches="tight"
... )
```

其输出结果如图 5-17 所示。

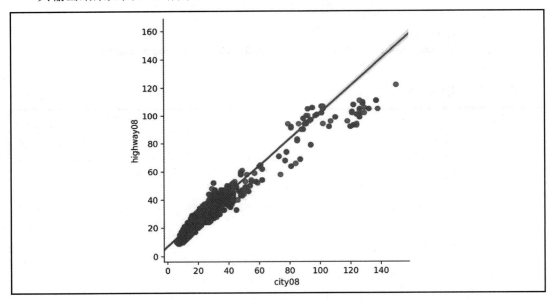

图 5-17 带有城市道路里程油耗和高速公路里程油耗关系回归线的 Seaborn 散点图

5.7.2 原理解释

皮尔逊相关系数告诉我们一个值对另一个值的影响方式。它为-1～1。在本示例中可以看到，城市道路里程油耗与高速公路里程油耗之间有很强的相关性。如果汽车在城市道路条件下有很好的里程油耗表现，那么在高速公路上的里程油耗指标也会很好。

协方差（covariance）让我们知道这些值是如何一起变化的。协方差对于比较那些具有相似相关性的多个连续列很有用。相关性不会随着比例的变化而变化，但是协方差则不然。例如，如果将 city08 与两倍的 Highway08 进行比较，则它们仍然具有相同的相关性，但是协方差则会发生变化。

```
>>> fueleco.city08.corr(fueleco.highway08 * 2)
0.932494506228495

>>> fueleco.city08.cov(fueleco.highway08 * 2)
92.6665204734725
```

热图是查看聚合相关性的好方法。我们可以寻找最蓝和最红的单元以找到最强的相关性。确保将 vmin 和 vmax 参数分别设置为-1 和 1，以便正确着色。

散点图是可视化连续变量之间关系的另一种方法。散点图使我们可以看到展现出来的趋势。笔者喜欢教给学生的一个技巧是确保将 alpha 参数设置为小于或等于 0.5 的值。这可以使点以半透明显示，与使用完全不透明标记的散点图相比，它的效果更好。

5.7.3　扩展知识

如果有更多需要比较的变量，则可以使用 Seaborn 向散点图添加更多维度。例如，使用 relplot 函数，可以按年份对点进行着色，并根据车辆消耗的汽油桶（barrel）数对点进行大小调整。这样我们就已经将散点图从两个维度变成了 4 个维度。

```
>>> res = sns.relplot(
...     x="city08",
...     y="highway08",
...     data=fueleco.assign(
...         cylinders=fueleco.cylinders.fillna(0)
...     ),
...     hue="year",
...     size="barrels08",
...     alpha=0.5,
...     height=8,
... )
>>> res.fig.savefig(
...     "c5-relplot2.png", dpi=300, bbox_inches="tight"
... )
```

其输出结果如图 5-18 所示。

请注意，我们还可以为 hue 添加分类维度，这样就可以按列查看分类值。

```
>>> res = sns.relplot(
...     x="city08",
...     y="highway08",
...     data=fueleco.assign(
...         cylinders=fueleco.cylinders.fillna(0)
```

```
...         ),
...         hue="year",
...         size="barrels08",
...         alpha=0.5,
...         height=8,
...         col="make",
...         col_order=["Ford", "Tesla"],
... )
>>> res.fig.savefig(
...         "c5-relplot3.png", dpi=300, bbox_inches="tight"
... )
```

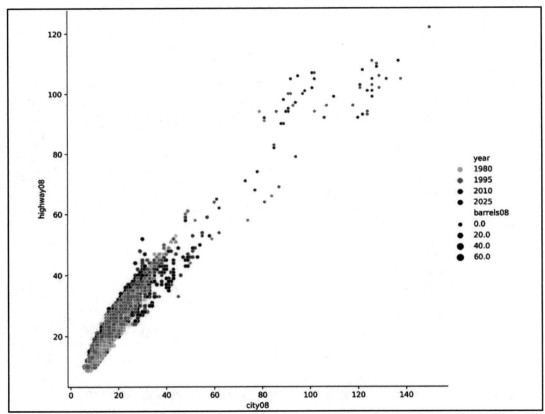

图 5-18　Seaborn 散点图，显示了按年份着色的里程油耗关系以及按汽车使用的汽油桶数显示的大小

其输出如图 5-19 所示。

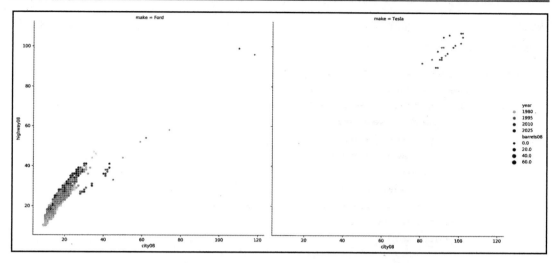

图 5-19 Seaborn 散点图,显示了里程油耗关系(按年份上色),
按汽车品牌分列,按使用的汽油桶数确定点的大小

皮尔逊相关旨在显示线性关系的强度。如果两个连续的列不具有线性关系,则另一种选择是使用 Spearman 相关(Spearman correlation)。此数字也从-1~1 不等。它测量关系是否为单调相关(并且不假定它是线性的),并使用每个数字的等级(rank)而不是数字本身。如果不确定各列之间是否存在线性关系,则使用此度量标准更好。

```
>>> fueleco.city08.corr(
...     fueleco.barrels08, method="spearman"
... )
-0.9743658646193255
```

5.8 使用分类值比较分类值

本节将重点介绍处理多个分类值。数据分析人员要记住的一件事是,可以通过对值进行分箱将连续的列转换为分类列。

本节将研究 make(品牌)和 vehicle(车辆)分类。

5.8.1 实战操作

(1)降低基数。对于车辆 VClass 列,限制为 6 个值(即 Car、SUV、Truck、Van、

Wagon 和 other）。而在品牌分类列 SClass 中，仅使用 Ford（福特）、Tesla（特斯拉）、BMW（宝马）和 Toyota（丰田）。

```
>>> def generalize(ser, match_name, default):
...     seen = None
...     for match, name in match_name:
...         mask = ser.str.contains(match)
...         if seen is None:
...             seen = mask
...         else:
...             seen |= mask
...         ser = ser.where(~mask, name)
...     ser = ser.where(seen, default)
...     return ser

>>> makes = ["Ford", "Tesla", "BMW", "Toyota"]
>>> data = fueleco[fueleco.make.isin(makes)].assign(
...     SClass=lambda df_: generalize(
...         df_.VClass,
...         [
...             ("Seaters", "Car"),
...             ("Car", "Car"),
...             ("Utility", "SUV"),
...             ("Truck", "Truck"),
...             ("Van", "Van"),
...             ("van", "Van"),
...             ("Wagon", "Wagon"),
...         ],
...         "other",
...     )
... )
```

（2）总结每个品牌的车辆分类数量。

```
>>> data.groupby(["make", "SClass"]).size().unstack()
SClass    Car      SUV   ...  Wagon   other
make                     ...
BMW     1557.0   158.0   ...   92.0     NaN
Ford    1075.0   372.0   ...  155.0   234.0
Tesla     36.0    10.0   ...    NaN     NaN
Toyota   773.0   376.0   ...  132.0   123.0
```

（3）使用 crosstab 函数而不是 Pandas 命令链。

```
>>> pd.crosstab(data.make, data.SClass)
SClass      Car    SUV  ...    Wagon    other
make                    ...
BMW        1557    158  ...       92        0
Ford       1075    372  ...      155      234
Tesla        36     10  ...        0        0
Toyota      773    376  ...      132      123
```

（4）添加更多维度。

```
>>> pd.crosstab(
...     [data.year, data.make], [data.SClass, data.VClass]
... )
SClass                    Car                   ...
other
VClass           Compact Cars    Large Cars     ...    Special Purpose Vehicle
4WD
year    make                                    ...
1984    BMW                 6             0     ...                          0
        Ford               33             3     ...                         21
        Toyota             13             0     ...                          3
1985    BMW                 7             0     ...                          0
        Ford               31             2     ...                          9
...                       ...           ...     ...                        ...
2017    Tesla               0             8     ...                          0
        Toyota              3             0     ...                          0
2018    BMW                37            12     ...                          0
        Ford                0             0     ...                          0
        Toyota              4             0     ...                          0
```

（5）使用 Cramér's V 测量表示分类的相关性。

💡 **提示：**
有关 Cramér's V 测量的详细信息，可访问以下网址。

https://stackoverflow.com/questions/46498455/categorical-features-correlation/46498792#46498792

```
>>> import scipy.stats as ss
>>> import numpy as np
>>> def cramers_v(x, y):
```

```
...         confusion_matrix = pd.crosstab(x, y)
...         chi2 = ss.chi2_contingency(confusion_matrix)[0]
...         n = confusion_matrix.sum().sum()
...         phi2 = chi2 / n
...         r, k = confusion_matrix.shape
...         phi2corr = max(
...             0, phi2 - ((k - 1) * (r - 1)) / (n - 1)
...         )
...         rcorr = r - ((r - 1) ** 2) / (n - 1)
...         kcorr = k - ((k - 1) ** 2) / (n - 1)
...         return np.sqrt(
...             phi2corr / min((kcorr - 1), (rcorr - 1))
...         )
>>> cramers_v(data.make, data.SClass)
0.2859720982171866
```

.corr 方法也是可调用的,因此另一种执行方式如下。

```
>>> data.make.corr(data.SClass, cramers_v)
0.2859720982171866
```

(6) 将交叉表可视化为条形图。

```
>>> fig, ax = plt.subplots(figsize=(10, 8))
>>> (
...     data.pipe(
...         lambda df_: pd.crosstab(df_.make, df_.SClass)
...     ).plot.bar(ax=ax)
... )
>>> fig.savefig("c5-bar.png", dpi=300, bbox_inches="tight")
```

其输出结果如图 5-20 所示。

(7) 使用 Seaborn 将交叉表可视化为条形图。

```
>>> res = sns.catplot(
...     kind="count", x="make", hue="SClass", data=data
... )
>>> res.fig.savefig(
...     "c5-barsns.png", dpi=300, bbox_inches="tight"
... )
```

其输出结果如图 5-21 所示。

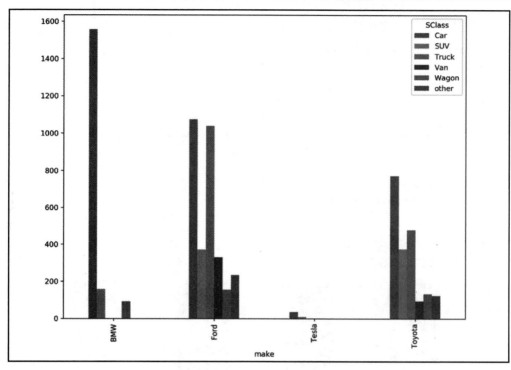

图 5-20　Pandas 条形图

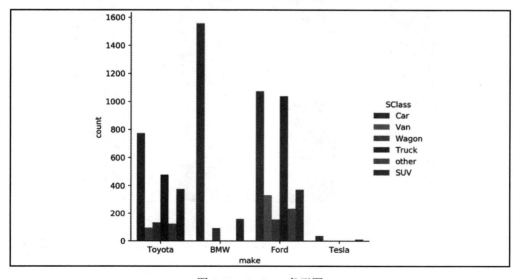

图 5-21　Seaborn 条形图

（8）通过归一化交叉表并制作堆叠条形图，以可视化组的相对大小。

```
>>> fig, ax = plt.subplots(figsize=(10, 8))
>>> (
...     data.pipe(
...         lambda df_: pd.crosstab(df_.make, df_.SClass)
...     )
...     .pipe(lambda df_: df_.div(df_.sum(axis=1), axis=0))
...     .plot.bar(stacked=True, ax=ax)
... )
>>> fig.savefig(
...     "c5-barstacked.png", dpi=300, bbox_inches="tight"
... )
```

其输出结果如图 5-22 所示。

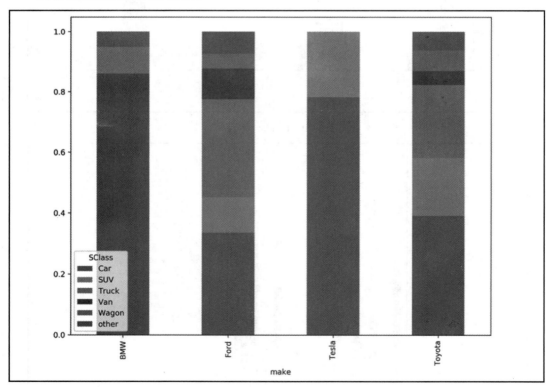

图 5-22　Pandas 条形图

5.8.2 原理解释

通过使用自创的 generalize 函数，我们减少了 VClass 列的基数。这样做是因为条形图需要间距——它们需要"呼吸"。一般来说，可以将条形图中的条形数量限制为少于 30 个。generalize 函数对于清理数据很有用，在你自己的数据分析案例中，也可以参考使用它。

可以通过创建交叉表（cross-tabulation）来汇总分类列的计数。你可以使用 group by 语义并分解结果来构建交叉表，或者利用 Pandas 中的内置函数 crosstab。请注意，crosstab 使用 0 填充缺失的数字，并将类型转换为整数。这是因为.unstack 方法可能会创建稀疏性（缺失值），而整数（int64 类型）又不支持缺少的值，因此，如果不填充 0 值的话，其类型将会被转换为浮点型。

可以在索引或列中添加任意深度，以在交叉表中创建层次结构。

可以使用一个数字 Cramér's V 量化两个分类列之间的关系。该数字的取值范围为 0～1。如果其值为 0，则一列中保存的值与另一列完全无关；如果其值为 1，则两列中的值将相对于彼此变化。

例如，如果将 make 列与 trany 列进行比较，则该值会显得比较大。

```
>>> cramers_v(data.make,data.trany)
0.6335899102918267
```

这告诉我们的是，随着 make（品牌）从福特变化到丰田，汽车的 trany（传动方式）列也应该会改变。我们可以做一个对比，用 make（品牌）与 model（车型）的值进行比较。从下面的计算可以看到，这两列之间的 Cramér's V 值非常接近 1。因此，就直觉而言，汽车品牌和派生车型的关系更有意义。

```
>>> cramers_v(data.make,data.model)
0.9542350243671587
```

最后，我们还可以使用各种条形图来查看计数或计数的相对大小。请注意，如果你使用的是 Seaborn，则可以通过设置 hue 或 col 来添加多个维度。

5.9 使用 Pandas 分析库

有一个第三方库 Pandas Profiling，其详细说明文档的网址如下。

https://pandas-profiling.github.io/pandas-profiling/docs/

Pandas Profiling 库可为每个列创建报告。这些报告与 .describe 方法的输出相似，但还包括图表和其他描述性统计信息。

本节将在燃油经济性数据上使用 Pandas Profiling 库。但是，在此之前你需要先使用以下命令安装该库。

```
pip install pandas-profiling
```

5.9.1 实战操作

运行 profile_report 函数以创建 HTML 报告。

```
>>> import pandas_profiling as pp
>>> pp.ProfileReport(fueleco)
```

其输出结果如图 5-23 和图 5-24 所示。

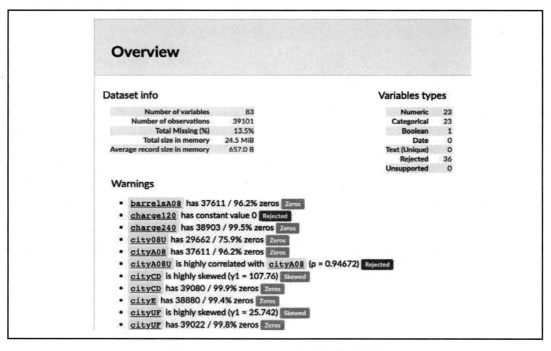

图 5-23 Pandas Profiling 创建的摘要报告

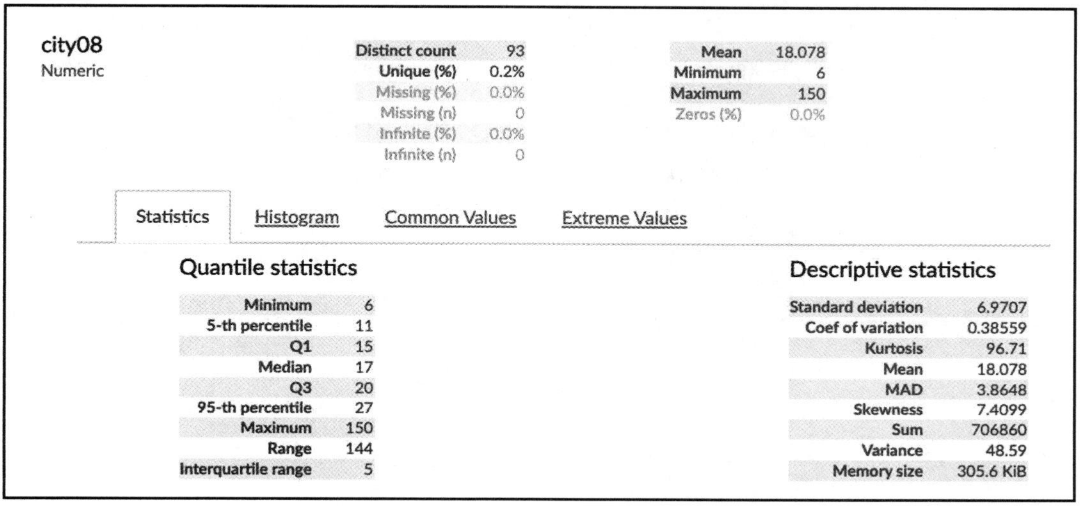

图 5-24 Pandas Profiling 创建的详细信息

5.9.2 原理解释

Pandas Profiling 库可以生成 HTML 格式的报告。如果你正在使用 Jupyter，那么它将以内联方式创建它。如果要将此报告保存到文件中（或者你没有使用 Jupyter），则可以使用.to_file 方法。

```
>>> report = pp.ProfileReport(fueleco)
>>> report.to_file("fuel.html")
```

Pandas Profiling 库非常适用于探索性数据分析，它可以确保你对数据有一个比较清晰的了解。当然，Pandas Profiling 库的大量输出信息也可能会使你感到有些难以招架，所以很可能会跳过该库，而不是深入研究该库。值得一提的是，尽管 Pandas Profiling 库非常适合探索性数据分析，但该库不会像本章中的某些示例那样进行列内比较（相关性除外），这也算是一个小小的缺憾吧。

第 6 章 选择数据子集

本章包含以下秘笈。
- 选择 Series 数据。
- 选择 DataFrame 行。
- 同时选择 DataFrame 行和列。
- 使用整数和标签选择数据。
- 按字典序切片。

6.1 介　　绍

Series 或 DataFrame 中数据的每个维都在索引（index）对象中标记。正是这个索引将 Pandas 数据结构与 NumPy 的 n 维数组区分开来。索引为数据的每行和每列均提供了有意义的标签，Pandas 用户可以通过使用这些标签来选择数据。此外，Pandas 还允许其用户根据行和列的位置选择数据。这种双重选择功能（一种使用名称，另一种使用位置）可谓有好有坏，好处是其功能强大，坏处则是在选择数据子集时往往会产生容易让人混淆的语法。

通过标签或位置选择数据并非 Pandas 独有。Python 字典和列表是内置的数据结构，它们就可以按这两种方式之一选择其数据。字典和列表均具有清晰的说明，并且限制了索引的用例。字典的键（也就是它的标签）必须是不可变的对象，如字符串、整数或元组。列表必须使用整数（位置）或切片（slice）对象进行选择。通过将键传递给索引操作符，字典一次只能选择一个对象。通过这种方式，Pandas 既可以像列表那样使用整数选择数据，又可以像字典那样使用标签来选择数据。

6.2 选择 Series 数据

Series 和 DataFrame 是复杂的数据容器，具有多个属性，这些属性可使用索引操作以不同的方式选择数据。除索引操作符本身外，.iloc 和 .loc 属性也是可用的，并以它们自己独特的方式使用索引运算符。

Series 和 DataFrame 允许按位置（就像 Python 列表一样）和标签（像 Python 字典一样）进行选择。.iloc 和 .loc 属性的区别如下。

- 当使用 .iloc 属性进行索引时，Pandas 仅按位置选择，并且其工作方式类似于 Python 列表。
- .loc 属性仅按索引标签选择数据，这类似于 Python 字典的工作方式。

.loc 和 .iloc 属性在 Series 和 DataFrame 上均可用。此秘笈展示了如何通过使用 .iloc 的位置和 .loc 的标签选择 Series 数据。这些索引器接收标量值、列表和切片。

我们之所以一再强调 .loc 和 .iloc 属性之间的区别，是因为这些术语和语法确实可能容易让人混淆。索引操作是在变量后加上一个方括号[]。例如，给定一个 Series s，你可以通过以下方式选择数据：s[item] 和 s.loc[item]。第一个是直接在 Series 上执行索引操作，而第二个则对 .loc 属性执行索引操作。

6.2.1 实战操作

（1）读取 college 数据集，该数据集以机构名称（institution name，INSTNM）作为索引，我们可以使用索引操作选择一列作为 Series。

```
>>> import pandas as pd
>>> import numpy as np
>>> college = pd.read_csv(
...     "data/college.csv", index_col="INSTNM"
... )
>>> city = college["CITY"]
>>> city
INSTNM
Alabama A & M University                    Normal
University of Alabama at Birmingham         Birmingham
Amridge University                          Montgomery
University of Alabama in Huntsville         Huntsville
Alabama State University                    Montgomery
                                            ...
SAE Institute of Technology                 San Francisco
Emeryville
Rasmussen College - Overland Park
```

```
Overland...
National Personal Training Institute of Cleveland
Highland...
Bay Area Medical Academy - San Jose Satellite Location
San Jose
Excel Learning Center-San Antonio South
San Antonio
Name: CITY, Length: 7535, dtype: object
```

（2）直接从 Series 中提取标量值。

```
>>> city["Alabama A & M University"]
'Normal'
```

（3）使用.loc 属性按名称提取出标量值。

```
>>> city.loc["Alabama A & M University"]
'Normal'
```

（4）使用.iloc 属性按位置提取出标量值。

```
>>> city.iloc[0]
'Normal'
```

（5）通过索引提取出若干个值。请注意，如果将列表传递给索引操作，则 Pandas 将返回一个 Series 而不是一个标量。

```
>>> city[
...     [
...         "Alabama A & M University",
...         "Alabama State University",
...     ]
... ]
INSTNM
Alabama A & M University        Normal
Alabama State University    Montgomery
Name: CITY, dtype: object
```

（6）使用.loc 重复上述步骤。

```
>>> city.loc[
...     [
...         "Alabama A & M University",
...         "Alabama State University",
...     ]
... ]
```

```
INSTNM
Alabama A & M University          Normal
Alabama State University      Montgomery
Name: CITY, dtype: object
```

（7）使用.iloc 重复上述步骤。

```
>>> city.iloc[[0, 4]]
INSTNM
Alabama A & M University          Normal
Alabama State University      Montgomery
Name: CITY, dtype: object
```

（8）使用切片提取出许多值。

```
>>> city[
...     "Alabama A & M University":"Alabama State University"
... ]
INSTNM
Alabama A & M University                  Normal
University of Alabama at Birmingham   Birmingham
Amridge University                    Montgomery
University of Alabama in Huntsville    Huntsville
Alabama State University              Montgomery
Name: CITY, dtype: object
```

（9）使用切片按位置提取出许多值。

```
>>> city[0:5]
INSTNM
Alabama A & M University                  Normal
University of Alabama at Birmingham   Birmingham
Amridge University                    Montgomery
University of Alabama in Huntsville    Huntsville
Alabama State University              Montgomery
Name: CITY, dtype: object
```

（10）使用切片通过.loc 提取许多值。

```
>>> city.loc[
...     "Alabama A & M University":"Alabama State University"
... ]
INSTNM
Alabama A & M University                  Normal
University of Alabama at Birmingham   Birmingham
```

```
Amridge University                           Montgomery
University of Alabama in Huntsville          Huntsville
Alabama State University                     Montgomery
Name: CITY, dtype: object
```

（11）使用切片通过 .iloc 提取许多值。

```
>>> city.iloc[0:5]
INSTNM
Alabama A & M University                     Normal
University of Alabama at Birmingham          Birmingham
Amridge University                           Montgomery
University of Alabama in Huntsville          Huntsville
Alabama State University                     Montgomery
Name: CITY, dtype: object
```

（12）使用布尔数组提取某些值。

```
>>> alabama_mask = city.isin(["Birmingham", "Montgomery"])
>>> city[alabama_mask]
INSTNM
University of Alabama at Birmingham          Birmingham
Amridge University                           Montgomery
Alabama State University                     Montgomery
Auburn University at Montgomery              Montgomery
Birmingham Southern College                  Birmingham
                                                ...
Fortis Institute-Birmingham                  Birmingham
Hair Academy                                 Montgomery
Brown Mackie College-Birmingham              Birmingham
Nunation School of Cosmetology               Birmingham
Troy University-Montgomery Campus            Montgomery
Name: CITY, Length: 26, dtype: object
```

6.2.2 原理解释

如果有 Series，则可以使用索引操作提取数据。根据使用的索引，可能会获得不同的类型作为输出。例如，如果在 Series 上使用标量索引，则将返回标量值；如果使用列表或切片索引，则将返回 Series。

仔细看看这些示例，似乎直接在 Series 上进行索引可以提供两全其美的效果，因为它既可以按位置又可以使用标签进行索引。但是，这里我们要告诫你的是，不要这样做。

记住，the zen of Python（Python 之禅）里面有一句是这么说的：Explicit is better than implicit（显式优于隐式）。.iloc 和.loc 都是显式的，但是直接在 Series 上索引则是隐式的。它要求我们考虑使用的是什么索引以及将会获得哪种类型的索引。

以一个简单的 Series 为例，它使用整数值作为索引。

```
>>> s = pd.Series([10, 20, 35, 28], index=[5, 2, 3, 1])
>>> s
5    10
2    20
3    35
1    28
dtype: int64

>>> s[0:4]
5    10
2    20
3    35
1    28
dtype: int64

>>> s[5]
10
>>> s[1]
28
```

当你直接在 Series 上使用切片进行索引（它使用的是位置），但在其他地方却使用标签时，可能会让未来的你和代码的其他读者感到困惑。请记住，针对代码可读性进行优化要比针对代码的易编写性进行优化更重要。因此，在这种情况下最简单便捷的解决方案就是使用.iloc 和.loc 索引器。

请记住，当按位置切片时，Pandas 会使用半开区间（half-open interval）。这个区间的概念你应该早就学习过（只是有可能忘记了）。半开区间包括第一个索引，但不包括结束索引。但是，当按标签切片时，Pandas 会使用封闭区间（closed interval），该区间同时包含开始和结束索引。此行为与 Python 不一致，但对于标签来说是实用的。

6.2.3 扩展知识

通过使用.loc 或.iloc，可以直接在原始 DataFrame 上执行本节中的所有示例。我们可以分别传入行和列标签或位置的元组（不带括号）。

```
>>> college.loc["Alabama A & M University", "CITY"]
'Normal'
```

```
>>> college.iloc[0, 0]
'Normal'

>>> college.loc[
...     [
...         "Alabama A & M University",
...         "Alabama State University",
...     ],
...     "CITY",
... ]
INSTNM
Alabama A & M University      Normal
Alabama State University      Montgomery
Name: CITY, dtype: object


>>> college.iloc[[0, 4], 0]
INSTNM
Alabama A & M University      Normal
Alabama State University      Montgomery
Name: CITY, dtype: object

>>> college.loc[
...     "Alabama A & M University":"Alabama State University",
...     "CITY",
... ]
INSTNM
Alabama A & M University               Normal
University of Alabama at Birmingham    Birmingham
Amridge University                     Montgomery
University of Alabama in Huntsville    Huntsville
Alabama State University               Montgomery
Name: CITY, dtype: object

>>> college.iloc[0:5, 0]
INSTNM
Alabama A & M University               Normal
University of Alabama at Birmingham    Birmingham
Amridge University                     Montgomery
University of Alabama in Huntsville    Huntsville
Alabama State University               Montgomery
Name: CITY, dtype: object
```

使用.loc 切片时应格外小心。如果起始索引出现在终止索引之后，那么将返回一个空的 Series，且不会抛出异常。

```
>>> city.loc[
...     "Reid State Technical College":"Alabama State University"
... ]
Series([], Name: CITY, dtype: object)
```

6.3　选择 DataFrame 行

选择 DataFrame 行的最明确的方法（也是首选方法）是使用.iloc 和.loc。.iloc 和.loc 都能够按行或按行和列进行选择。

6.3.1　实战操作

此秘笈将展示如何使用.iloc 和.loc 索引器从 DataFrame 中选择行。

（1）读取 college 数据集，并将索引设置为机构名称。

```
>>> college = pd.read_csv(
...     "data/college.csv", index_col="INSTNM"
... )
>>> college.sample(5, random_state=42)
                CITY     STABBR  ...  MD_EARN_WNE_P10  GRAD_DEBT_MDN_SUPP
INSTNM                           ...
Career Po...    San Antonio   TX   ...       20700              14977
Ner Israe...    Baltimore     MD   ...    PrivacyS...        PrivacyS...
Reflectio...    Decatur       IL   ...         NaN         PrivacyS...
Capital A...    Baton Rouge   LA   ...       26400         PrivacyS...
West Virg...    Montgomery    WV   ...       43400              23969
<BLANKLINE>
[5 rows x 26 columns]
```

（2）要选择整行，可以将一个整数传递给.iloc。

```
>>> college.iloc[60]
CITY                 Anchorage
STABBR                      AK
HBCU                         0
MENONLY                      0
WOMENONLY                    0
```

```
                          ...
PCTPELL                0.2385
PCTFLOAN               0.2647
UG25ABV                0.4386
MD_EARN_WNE_P10         42500
GRAD_DEBT_MDN_SUPP    19449.5
Name: University of Alaska Anchorage, Length: 26, dtype: object
```

因为 Python 索引是从 0 开始的,所以上述代码实际上选择的是第 61 行。请注意,Pandas 将此行表示为 Series。

(3)要获取与步骤(2)相同的行,可以将索引标签传递给 .loc。

```
>>> college.loc["University of Alaska Anchorage"]
CITY                 Anchorage
STABBR                      AK
HBCU                         0
MENONLY                      0
WOMENONLY                    0
                          ...
PCTPELL                0.2385
PCTFLOAN               0.2647
UG25ABV                0.4386
MD_EARN_WNE_P10         42500
GRAD_DEBT_MDN_SUPP    19449.5
Name: University of Alaska Anchorage, Length: 26, dtype: object
```

(4)要将一组不连续的行选择为 DataFrame,则可以将一个整数列表传递给 .iloc。

```
>>> college.iloc[[60, 99, 3]]
                  CITY STABBR ... MD_EARN_WNE_P10 GRAD_DEBT_MDN_SUPP
INSTNM                      ...
Universit...  Anchorage     AK ...           42500            19449.5
Internati...      Tempe     AZ ...           22200             10556
Universit... Huntsville     AL ...           45500             24097
<BLANKLINE>
[3 rows x 26 columns]
```

因为传入的是行位置的列表,所以这里将返回一个 DataFrame。

(5)通过向 .loc 中传递机构名称列表,可以使用 .loc 复制与步骤(4)相同的 DataFrame。

```
>>> labels = [
...     "University of Alaska Anchorage",
...     "International Academy of Hair Design",
```

```
...         "University of Alabama in Huntsville",
... ]
>>> college.loc[labels]
                    CITY STABBR ...  MD_EARN_WNE_P10  GRAD_DEBT_MDN_SUPP
INSTNM                          ...
Universit...   Anchorage     AK ...            42500             19449.5
Internati...       Tempe     AZ ...            22200               10556
Universit...   Huntsville    AL ...            45500               24097
<BLANKLINE>
[3 rows x 26 columns]
```

（6）将切片符号与 .iloc 一起使用以选择数据的连续行。

```
>>> college.iloc[99:102]
                    CITY STABBR ...  MD_EARN_WNE_P10  GRAD_DEBT_MDN_SUPP
INSTNM                          ...
Internati...       Tempe     AZ ...            22200               10556
GateWay C...     Phoenix     AZ ...            29800                7283
Mesa Comm...        Mesa     AZ ...            35200                8000
<BLANKLINE>
[3 rows x 26 columns]
```

（7）切片符号也可以与 .loc 一起使用，但要注意的是，它使用的是一个封闭区间（即包括开始标签和停止标签）。

```
>>> start = "International Academy of Hair Design"
>>> stop = "Mesa Community College"
>>> college.loc[start:stop]
                    CITY STABBR ...  MD_EARN_WNE_P10  GRAD_DEBT_MDN_SUPP
INSTNM                          ...
Internati...       Tempe     AZ ...            22200               10556
GateWay C...     Phoenix     AZ ...            29800                7283
Mesa Comm...        Mesa     AZ ...            35200                8000
<BLANKLINE>
[3 rows x 26 columns]
```

6.3.2 原理解释

当我们将标量值、标量列表或切片传递给 .iloc 或 .loc 时，这将导致 Pandas 扫描索引以查找适当的行并返回它们。如果传递单个标量值，则返回一个 Series；如果传递了列表或切片，则将返回 DataFrame。

6.3.3 扩展知识

在步骤（5）中，可以直接从步骤（4）返回的 DataFrame 中选择索引标签的列表，而无须复制和粘贴。

```
>>> college.iloc[[60, 99, 3]].index.tolist()
['University of Alaska Anchorage', 'International Academy of Hair
Design', 'University of Alabama in Huntsville']
```

6.4 同时选择 DataFrame 行和列

Pandas 有很多选择行和列的方法。要从 DataFrame 中选择一个或多个列，最简单的方法是从 DataFrame 中索引。但是，这种方法有局限性。直接在 DataFrame 中建立索引不允许同时选择行和列。因此，要同时选择行和列，你需要将有效的行和列选择以逗号分隔的形式传递给.iloc 或.loc。

选择行和列的通用形式将类似于以下代码。

```
df.iloc[row_idxs, column_idxs]
df.loc[row_names, column_names]
```

其中，row_idxs 和 column_idxs 可以是标量整数、整数列表或整数切片。虽然 row_names 和 column_names 可以是标量名称、名称列表或名称切片，但 row_names 也可以是布尔数组。

在此秘笈中，每个步骤都将显示使用.iloc 和.loc 同时选择行和列。

6.4.1 实战操作

（1）读取 college 数据集，并将索引设置为机构名称列（INSTNM）。使用切片符号选择前 3 行和前 4 列。

```
>>> college = pd.read_csv(
...     "data/college.csv", index_col="INSTNM"
... )
>>> college.iloc[:3, :4]
                CITY    STABBR   HBCU   MENONLY
INSTNM
Alabama A...    Normal  AL       1.0    0.0
```

```
Universit...    Birmingham     AL    0.0         0.0
Amridge U...    Montgomery     AL    0.0         0.0

>>> college.loc[:"Amridge University", :"MENONLY"]
                CITY       STABBR   HBCU    MENONLY
INSTNM
Alabama A...    Normal         AL    1.0         0.0
Universit...    Birmingham     AL    0.0         0.0
Amridge U...    Montgomery     AL    0.0         0.0
```

（2）选择两个不同列的所有行。

```
>>> college.iloc[:, [4, 6]].head()
                                        WOMENONLY    SATVRMID
INSTNM
Alabama A & M University                      0.0       424.0
University of Alabama at Birmingham           0.0       570.0
Amridge University                            0.0         NaN
University of Alabama in Huntsville           0.0       595.0
Alabama State University                      0.0       425.0

>>> college.loc[:, ["WOMENONLY", "SATVRMID"]].head()
                                        WOMENONLY    SATVRMID
INSTNM
Alabama A & M University                      0.0       424.0
University of Alabama at Birmingham           0.0       570.0
Amridge University                            0.0         NaN
University of Alabama in Huntsville           0.0       595.0
Alabama State University                      0.0       425.0
```

（3）选择不相交的行和列。

```
>>> college.iloc[[100, 200], [7, 15]]
                                        SATMTMID    UGDS_NHPI
INSTNM
GateWay Community College                    NaN       0.0029
American Baptist Seminary of the West        NaN          NaN

>>> rows = [
...     "GateWay Community College",
...     "American Baptist Seminary of the West",
... ]
>>> columns = ["SATMTMID", "UGDS_NHPI"]
>>> college.loc[rows, columns]
```

	SATMTMID	UGDS_NHPI
INSTNM		
GateWay Community College	NaN	0.0029
American Baptist Seminary of the West	NaN	NaN

（4）选择一个标量值。

```
>>> college.iloc[5, -4]
0.401
>>> college.loc["The University of Alabama", "PCTFLOAN"]
0.401
```

（5）切片行并选择一个列。

```
>>> college.iloc[90:80:-2, 5]
INSTNM
Empire Beauty School-Flagstaff       0
Charles of Italy Beauty College      0
Central Arizona College              0
University of Arizona                0
Arizona State University-Tempe       0
Name: RELAFFIL, dtype: int64

>>> start = "Empire Beauty School-Flagstaff"
>>> stop = "Arizona State University-Tempe"
>>> college.loc[start:stop:-2, "RELAFFIL"]
INSTNM
Empire Beauty School-Flagstaff       0
Charles of Italy Beauty College      0
Central Arizona College              0
University of Arizona                0
Arizona State University-Tempe       0
Name: RELAFFIL, dtype: int64
```

6.4.2 原理解释

同时选择行和列的关键之一是了解方括号中逗号的用法。逗号左侧的选择始终是根据行索引选择行，逗号右边的选择始终是根据列索引选择列。

不必同时选择行和列。步骤（2）显示了如何选择所有行和列的子集。冒号（:）表示一个切片对象，该对象返回该维的所有值。

6.4.3 扩展知识

要仅选择行（以及所有列），则不必在逗号后使用冒号。如果没有逗号，则默认行为是选择所有列。

先前的秘笈正是以这种方式选择了行。当然，你也可以使用冒号表示所有列的切片。以下代码行是等效的。

```
college.iloc [:10]
college.iloc [:10, :]
```

6.5 使用整数和标签选择数据

有时，你可能会希望同时使用.iloc 和.loc 的功能，即可以同时通过位置和标签来选择数据。在 Pandas 的早期版本中，.ix 可以同时通过位置和标签选择数据。尽管这在某些特定情况下很方便，但是它本质上是模棱两可的，并且容易让许多 Pandas 用户感到困惑。.ix 索引器随后被弃用，因此应避免使用。

在弃用.ix 之前，可以使用 college.ix[:5, 'UGDS_WHITE':'UGDS_UNKN']从 UGDS_WHITE～UGDS_UNKN 选择 college 数据集的前 5 行和列。现在不可能直接使用.loc 或.iloc 执行此操作。

以下秘笈将显示如何查找列的整数位置，然后使用.iloc 完成选择。

6.5.1 实战操作

（1）读取 college 数据集，并将机构名称（INSTNM）分配为索引。

```
>>> college = pd.read_csv(
...     "data/college.csv", index_col="INSTNM"
... )
```

（2）使用 Index 方法.get_loc 查找所需列的整数位置。

```
>>> col_start = college.columns.get_loc("UGDS_WHITE")
>>> col_end = college.columns.get_loc("UGDS_UNKN") + 1
>>> col_start, col_end
(10, 19)
```

（3）使用 col_start 和 col_end，通过.iloc 按位置选择列。

第 6 章 选择数据子集

```
>>> college.iloc[:5, col_start:col_end]
             UGDS_WHITE  UGDS_BLACK  ...  UGDS_NRA  UGDS_UNKN
INSTNM                               ...
Alabama A...     0.0333      0.9353  ...    0.0059     0.0138
Universit...     0.5922      0.2600  ...    0.0179     0.0100
Amridge U...     0.2990      0.4192  ...    0.0000     0.2715
Universit...     0.6988      0.1255  ...    0.0332     0.0350
Alabama S...     0.0158      0.9208  ...    0.0243     0.0137
<BLANKLINE>
[5 rows x 9 columns]
```

6.5.2 原理解释

步骤（2）首先通过.columns 属性检索列索引。索引具有.get_loc 方法，该方法接收索引标签并返回其整数位置。我们找到要切片的列的开始和结束整数位置。这里加 1 是因为使用.iloc 进行切片时使用的是半开区间，并且不包括最后一项。步骤（3）则是将切片符号与行和列位置一起使用。

6.5.3 扩展知识

我们可以执行一个非常类似的操作来使用位置获取标签，以便让.loc 工作。下面显示了如何选择第 10 行～第 15 行（包括第 15 行）以及列 UGDS_WHITE～UGDS_UNKN。

```
>>> row_start = college.index[10]
>>> row_end = college.index[15]
>>> college.loc[row_start:row_end, "UGDS_WHITE":"UGDS_UNKN"]
             UGDS_WHITE  UGDS_BLACK  ...  UGDS_NRA  UGDS_UNKN
INSTNM                               ...
Birmingha...     0.7983      0.1102  ...    0.0000     0.0051
Chattahoo...     0.4661      0.4372  ...    0.0000     0.0139
Concordia...     0.0280      0.8758  ...    0.0466     0.0000
South Uni...     0.3046      0.6054  ...    0.0019     0.0326
Enterpris...     0.6408      0.2435  ...    0.0012     0.0069
James H F...     0.6979      0.2259  ...    0.0007     0.0009
<BLANKLINE>
[6 rows x 9 columns]
```

使用.ix（它已从 Pandas 1.0 中被删除，因此请不要执行此操作）时，看起来如下所示（在 1.0 之前的版本中）。

```
>>> college.ix[10:16, "UGDS_WHITE":"UGDS_UNKN"]
```

```
           UGDS_WHITE    UGDS_BLACK    ...    UGDS_NRA    UGDS_UNKN
INSTNM                                 ...
Birmingha...   0.7983       0.1102     ...     0.0000      0.0051
Chattahoo...   0.4661       0.4372     ...     0.0000      0.0139
Concordia...   0.0280       0.8758     ...     0.0466      0.0000
South Uni...   0.3046       0.6054     ...     0.0019      0.0326
Enterpris...   0.6408       0.2435     ...     0.0012      0.0069
James H F...   0.6979       0.2259     ...     0.0007      0.0009
<BLANKLINE>
[6 rows x 9 columns]
```

通过将.loc 和.iloc 链接在一起可以实现相同的结果，但是链接索引器通常是一个坏主意。它可能会更慢，并且还不确定返回的是视图还是副本（查看数据时这没有问题，但在更新数据时就可能有问题，你可能会看到臭名昭著的 SettingWithCopyWarning 警告）。

```
>>> college.iloc[10:16].loc[:, "UGDS_WHITE":"UGDS_UNKN"]
           UGDS_WHITE    UGDS_BLACK    ...    UGDS_NRA    UGDS_UNKN
INSTNM                                 ...
Birmingha...   0.7983       0.1102     ...     0.0000      0.0051
Chattahoo...   0.4661       0.4372     ...     0.0000      0.0139
Concordia...   0.0280       0.8758     ...     0.0466      0.0000
South Uni...   0.3046       0.6054     ...     0.0019      0.0326
Enterpris...   0.6408       0.2435     ...     0.0012      0.0069
James H F...   0.6979       0.2259     ...     0.0007      0.0009
<BLANKLINE>
[6 rows x 9 columns]
```

6.6 按字典序切片

.loc 属性通常会根据索引的确切字符串标签选择数据。但是，它也允许分析人员根据索引中值的字典序（lexicographic order）选择数据。具体来说，就是.loc 允许使用切片符号按字典序选择具有索引的所有行，该操作仅在索引已排序时才有效。

在本秘笈中，我们将首先对索引进行排序，然后在.loc 索引器中使用切片符号（:）选择两个字符串之间的所有行。

6.6.1 实战操作

（1）读取 college 数据集，并将机构名称（INSTNM）列设置为索引。

```
>>> college = pd.read_csv(
...     "data/college.csv", index_col="INSTNM"
... )
```

（2）尝试选择所有名称的字典序为 Sp～Su 的大学。

```
>>> college.loc["Sp":"Su"]
Traceback (most recent call last):
...
ValueError: index must be monotonic increasing or decreasing

During handling of the above exception, another exception occurred:

Traceback (most recent call last):
...
KeyError: 'Sp'
```

（3）可以看到，由于未对索引进行排序，上面的命令失败了。因此，接下来我们需要首先对索引进行排序。

```
>>> college = college.sort_index()
```

（4）现在重新运行步骤（2）中的命令。

```
>>> college.loc["Sp":"Su"]
                 CITY STABBR ...  MD_EARN_WNE_P10  GRAD_DEBT_MDN_SUPP
INSTNM                       ...
Spa Tech...    Ipswich     MA ...            21500                6333
Spa Tech...   Plymouth     MA ...            21500                6333
Spa Tech...   Westboro     MA ...            21500                6333
Spa Tech...  Westbrook     ME ...            21500                6333
Spalding...  Louisville    KY ...            41700               25000
...                ...    ... ...              ...                 ...
Studio Ac... Chandler     AZ ...              NaN                6333
Studio Je... New York     NY ...       PrivacyS...         PrivacyS...
Stylemast... Longview     WA ...            17000               13320
Styles an...   Selmer     TN ...       PrivacyS...         PrivacyS...
Styletren... Rock Hill    SC ...       PrivacyS...              9495.5
<BLANKLINE>
[201 rows x 26 columns]
```

6.6.2 原理解释

.loc 的正常行为是根据传递给它的确切标签来选择数据。当在索引中找不到这些标签时，就会抛出 KeyError。但是，这里有一个特殊的例外情况，那就是只要索引是按字典序排序的，并且将切片传递给该索引，那么，即使在索引中找不到这些值，也可以在切片的开始和结束标签之间进行选择。

6.6.3 扩展知识

通过此秘笈可以轻松地选择名称在两个字母之间的大学。例如，要选择所有以字母 D~S 开头的大学，可以使用以下代码。

```
college.loc ['D':'T']
```

像这样的切片使用的是封闭区间，并包含了最后一个索引，因此从技术上讲，这将返回名称以 T 开头的大学。

当索引按相反方向排序时，这种切片方式也是适用的。分析人员可以使用 .is_monotonic_increasing 或 .is_monotonic_decreasing 索引属性来确定索引的排序方向。为了使字典序切片能够正常工作，这两个索引属性必须有一个为 True。例如，以下代码将按字典序对索引从 Z~A 进行排序。

```
>>> college = college.sort_index(ascending=False)
>>> college.index.is_monotonic_decreasing
True
>>> college.loc["E":"B"]
                                              CITY    ...
INSTNM                                                ...
Dyersburg State Community College        Dyersburg    ...
Dutchess Community College              Poughkeepsie  ...
Dutchess BOCES-Practical Nursing Program Poughkeepsie ...
Durham Technical Community College         Durham     ...
Durham Beauty Academy                      Durham     ...
...                                          ...      ...
Bacone College                            Muskogee    ...
Babson College                            Wellesley   ...
BJ's Beauty & Barber College               Auburn     ...
BIR Training Center                       Chicago     ...
B M Spurr School of Practical Nursing    Glen Dale    ...
```

第 7 章 过 滤 行

本章包含以下秘笈。
- 计算布尔统计信息。
- 构造多个布尔条件。
- 用布尔数组过滤。
- 比较行过滤和索引过滤。
- 使用唯一索引和排序索引进行选择。
- 转换 SQL WHERE 子句。
- 使用查询方法提高布尔索引的可读性。
- 使用.where 方法保留 Series 大小。
- 屏蔽 DataFrame 行。
- 使用布尔值、整数位置和标签进行选择。

7.1 介　　绍

过滤数据集中的数据是最常见的基本操作之一。有许多方法可以使用布尔索引来过滤 Pandas 中的数据（或创建子集）。布尔索引（boolean indexing）也称为布尔选择（boolean selection），这可能是一个令人困惑的术语，但是在 Pandas 领域中，它指的是通过提供布尔数组（boolean array）来选择行。布尔数组是一个具有相同索引的 Pandas Series，只不过它每行的值均为 True 或 False。该名称来自 NumPy（在 NumPy 中就是以类似方式处理过滤逻辑的），因此虽然它实际上是一个包含布尔值的 Series，但也称为布尔数组。

本章将首先创建布尔 Series，并计算它们的统计数据，然后继续创建更复杂的条件，再以多种方式使用布尔索引来过滤数据。

7.2　计算布尔统计信息

计算布尔数组的基本摘要统计信息可能会很有帮助。布尔数组的每个值（True 或 False）的求值分别为 1 或 0，因此所有使用数值的 Series 方法也适用于布尔值。

在此秘笈中，我们通过将条件应用于数据列来创建布尔数组，然后从布尔数组中计算汇总统计信息。

7.2.1 实战操作

（1）读取 movie 数据集，将索引设置为片名（movie_title）列，然后检查 duration（放映时长）列的前几行。

```
>>> import pandas as pd
>>> import numpy as np
>>> movie = pd.read_csv(
...     "data/movie.csv", index_col="movie_title"
... )
>>> movie[["duration"]].head()
                                            Duration
movie_title
Avatar                                         178.0
Pirates of the Caribbean: At World's End       169.0
Spectre                                        148.0
The Dark Knight Rises                          164.0
Star Wars: Episode VII - The Force Awakens       NaN
```

（2）对 duration（放映时长）列使用大于（>）比较运算符，以确定每部电影的放映时长是否大于 2h。

```
>>> movie_2_hours = movie["duration"] > 120
>>> movie_2_hours.head(10)
movie_title
Avatar                                        True
Pirates of the Caribbean: At World's End      True
Spectre                                       True
The Dark Knight Rises                         True
Star Wars: Episode VII - The Force Awakens    False
John Carter                                   True
Spider-Man 3                                  True
Tangled                                       False
Avengers: Age of Ultron                       True
Harry Potter and the Half-Blood Prince        True
Name: duration, dtype: bool
```

（3）现在可以使用该 Series 来确定放映时长超过 2h 的电影数量。

```
>>> movie_2_hours.sum()
1039
```

（4）要查找数据集中放映时长超过 2h 的电影的百分比，可使用.mean 方法。

```
>>> movie_2_hours.mean()* 100
21.13506916192026
```

（5）遗憾的是，步骤（4）的输出具有误导性，因为 duration（放映时长）列包含了一些缺失值。如果回头看步骤（1）的 DataFrame 输出，就可以看到 duration（放映时长）列最后一行包含了缺失值。步骤（2）中的布尔条件也因此而返回 False。要解决该问题，可以先删除缺失值，然后评估条件并取均值。

```
>>> movie["duration"].dropna().gt(120).mean() * 100
21.199755152009794
```

（6）使用.describe 方法在布尔数组上输出摘要统计信息。

```
>>> movie_2_hours.describe()
count         4916
unique           2
top          False
freq          3877
Name: duration, dtype: object
```

7.2.2 原理解释

大多数 DataFrame 都不会像 movie 数据集那样具有布尔值列。产生布尔数组的最直接方法是将条件运算符应用于其中一列。在步骤（2）中，我们使用了大于（>）比较运算符来测试每部电影的放映时长是否超过 120min。步骤（3）和步骤（4）均通过布尔 Series 计算了两个重要的量，即 sum 值和 mean 值。这些方法之所以可行，是因为 Python 将 False 值和 True 值分别评估为 0 和 1。

你可以自己证明布尔数组的均值代表 True 值的百分比。要进行这种证明，可以使用.value_counts 方法，并且将 normalize 参数设置为 True，然后进行计数以获取其分布。

```
>>> movie_2_hours.value_counts(normalize=True)
False    0.788649
True     0.211351
Name: duration, dtype: float64
```

步骤（5）提醒我们，步骤（4）的结果不正确。即使 duration 列中包含了缺失值，布尔条件也会将所有这些与缺失值的比较一起评估为 False。删除这些缺失值使我们能够计算出正确的统计量。通过方法链，此操作仅需一步即可完成。

> **注意：**
> 在计算之前要确保已经处理了缺失值。

步骤（6）显示了 Pandas 通过显示频率信息的方式，将.describe 方法应用于布尔数组，该方式与将其应用于对象或字符串列的方式相同。这是考虑布尔数组而不是显示分位数的自然方法。

如果需要分位数信息，则可以将该 Series 转换为整数。

```
>>> movie_2_hours.astype(int).describe()
count    4916.000000
mean        0.211351
std         0.408308
min         0.000000
25%         0.000000
50%         0.000000
75%         0.000000
max         1.000000
Name: duration, dtype: float64
```

7.2.3 扩展知识

可以比较来自同一 DataFrame 的两列以生成布尔系列。例如，我们可以计算演员 1 在 Facebook 平台上获得的喜欢/点赞数多于演员 2 在该平台上获得的喜欢/点赞数的电影的百分比。要做到这一点，我们可以选择这两个列，然后删除任何包含缺失值的行。最后进行比较并计算均值。

```
>>> actors = movie[
...     ["actor_1_facebook_likes", "actor_2_facebook_likes"]
... ].dropna()
>>> (
...     actors["actor_1_facebook_likes"]
...     > actors["actor_2_facebook_likes"]
... ).mean()
0.9777687130328371
```

7.3 构造多个布尔条件

在 Python 中，布尔表达式使用的是内置的逻辑运算符 and、or 和 not。这些关键字不适用于 Pandas 中的布尔索引，而是分别用&、|和~替换。此外，在组合表达式时，每个

表达式必须用括号括起来；否则将（由于运算符的优先级问题）引发错误。

为数据集构建过滤器时，可能需要将多个布尔表达式组合在一起以提取所需的行。在本秘笈中，我们将构造多个布尔表达式，然后将它们组合在一起，以查找 imdb_score（互联网电影数据库评分）大于 8，content_rating（内容分级）为 PG-13，并且 title_year（影片发行年份）为 2000 年之前或 2009 年之后。

7.3.1 实战操作

（1）读取 movie 数据集并将片名（movie_title）列设置为索引。

```
>>> movie = pd.read_csv(
...     "data/movie.csv", index_col="movie_title"
... )
```

（2）创建一个变量以将每个过滤器保存为布尔数组。

```
>>> criteria1 = movie.imdb_score > 8
>>> criteria2 = movie.content_rating == "PG-13"
>>> criteria3 = (movie.title_year < 2000) | (
...     movie.title_year > 2009
... )
```

（3）将所有过滤器组合成一个布尔数组。

```
>>> criteria_final = criteria1 & criteria2 & criteria3
>>> criteria_final.head()
movie_title
Avatar                                          False
Pirates of the Caribbean: At World's End        False
Spectre                                         False
The Dark Knight Rises                            True
Star Wars: Episode VII - The Force Awakens      False
dtype: bool
```

7.3.2 原理解释

可以使用标准比较运算符（<、>、==、!=、<=和>=）将 Series 中的所有值与标量值进行比较。表达式movie.imdb_score>8将生成一个布尔数组，其中所有大于 8 的 imdb_score 值均为 True，小于或等于 8 的 imdb_score 值为 False。此布尔数组的索引与 movie DataFrame 具有相同的索引。

可以通过组合两个布尔数组来创建 criteria3 变量；每个表达式必须用括号括起来才能正常运行；管道字符（|）用于在两个 Series 的每个值之间创建逻辑或条件。

这 3 个条件都必须为 True 才能满足秘笈要求。它们都通过与号（&）被组合，从而在每个 Series 值之间创建逻辑和（and）条件。

7.3.3 扩展知识

Pandas 对逻辑运算符使用不同语法的结果是，运算符的优先级不再相同。比较运算符的优先级高于 and、or 和 not。但是，Pandas 使用的运算符（按位运算符&、|和~）比那些比较运算符具有更高的优先级，因此需要使用括号括起来。

可以通过一个示例来解释清楚这个问题。来看以下表达式。

```
>>> 5 < 10 and 3 > 4
False
```

在上面的表达式中，首先计算 5 <10，然后计算 3 > 4，最后评估 and 的结果。Python 按以下过程评估表达式。

```
>>> 5 < 10 and 3 > 4
False
>>> True and 3 > 4
False
>>> True and False
False
>>> False
False
```

现在来看看，如果在 criteria3 的表达式中按如下方式编写会发生什么。

```
>>> movie.title_year < 2000 | movie.title_year > 2009
Traceback (most recent call last):
    ...
TypeError: ufunc 'bitwise_or' not supported for the input types, and the
inputs could not be safely coerced to any supported types according to
the casting rule ''safe''

During handling of the above exception, another exception occurred:

Traceback (most recent call last):
    ...
TypeError: cannot compare a dtyped [float64] array with a scalar of type
[bool]
```

由于按位运算符的优先级高于比较运算符，因此 2000 | movie.title_year 首先被评估，这是荒谬的，并且会引发错误。因此，需要使用括号来强制运算符优先级。

为何 Pandas 不能使用 and、or 和 not？当评估这些关键字时，Python 会尝试查找对象作为一个整体的真实性。由于将 Series 视为一个整体并评估其 True 或 False（仅每个元素）没有意义，因此 Pandas 会引发错误。

Python 中的所有对象都具有布尔表示形式，通常被称为真实性（truthiness）。例如，除 0 以外的所有整数均被视为 True；除空字符串外，所有字符串均为 True，所有非空集、元组、字典和列表均为 True；一般来说，要评估 Python 对象的真实性，可将其传递给 bool 函数。空的 DataFrame 或 Series 不会评估出 True 或 False，而是会引发错误。

7.4 用布尔数组过滤

Series 和 DataFrame 都可以使用布尔数组进行过滤。数据分析人员可以直接通过对象或 .loc 属性建立索引。

本秘笈为电影的不同行构造了两个复杂的过滤器。

- 第一个过滤器使用了以下条件过滤电影：imdb_score（互联网电影数据库评分）大于 8，content_rating（内容分级）为 PG-13，并且 title_year（影片发行年份）为 2000 年之前或 2009 年之后。
- 第二个过滤器使用了以下条件过滤电影：imdb_score（互联网电影数据库评分）小于 5，content_rating（内容分级）为 R，并且 title_year（影片发行年份）为 2000—2010 年。

最后，我们将结合使用这些过滤器。

7.4.1 实战操作

（1）读取 movie 数据集，将索引设置为 movie_title（片名）列，然后创建第一组条件。

```
>>> movie = pd.read_csv(
...     "data/movie.csv", index_col="movie_title"
... )
>>> crit_a1 = movie.imdb_score > 8
>>> crit_a2 = movie.content_rating == "PG-13"
>>> crit_a3 = (movie.title_year < 2000) | (
...     movie.title_year > 2009
... )
```

```
>>> final_crit_a = crit_a1 & crit_a2 & crit_a3
```

（2）创建第二组条件。

```
>>> crit_b1 = movie.imdb_score < 5
>>> crit_b2 = movie.content_rating == "R"
>>> crit_b3 = (movie.title_year >= 2000) & (
...         movie.title_year <= 2010
... )
>>> final_crit_b = crit_b1 & crit_b2 & crit_b3
```

（3）使用 Pandas 或运算符组合两组标准。这会产生一个布尔数组，其中所有影片都是这两个集合的成员。

```
>>> final_crit_all = final_crit_a | final_crit_b
>>> final_crit_all.head()
movie_title
Avatar                                          False
Pirates of the Caribbean: At World's End        False
Spectre                                         False
The Dark Knight Rises                           True
Star Wars: Episode VII - The Force Awakens      False
dtype: bool
```

（4）拥有布尔数组后，即可将其传递给索引运算符以过滤数据。

```
>>> movie[final_crit_all].head()
                              color    ...    movie/likes
movie_title                            ...
The Dark Knight Rises         Color    ...        164000
The Avengers                  Color    ...        123000
Captain America: Civil War    Color    ...         72000
Guardians of the Galaxy       Color    ...         96000
Interstellar                  Color    ...        349000
```

（5）我们还可以过滤掉.loc 属性。

```
>>> movie.loc[final_crit_all].head()
                              color    ...    movie/likes
movie_title                            ...
The Dark Knight Rises         Color    ...        164000
The Avengers                  Color    ...        123000
Captain America: Civil War    Color    ...         72000
Guardians of the Galaxy       Color    ...         96000
Interstellar                  Color    ...        349000
```

（6）另外，还可以使用.loc 属性指定要选择的列。

```
>>> cols = ["imdb_score", "content_rating", "title_year"]
>>> movie_filtered = movie.loc[final_crit_all, cols]
>>> movie_filtered.head(10)
              imdb_score   content_rating   title_year
movie_title
The Dark ...     8.5          PG-13          2012.0
The Avengers     8.1          PG-13          2012.0
Captain A...     8.2          PG-13          2016.0
Guardians...     8.1          PG-13          2014.0
Interstellar     8.6          PG-13          2014.0
Inception        8.8          PG-13          2010.0
The Martian      8.1          PG-13          2015.0
Town & Co...     4.4           R             2001.0
Sex and t...     4.3           R             2010.0
Rollerball       3.0           R             2002.0
```

7.4.2 原理解释

在步骤（1）和步骤（2）中，每组条件都是从更简单的布尔数组构建的。虽然不必像此处所做的那样为每个布尔表达式创建一个不同的变量，但这确实使读取和调试任何逻辑错误变得容易得多。当我们需要这两组电影时，步骤（3）即使用了 Pandas 逻辑或（or）运算符将它们组合在一起。

在步骤（4）中，我们将步骤（3）中创建的布尔 Series 直接传递给索引运算符。只有 final_crit_all 为 True 值的电影才会被选择。

如步骤（6）所示，通过同时选择行（使用布尔数组）和列，这种过滤还可以与.loc 属性一起使用。精简后的 DataFrame 可轻松手动检查逻辑是否正确实现。

请注意，.iloc 属性不支持布尔数组！如果将布尔 Series 传递给它，则会引发异常。当然，.loc 属性可与 NumPy 数组协同工作，因此，如果调用.to_numpy()方法，则可以对其进行过滤。

```
>>> movie.iloc[final_crit_all]
Traceback (most recent call last):
    ...
ValueError: iLocation based boolean indexing cannot use an indexable
as a mask

>>> movie.iloc[final_crit_all.to_numpy()]
```

```
                              color    ...  movie/likes
movie_title                            ...
The Dark Knight Rises         Color    ...  164000
The Avengers                  Color    ...  123000
Captain America: Civil War    Color    ...  72000
Guardians of the Galaxy       Color    ...  96000
Interstellar                  Color    ...  349000
...                           ...      ...  ...
The Young Unknowns            Color    ...  4
Bled                          Color    ...  128
Hoop Dreams                   Color    ...  0
Death Calls                   Color    ...  16
The Legend of God's Gun       Color    ...  13
```

7.4.3 扩展知识

如前文所述，我们可以使用一个长布尔表达式代替其他几个较短的布尔表达式。例如，要使用一个较长的代码复制步骤（1）中的 final_crit_a 变量，可以执行以下操作。

```
>>> final_crit_a2 = (
...     (movie.imdb_score > 8)
...     & (movie.content_rating == "PG-13")
...     & (
...         (movie.title_year < 2000)
...         | (movie.title_year > 2009)
...     )
... )
>>> final_crit_a2.equals(final_crit_a)
True
```

7.5 比较行过滤和索引过滤

通过利用索引，可以复制布尔选择的特定情况。

在本秘笈中，我们将使用 college 数据集，分别通过布尔索引和索引选择方式，选择来自特定州（得克萨斯州）的所有机构，然后对它们各自的性能进行相互比较。

笔者更喜欢按列（使用布尔数组）进行过滤，而不是按索引进行过滤。列过滤功能更强大，因为它可以使用其他逻辑运算符并在多个列上进行过滤。

7.5.1 实战操作

(1) 读取 college 数据集,并使用布尔索引选择来自得克萨斯州(TX)的所有机构。

```
>>> college = pd.read_csv("data/college.csv")
>>> college[college["STABBR"] == "TX"].head()
                          INSTNM   ...   GRAD_/_SUPP
3610     Abilene Christian University   ...         25985
3611        Alvin Community College   ...          6750
3612              Amarillo College   ...         10950
3613              Angelina College   ...   PrivacySuppressed
3614         Angelo State University   ...         21319.5
```

(2) 要使用索引选择方式重复此操作,可以将 STABBR 列移入索引中。然后在 .loc 索引器中使用基于标签的选择。

```
>>> college2 = college.set_index("STABBR")
>>> college2.loc["TX"].head()
                          INSTNM   ...   GRAD_/_SUPP
3610     Abilene Christian University   ...         25985
3611        Alvin Community College   ...          6750
3612              Amarillo College   ...         10950
3613              Angelina College   ...   PrivacySuppressed
3614         Angelo State University   ...         21319.5
```

(3) 现在可以比较这两种方法的速度。

```
>>> %timeit college[college['STABBR'] == 'TX']
1.75 ms ± 187 µs per loop (mean ± std. dev. of 7 runs, 1000 loops each)

>>> %timeit college2.loc['TX']
882 µs ± 69.3 µs per loop (mean ± std. dev. of 7 runs, 1000 loops each)
```

(4) 可以看到,布尔索引的时间是索引选择的两倍。由于设置索引并不是没有成本的,因此我们也可以对该操作计时。

```
>>> %timeit college2 = college.set_index('STABBR')
2.01 ms ± 107 µs per loop (mean ± std. dev. of 7 runs, 100 loops each)
```

7.5.2 原理解释

步骤(1)通过确定哪些数据行的 STABBR 等于得克萨斯州(TX)来创建布尔 Series。

该 Series 将传递给索引操作符，后者将选择数据。

要复制此过程，可以将同一列移到索引中，并使用基于基本标签的索引和 .loc 来选择。通过索引选择比布尔选择快得多。

当然，如果需要在多个列上进行过滤，则重复切换索引同样将产生开销（以及令人困惑的代码）。我们的建议是不要切换索引，就使用它过滤即可。

7.5.3 扩展知识

此秘笈仅选择了一个州，其实也可以通过布尔选择和索引选择方式来选择多个州。例如，我们可以选择得克萨斯（TX）、加利福尼亚（CA）和纽约（NY）。使用布尔选择方式时，可以使用 .isin 方法；但是使用索引时，只需将列表传递给 .loc。

```
>>> states = ["TX", "CA", "NY"]
>>> college[college["STABBR"].isin(states)]
           INSTNM          CITY   ...   MD_EARN_WNE_P10   GRAD_DEBT_MDN_SUPP
192     Academy ...    San Fran...  ...            36000                35093
193     ITT Tech...    Rancho C...  ...            38800              25827.5
194     Academy ...       Oakland   ...              NaN            PrivacyS...
195     The Acad...    Huntingt...  ...            28400                 9500
196     Avalon S...       Alameda   ...            21600                 9860
...           ...           ...    ...              ...                  ...
7528    WestMed ...        Merced   ...              NaN              15623.5
7529    Vantage ...       El Paso   ...              NaN                 9500
7530    SAE Inst...    Emeryville   ...              NaN                 9500
7533    Bay Area...      San Jose   ...              NaN            PrivacyS...
7534    Excel Le...   San Antonio   ...              NaN                12125

>>> college2.loc[states]
            INSTNM          CITY  ...   MD_EARN_WNE_P10   GRAD_DEBT_MDN_SUPP
STABBR                             ...
TX         Abilene ...     Abilene  ...            40200                25985
TX         Alvin Co...       Alvin  ...            34500                 6750
TX         Amarillo...    Amarillo  ...            31700                10950
TX         Angelina...      Lufkin  ...            26900            PrivacyS...
TX         Angelo S...  San Angelo  ...            37700              21319.5
...             ...           ...  ...              ...                  ...
NY         Briarcli...    Patchogue ...            38200              28720.5
NY         Jamestow...    Salamanca ...              NaN                12050
NY         Pratt Ma...     New York ...            40900                26691
NY         Saint Jo...    Patchogue ...            52000              22143.5
NY         Franklin...     Brooklyn ...            20000            PrivacyS...
```

上述技巧已经超出了此秘笈的解释范围。Pandas 可以根据索引是唯一索引还是排序索引来实现不同方式的索引。有关详细信息，请参见 7.6 节"使用唯一索引和排序索引进行选择"秘笈。

7.6 使用唯一索引和排序索引进行选择

当索引是唯一的或已排序时，索引选择的性能会大大提高。先前的秘笈使用了包含重复项的未排序索引，这使得选择相对较慢。

在本秘笈中，我们使用 college 数据集来形成唯一索引或排序索引，以提高索引选择的性能。我们还将继续使用 college 数据集与布尔索引的性能进行比较。

如果你仅需要从单个列中进行选择，并且这对于你来说是一个瓶颈，那么此秘笈可以为你节省 10 倍的工作量。

7.6.1 实战操作

（1）读取 college 数据集，使用 STABBR 作为索引创建一个单独的 DataFrame，然后检查索引是否已排序。

```
>>> college = pd.read_csv("data/college.csv")
>>> college2 = college.set_index("STABBR")
>>> college2.index.is_monotonic
False
```

（2）对来自 college2 的索引进行排序，并将其存储为另一个对象。

```
>>> college3 = college2.sort_index()
>>> college3.index.is_monotonic
True
```

（3）对于从 3 个 DataFrame 中选择得克萨斯州（TX）的操作进行计时。

```
>>> %timeit college[college['STABBR'] == 'TX']
1.75 ms ± 187 µs per loop (mean ± std. dev. of 7 runs, 1000 loops each)

>>> %timeit college2.loc['TX']
1.09 ms ± 232 µs per loop (mean ± std. dev. of 7 runs, 1000 loops each)

>>> %timeit college3.loc['TX']
304 µs ± 17.8 µs per loop (mean ± std. dev. of 7 runs, 1000 loops each)
```

（4）可以看到，排序索引的执行速度比布尔选择快近一个数量级。现在我们转向唯一索引。为此，可以使用机构名称（INSTNM）作为索引。

```
>>> college_unique = college.set_index("INSTNM")
>>> college_unique.index.is_unique
True
```

（5）使用布尔索引方式选择斯坦福大学。请注意，这将返回一个 DataFrame。

```
>>> college[college["INSTNM"] == "Stanford University"]
            INSTNM      CITY    ...    MD_EARN_WNE_P10   GRAD_DEBT_MDN_SUPP
4217    Stanford...   Stanford  ...         86000                12782
```

（6）使用索引选择方式选择斯坦福大学。请注意，这将返回一个 Series。

```
>>> college_unique.loc["Stanford University"]
CITY                    Stanford
STABBR                  CA
HBCU                    0
MENONLY                 0
WOMENONLY               0
                        ...
PCTPELL                 0.1556
PCTFLOAN                0.1256
UG25ABV                 0.0401
MD_EARN_WNE_P10         86000
GRAD_DEBT_MDN_SUPP      12782
Name: Stanford University, Length: 26, dtype: object
```

（7）如果要使用 DataFrame 而不是 Series，则需要将索引值列表传递到 .loc 中。

```
>>> college_unique.loc[["Stanford University"]]
            INSTNM      CITY    ...    MD_EARN_WNE_P10   GRAD_DEBT_MDN_SUPP
4217    Stanford...   Stanford  ...         86000                12782
```

（8）布尔索引和索引选择方式可以产生相同的数据，只是对象不同。现在来对这两种操作方法计时。

```
>>> %timeit college[college['INSTNM'] == 'Stanford University']
1.92 ms ± 396 µs per loop (mean ± std. dev. of 7 runs, 1000 loops each)

>>> %timeit college_unique.loc[['Stanford University']]
988 µs ± 122 µs per loop (mean ± std. dev. of 7 runs, 1000 loops each)
```

7.6.2 原理解释

当索引未被排序并且包含重复项时（如使用 college2），Pandas 将需要检查索引中的每个值以进行正确选择；当索引被排序时（如使用 college3），Pandas 将利用二分搜索（binary search）算法提高搜索性能。

在秘笈的后半部分，我们使用了唯一列作为索引。Pandas 可通过哈希表实现唯一索引，从而使选择速度更快。每个索引位置几乎可以在同一时间查找到，而不管其长度如何。

7.6.3 扩展知识

布尔选择比索引选择具有更大的灵活性，因为它可以对任意数量的列进行条件调整。在此秘笈中，我们使用单列作为索引，但其实也可以将多个列连接在一起以形成索引。例如，在以下代码中，我们就将索引设置为等于 city（城市）和 state（州）列的连接。

```
>>> college.index = (
...     college["CITY"] + ", " + college["STABBR"]
... )
>>> college = college.sort_index()
>>> college.head()
                    INSTNM         CITY ... MD_EARN_WNE_P10 GRAD_DEBT_MDN_SUPP
ARTESIA, CA      Angeles ...     ARTESIA ...             NaN              16850
Aberdeen, SD     Presenta... Aberdeen ...           35900              25000
Aberdeen, SD     Northern... Aberdeen ...           33600              24847
Aberdeen, WA     Grays Ha... Aberdeen ...           27000              11490
Abilene, TX      Hardin-S... Abilene  ...           38700              25864
```

以此为基础，我们可以从特定城市和州的组合中选择所有大学，而无须布尔索引。例如，我们可以选择佛罗里达州迈阿密（Miami，FL）市的所有大学。

```
>>> college.loc["Miami, FL"].head()
                INSTNM       CITY ... MD_EARN_WNE_P10 GRAD_DEBT_MDN_SUPP
Miami, FL   New Prof...    Miami ...           18700               8682
Miami, FL   Manageme...    Miami ...        PrivacyS...            12182
Miami, FL   Strayer  ...   Miami ...           49200             36173.5
Miami, FL   Keiser U...    Miami ...           29700              26063
Miami, FL   George T...    Miami ...           38600          PrivacyS...
```

现在将这种复合索引选择方式与布尔索引的速度进行比较。可以看到，它们几乎有一个数量级的差异。

```
>>> %%timeit
>>> crit1 = college["CITY"] == "Miami"
>>> crit2 = college["STABBR"] == "FL"
>>> college[crit1 & crit2]
3.05 ms ± 66.4 µs per loop (mean ± std. dev. of 7 runs, 100 loops each)

>>> %timeit college.loc['Miami, FL']
369 µs ± 130 µs per loop (mean ± std. dev. of 7 runs, 1000 loops each)
```

7.7 转换 SQL WHERE 子句

许多 Pandas 用户都拥有使用结构化查询语言（structured query language，SQL）与数据库进行交互的经验。SQL 是定义、操作和控制存储在数据库中的数据的标准。

SQL 是数据科学家需要了解的重要语言。世界上有很多数据都被存储在需要 SQL 来检索和操作的数据库中，SQL 语法非常简单易学。Oracle、Microsoft 和 IBM 等公司提供了许多不同的 SQL 实现。

在 SQL SELECT 语句中，WHERE 子句很常见，该子句用于过滤数据。该秘笈将编写与选择雇员数据集的某个子集的 SQL 查询等效的 Pandas 代码。

本节假设有一项任务是找到所有在警察或消防部门工作的、基本薪水为 8 万～12 万美元的女性雇员。

以下 SQL 语句将为我们回答此查询。

```
SELECT
    UNIQUE_ID,
    DEPARTMENT,
    GENDER,
    BASE_SALARY
FROM
    EMPLOYEE
WHERE
    DEPARTMENT IN ( 'Houston Police Department-HPD',
                    'Houston Fire Department (HFD)') AND
    GENDER = 'Female' AND
    BASE_SALARY BETWEEN 80000 AND 120000;
```

本秘笈假定你已经在 CSV 文件中转储了 EMPLOYEE 数据库，并且想要使用 Pandas 复制上述查询。

7.7.1 实战操作

(1) 读取 employee 数据集作为 DataFrame。

```
>>> employee = pd.read_csv("data/employee.csv")
```

(2) 在过滤数据之前,可以对每个要过滤的列进行一些手动检查,以了解将在过滤器中使用的确切值。

```
>>> employee.dtypes
UNIQUE_ID                int64
POSITION_TITLE          object
DEPARTMENT              object
BASE_SALARY            float64
RACE                    object
EMPLOYMENT_TYPE         object
GENDER                  object
EMPLOYMENT_STATUS       object
HIRE_DATE               object
JOB_DATE                object
dtype: object

>>> employee.DEPARTMENT.value_counts().head()
Houston Police Department-HPD      638
Houston Fire Department (HFD)      384
Public Works & Engineering-PWE     343
Health & Human Services            110
Houston Airport System (HAS)       106
Name: DEPARTMENT, dtype: int64

>>> employee.GENDER.value_counts()
Male      1397
Female     603
Name: GENDER, dtype: int64

>>> employee.BASE_SALARY.describe()
count     1886.000000
mean     55767.931601
std      21693.706679
min      24960.000000
25%      40170.000000
50%      54461.000000
```

```
75%        66614.000000
max       275000.000000
Name: BASE_SALARY, dtype: float64
```

（3）为每个条件编写一个语句。使用 isin 方法检查雇员是否在警察或消防部工作（是否等于 depts 中的两个值之一）。

```
>>> depts = [
...     "Houston Police Department-HPD",
...     "Houston Fire Department (HFD)",
... ]
>>> criteria_dept = employee.DEPARTMENT.isin(depts)
>>> criteria_gender = employee.GENDER == "Female"
>>> criteria_sal = (employee.BASE_SALARY >= 80000) & (
...     employee.BASE_SALARY <= 120000
... )
```

（4）合并所有布尔数组。

```
>>> criteria_final = (
...     criteria_dept & criteria_gender & criteria_sal
... )
```

（5）使用布尔索引选择仅符合最终条件的行。

```
>>> select_columns = [
...     "UNIQUE_ID",
...     "DEPARTMENT",
...     "GENDER",
...     "BASE_SALARY",
... ]
>>> employee.loc[criteria_final, select_columns].head()
     UNIQUE_ID   DEPARTMENT  GENDER   BASE_SALARY
61          61     Houston ... Female      96668.0
136        136     Houston ... Female      81239.0
367        367     Houston ... Female      86534.0
474        474     Houston ... Female      91181.0
513        513     Houston ... Female      81239.0
```

7.7.2 原理解释

在进行任何过滤之前，你都需要知道将用来过滤的确切字符串名称。.value_counts 就是一种能够获取确切的字符串名称和字符串值出现次数的方法。

.isin 方法等效于 SQL 的 IN 操作符，并接收要保留的所有可能值的列表。可以使用一系列 OR 条件来复制该表达式，但效率或惯用性都不那么高。

薪水标准（criterias_sal）是通过组合两个简单的不等式（大于或等于 8 万或小于等于 12 万美元）而形成的。将所有这些条件使用 Pandas 的 AND 运算符（&）组合在一起，即可生成单个布尔数组作为过滤器。

7.7.3 扩展知识

对于许多操作而言，Pandas 可以通过多种方式来实现同一目的。在前面的秘笈中，薪水标准使用了两个单独的布尔表达式。与 SQL 类似，Series 也有一个 .between 方法，可以实现上述薪水标准的等效编写。我们将在硬编码数字中使用下画线，以帮助提高可读性。

```
>>> criteria_sal = employee.BASE_SALARY.between(
...        80_000, 120_000
... )
```

.isin 方法的另一个非常实用的应用是提供由其他一些 Pandas 语句自动生成的值序列，这样，在查找要存储在列表中的确切字符串名称时，就不必进行任何手动查看操作。

例如，我们可以尝试从最经常出现的 5 个部门中排除行。

```
>>> top_5_depts = employee.DEPARTMENT.value_counts().index[
...        :5
... ]
>>> criteria = ~employee.DEPARTMENT.isin(top_5_depts)
>>> employee[criteria]
      UNIQUE_ID  POSITION_TITLE   ...  HIRE_DATE   JOB_DATE
0             0  ASSISTAN...      ...  2006-06-12  2012-10-13
1             1  LIBRARY ...      ...  2000-07-19  2010-09-18
4             4  ELECTRICIAN      ...  1989-06-19  1994-10-22
18           18  MAINTENA...      ...  2008-12-29  2008-12-29
32           32  SENIOR A...      ...  1991-02-11  2016-02-13
...         ...  ...              ...  ...         ...
1976       1976  SENIOR S...      ...  2015-07-20  2016-01-30
1983       1983  ADMINIST...     ...  2006-10-16  2006-10-16
1985       1985  TRUCK DR...      ...  2013-06-10  2015-08-01
1988       1988  SENIOR A...      ...  2013-01-23  2013-03-02
1990       1990  BUILDING...     ...  1995-10-14  2010-03-20
```

上述操作的 SQL 等效语句如下。

```
SELECT *
    FROM
        EMPLOYEE
    WHERE
        DEPARTMENT not in
        (
            SELECT
                DEPARTMENT
FROM ( SELECT
DEPARTMENT,
                COUNT(1) as CT
            FROM
                EMPLOYEE
            GROUP BY
                DEPARTMENT
            ORDER BY
                CT DESC
            LIMIT 5
) );
```

可以看到，在上述示例中使用了 Pandas 的 not 运算符（~），它否定了 Series 的所有布尔值。

7.8　使用查询方法提高布尔索引的可读性

布尔索引不一定是最令人愉快的读取或写入语法，尤其是在需要使用单行编写复杂的过滤器时。Pandas 有一种替代性的基于字符串的语法，它可以通过 DataFrame 查询方法提供更好的代码可读性。

此秘笈重复了前面 7.7 节"转换 SQL WHERE 子句"秘笈中的操作，但利用的却是 DataFrame 的 .query 方法。

和 7.7 节一样，我们的任务是找到所有在警察或消防部门工作的、基本薪水在 8 万～12 万美元的女性雇员。

7.8.1　实战操作

（1）读取 employee 数据，分配选定的部门，并将列导入变量中。

```
>>> employee = pd.read_csv("data/employee.csv")
>>> depts = [
...     "Houston Police Department-HPD",
```

```
...         "Houston Fire Department (HFD)",
...     ]
>>> select_columns = [
...         "UNIQUE_ID",
...         "DEPARTMENT",
...         "GENDER",
...         "BASE_SALARY",
...     ]
```

（2）构建查询字符串并执行该方法。请注意，.query 方法不喜欢使用跨越多行的三重引号将字符串引起来，因此连接起来很不方便。

```
>>> qs = (
...         "DEPARTMENT in @depts "
...         " and GENDER == 'Female' "
...         " and 80000 <= BASE_SALARY <= 120000"
...     )
>>> emp_filtered = employee.query(qs)
>>> emp_filtered[select_columns].head()
     UNIQUE_ID  DEPARTMENT       GENDER  BASE_SALARY
61          61     Houston  ...  Female      96668.0
136        136     Houston  ...  Female      81239.0
367        367     Houston  ...  Female      86534.0
474        474     Houston  ...  Female      91181.0
513        513     Houston  ...  Female      81239.0
```

7.8.2 原理解释

传递给 .query 方法的字符串看起来比普通的 Pandas 代码更像普通英语。与 depts 一样，可以使用 at 符号（@）来引用 Python 变量。所有 DataFrame 列名称都可以通过引用其名称在查询名称空间中使用，而无须额外的引号。如果需要一个字符串（如 Female），则需要用引号将其引起来。

查询语法的另一个很好用的功能是可以使用 and、or 和 not 组合布尔运算符。

7.8.3 扩展知识

不用手动输入部门名称列表，我们可以按编程方式创建它。例如，如果要按频率查找不是排名前 10 部门成员的所有女性雇员，则可以运行以下代码。

```
>>> top10_depts = (
...     employee.DEPARTMENT.value_counts()
```

```
...         .index[:10]
...         .tolist()
... )
>>> qs = "DEPARTMENT not in @top10_depts and GENDER == 'Female'"
>>> employee_filtered2 = employee.query(qs)
>>> employee_filtered2.head()
     UNIQUE_ID    POSITION_TITLE   ...   HIRE_DATE     JOB_DATE
0            0         ASSISTAN... ...   2006-06-12   2012-10-13
73          73         ADMINIST... ...   2011-12-19   2013-11-23
96          96         ASSISTAN... ...   2013-06-10   2013-06-10
117        117         SENIOR A... ...   1998-03-20   2012-07-21
146        146         SENIOR S... ...   2014-03-17   2014-03-17
```

7.9 使用 .where 方法保留 Series 大小

在使用布尔数组进行过滤时，获得的 Series 或 DataFrame 通常较小。.where 方法将保留 Series 或 DataFrame 的大小，并将不符合条件的值设置为缺失值或将其替换为其他值。除了丢弃所有这些值，也可以考虑保留它们。

当你将此功能与 other 参数结合使用时，则可以创建类似于合并（coalesce）的功能，后者可以在数据库中找到。

在此秘笈中，我们将给 .where 方法传递布尔条件，以便对 movie 数据集中演员 1 在 Facebook 平台上获得的最小和最大喜欢/点赞数设置上下限。

7.9.1 实战操作

（1）读取 movie 数据集，将片名（movie_title）列设置为索引，然后在 actor_1_facebook_likes 列中选择所有未缺失的值。

```
>>> movie = pd.read_csv(
...     "data/movie.csv", index_col="movie_title"
... )
>>> fb_likes = movie["actor_1_facebook_likes"].dropna()
>>> fb_likes.head()
movie_title
Avatar                                         1000.0
Pirates of the Caribbean: At World's End      40000.0
Spectre                                       11000.0
The Dark Knight Rises                         27000.0
Star Wars: Episode VII - The Force Awakens      131.0
Name: actor_1_facebook_likes, dtype: float64
```

(2)使用.describe 方法了解分布情况。

```
>>> fb_likes.describe()
count      4909.000000
mean       6494.488491
std       15106.986884
min           0.000000
25%         607.000000
50%         982.000000
75%       11000.000000
max      640000.000000
Name: actor_1_facebook_likes, dtype: float64
```

(3)此外,我们还可以绘制该 Series 的直方图以直观地检查分布。下面的代码调用了 plt.subplots 来指定图形尺寸,但一般来说是不需要的。

```
>>> import matplotlib.pyplot as plt
>>> fig, ax = plt.subplots(figsize=(10, 8))
>>> fb_likes.hist(ax=ax)
>>> fig.savefig(
...     "c7-hist.png", dpi=300
... )
```

其输出结果如图 7-1 所示。

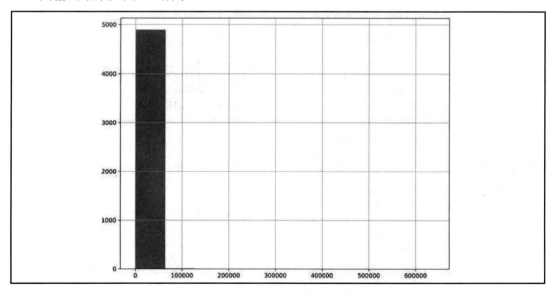

图 7-1　默认 Pandas 直方图

（4）这种可视化使得我们很难了解分布情况。另外，步骤（2）的汇总统计信息似乎在告诉我们，在进行一些非常大的观察时，数据高度偏向右侧（比中位数大一个数量级）。因此，可以创建标准来测试喜欢/点赞次数是否少于 20000。

```
>>> criteria_high = fb_likes < 20_000
>>> criteria_high.mean().round(2)
0.91
```

（5）大约 91%的电影其演员 1 所获得的喜欢/点赞数都少于 20000 个。现在，我们将使用.where 方法，该方法接收一个布尔数组。默认行为是返回与原始大小相同的 Series，但是将所有 False 位置替换为缺失值。

```
>>> fb_likes.where(criteria_high).head()
movie_title
Avatar                                         1000.0
Pirates of the Caribbean: At World's End          NaN
Spectre                                       11000.0
The Dark Knight Rises                             NaN
Star Wars: Episode VII - The Force Awakens      131.0
Name: actor_1_facebook_likes, dtype: float64
```

（6）.where 方法的第二个参数（other）允许你控制替换值。让我们将所有缺失值更改为 20000。

```
>>> fb_likes.where(criteria_high, other=20000).head()
movie_title
Avatar                                         1000.0
Pirates of the Caribbean: At World's End      20000.0
Spectre                                       11000.0
The Dark Knight Rises                         20000.0
Star Wars: Episode VII - The Force Awakens      131.0
Name: actor_1_facebook_likes, dtype: float64
```

（7）同样地，我们还可以创建条件以对最小的喜欢/点赞次数设置下限。在这里，我们将链接另一个.where 方法，并将不满足条件的值替换为 300。

```
>>> criteria_low = fb_likes > 300
>>> fb_likes_cap = fb_likes.where(
...     criteria_high, other=20_000
... ).where(criteria_low, 300)
>>> fb_likes_cap.head()
movie_title
Avatar                                         1000.0
Pirates of the Caribbean: At World's End      20000.0
```

```
Spectre                                       11000.0
The Dark Knight Rises                         20000.0
Star Wars: Episode VII - The Force Awakens      300.0
Name: actor_1_facebook_likes, dtype: float64
```

（8）原始 Series 和经修改的 Series 的长度相同。

```
>>> len(fb_likes), len(fb_likes_cap)
(4909, 4909)
```

（9）使用修改后的 Series 创建直方图。当数据范围更窄时，该 Series 应产生更好的图。

```
>>> fig, ax = plt.subplots(figsize=(10, 8))
>>> fb_likes_cap.hist(ax=ax)
>>> fig.savefig(
...     "c7-hist2.png", dpi=300
... )
```

其输出结果如图 7-2 所示。

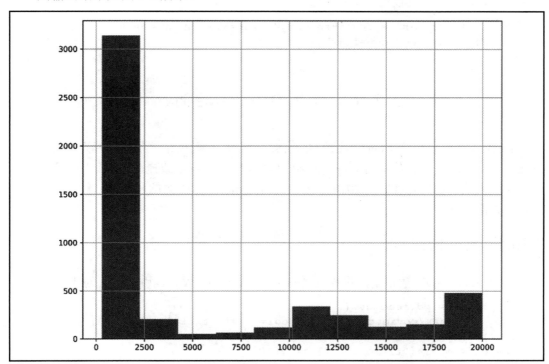

图 7-2　范围更小的 Pandas 直方图

7.9.2 原理解释

.where 方法再次保留了调用对象的大小和形状，并且在传递的布尔值为 True 的情况下，不修改该值。在步骤（1）中删除缺失值非常重要，因为.where 方法最终会在以后的步骤中将这些缺失值替换为有效的数字。

步骤（2）中的汇总统计信息使我们对限制数据的上限有了一些了解。另外，步骤（3）中的直方图似乎会将所有数据聚集到一个箱子中。对于纯直方图，数据有太多离群值，因此无法绘制出正确的图。.where 方法允许我们在数据上分别设置一个上限和下限，从而产生方差较小的直方图。

7.9.3 扩展知识

Pandas 实际上具有复制此操作的内置方法，即.clip、.clip_lower 和.clip_upper。其中，.clip 方法可以同时设置上限和下限。

```
>>> fb_likes_cap2 = fb_likes.clip(lower=300, upper=20000)
>>> fb_likes_cap2.equals(fb_likes_cap)
True
```

7.10 屏蔽 DataFrame 行

.mask 方法可以执行.where 方法的补充。默认情况下，无论布尔条件是否为 True，.mask 方法都会创建缺失值。从本质上讲，.mask 方法实际上是屏蔽或掩盖数据集中的值。

在此秘笈中，我们将屏蔽 2010 年之后制作的 movie 数据集的所有行，然后过滤所有包含缺失值的行。

7.10.1 实战操作

（1）读取 movie 数据集，将片名（movie_title）列设置为索引，然后创建条件。

```
>>> movie = pd.read_csv(
...     "data/movie.csv", index_col="movie_title"
... )
>>> c1 = movie["title_year"] >= 2010
>>> c2 = movie["title_year"].isna()
>>> criteria = c1 | c2
```

(2) 在 DataFrame 上使用 .mask 方法，以删除包含 2010 年以来所制作电影的行中的所有值。任何最初具有 title_year 缺失值的电影也将被屏蔽。

```
>>> movie.mask(criteria).head()
                                            color  ...
movie_title                                        ...
Avatar                                      Color  ...
Pirates of the Caribbean: At World's End    Color  ...
Spectre                                       NaN  ...
The Dark Knight Rises                         NaN  ...
Star Wars: Episode VII - The Force Awakens    NaN  ...
```

(3) 请注意，前面的 DataFrame 中的第三、第四和第五行中的所有值都丢失了。链接 .dropna 方法以删除包含缺失值的行。

```
>>> movie_mask = movie.mask(criteria).dropna(how="all")
>>> movie_mask.head()
                                            color  ...
movie_title                                        ...
Avatar                                      Color  ...
Pirates of the Caribbean: At World's End    Color  ...
Spider-Man 3                                Color  ...
Harry Potter and the Half-Blood Prince      Color  ...
Superman Returns                            Color  ...
```

(4) 步骤（3）中的操作只是进行基本布尔索引的一种复杂方法。我们可以检查 .mask 和 .dropna 方法是否产生相同的 DataFrame。

```
>>> movie_boolean = movie[movie["title_year"] < 2010]
>>> movie_mask.equals(movie_boolean)
False
```

(5) .equals 方法告知我们它们不相等。这里出了点问题。让我们进行一些完整性检查，看看它们是否相同。

```
>>> movie_mask.shape == movie_boolean.shape
True
```

(6) 当我们使用前面的 .mask 方法时，该方法创建了许多缺失值。缺失值是 float 数据类型，因此任何具有缺失值的 integer 类型列都将被转换为 float 类型。如果列的数据类型不同，即使值相同，那么 .equals 方法也将返回 False。

现在来检查数据类型的相等性，以查看是否发生了这种情况。

```
>>> movie_mask.dtypes == movie_boolean.dtypes
color                         True
director_name                 True
num_critic_for_reviews        True
duration                      True
director_facebook_likes       True
                              ...
title_year                    True
actor_2_facebook_likes        True
imdb_score                    True
aspect_ratio                  True
movie_facebook_likes          False
Length: 27, dtype: bool
```

（7）事实证明，有几列没有相同的数据类型。Pandas 可以替代这些情况。在主要由开发人员使用的 testing 模块中，有一个函数 assert_frame_equal，该函数可以检查 Series 和 DataFrame 的相等性，而无须同时检查数据类型的相等性。

```
>>> from pandas.testing import assert_frame_equal
>>> assert_frame_equal(
...     movie_boolean, movie_mask, check_dtype=False
... )
```

7.10.2　原理解释

默认情况下，.mask 方法可使用 NaN 填充布尔数组为 True 的行。.mask 方法的第一个参数是布尔数组。因为.mask 方法是从 DataFrame 中调用的，所以条件为 True 的每行中的所有值都将被更改为 missing。步骤（3）使用此屏蔽的 DataFrame 删除包含所有缺失值的行。步骤（4）显示了如何使用索引操作执行相同的过程。

在数据分析过程中，持续验证结果很重要。检查 Series 和 DataFrame 的相等性是一种验证方法。我们在步骤（4）中的首次尝试产生了意外结果。在深入研究之前，一些基本的健全性检查（如确保行和列的数目相同或行和列的名称相同）是很好的检查。

步骤（6）比较了两个 Series 的数据类型。在这里，我们揭示了 DataFrame 不相等的原因。.equals 方法检查值和数据类型是否相同。步骤（7）中的 assert_frame_equal 函数具有许多可用参数，可以通过各种方式测试相等性。请注意，在调用 assert_frame_equal 函数之后没有任何输出。当两个 DataFrame 相等时，此方法返回 None；否则，将引发错误。

7.10.3 扩展知识

现在可以比较屏蔽和删除缺失值行与使用布尔数组过滤之间的速度差异。在这种情况下，过滤大约快一个数量级。

```
>>> %timeit movie.mask(criteria).dropna(how='all')
11.2 ms ± 144 µs per loop (mean ± std. dev. of 7 runs, 100 loops each)

>>> %timeit movie[movie['title_year'] < 2010]
ms ± 34.9 µs per loop (mean ± std. dev. of 7 runs, 1000 loops each)
```

7.11 使用布尔值、整数位置和标签进行选择

前文介绍了有关通过.iloc 和.loc 属性选择不同数据子集的各种方法。这些都通过整数位置或标签同时选择行和列。

在此秘笈中，我们将使用.iloc 和.loc 属性过滤行和列。

7.11.1 实战操作

（1）读取 movie 数据集，将索引设置为 movie_title（片名）列，然后创建一个布尔数组，该布尔数组匹配所有内容分级为 G 且 IMDB 分数小于 4 的电影。

```
>>> movie = pd.read_csv(
...     "data/movie.csv", index_col="movie_title"
... )
>>> c1 = movie["content_rating"] == "G"
>>> c2 = movie["imdb_score"] < 4
>>> criteria = c1 & c2
```

（2）将这些条件传递给.loc 以过滤行。

```
>>> movie_loc = movie.loc[criteria]
>>> movie_loc.head()
                                  color  ...  movie/likes
movie_title                              ...
The True Story of Puss'N Boots    Color  ...           90
Doogal                            Color  ...          346
Thomas and the Magic Railroad     Color  ...          663
```

```
Barney's Great Adventure         Color    ...      436
Justin Bieber: Never Say Never   Color    ...      62000
```

(3)检查此 DataFrame 是否完全等于直接从索引操作符生成的 DataFrame。

```
>>> movie_loc.equals(movie[criteria])
True
```

(4)尝试使用.iloc 索引器进行相同的布尔索引。

```
>>> movie_iloc = movie.iloc[criteria]
Traceback (most recent call last):
...
ValueError: iLocation based boolean indexing cannot use an
indexable as a mask
```

(5)事实证明,由于索引的关系,我们不能直接使用布尔值的 Series。但是,我们可以使用布尔值的 ndarray。要获取该数组,可使用.to_numpy()方法。

```
>>> movie_iloc = movie.iloc[criteria.to_numpy()]
>>> movie_iloc.equals(movie_loc)
True
```

(6)尽管不是很常见,但可以进行布尔索引来选择特定的列。在这里,我们选择所有数据类型为 64 位整数的列。

```
>>> criteria_col = movie.dtypes == np.int64
>>> criteria_col.head()
color                         False
director_name                 False
num_critic_for_reviews        False
duration                      False
director_facebook_likes       False
dtype: bool

>>> movie.loc[:, criteria_col].head()
             num_voted_users  cast_total_facebook_likes  movie_facebook_likes
movie_title
Avatar               886204                       4834                 33000
Pirates o...         471220                      48350                     0
Spectre              275868                      11700                 85000
The Dark ...        1144337                     106759                164000
Star Wars...              8                        143                     0
```

(7)由于 criteria_col 是一个始终有索引的 Series,因此必须使用底层 ndarray 使其能够与.iloc协同工作。以下代码将产生与步骤（6）相同的结果。

```
>>> movie.iloc[ :, criteria_col.to_numpy()].head()
              num_voted_users cast_total_facebook_likes  movie_facebook_likes
movie_title
Avatar                 886204                      4834                 33000
Pirates o...           471220                     48350                     0
Spectre                275868                     11700                 85000
The Dark ...          1144337                    106759                164000
Star Wars...                8                       143                     0
```

（8）使用.loc时，可以使用布尔数组选择行，并使用标签列表指定所需的列。请记住，你需要在行和列选择之间加上逗号。让我们保持相同的行条件，然后选择content_rating、imdb_score、title_year和gross列。

```
>>> cols = [
...     "content_rating",
...     "imdb_score",
...     "title_year",
...     "gross",
... ]
>>> movie.loc[criteria, cols].sort_values("imdb_score")
              content_rating  imdb_score  title_year       gross
movie_title
Justin Bi...               G         1.6      2011.0  73000942.0
Sunday Sc...               G         2.5      2008.0         NaN
Doogal                     G         2.8      2006.0   7382993.0
Barney's ...               G         2.8      1998.0  11144518.0
The True ...               G         2.9      2009.0         NaN
Thomas an...               G         3.6      2000.0  15911333.0
```

（9）可以使用.iloc创建相同的操作，但是需要指定列的位置。

```
>>> col_index = [movie.columns.get_loc(col) for col in cols]
>>> col_index
[20, 24, 22, 8]
>>> movie.iloc[criteria.to_numpy(), col_index].sort_values(
...     "imdb_score"
... )
              content_rating  imdb_score  title_year       gross
movie_title
Justin Bi...               G         1.6      2011.0  73000942.0
Sunday Sc...               G         2.5      2008.0         NaN
Doogal                     G         2.8      2006.0   7382993.0
```

Barney's ...	G	2.8	1998.0	11144518.0
The True ...	G	2.9	2009.0	NaN
Thomas an...	G	3.6	2000.0	15911333.0

7.11.2 原理解释

.iloc 和.loc 属性都具有对布尔数组的过滤支持（需要注意的是，不能将.iloc 传递给 Series，但可以传递底层的 ndarray）。让我们来看看一维 ndarray 的基础 criteria。

```
>>> a = criteria.to_numpy()
>>> a[:5]
array([False, False, False, False, False])
>>> len(a), len(criteria)
(4916, 4916)
```

数组的长度与 Series 的长度相同，而 Series 与电影的 DataFrame 长度相同。布尔数组的整数位置与 DataFrame 的整数位置对齐，并且过滤将按预期进行。这些数组也可以与.loc 属性一起使用，但对于.iloc 来说则是必须的。

步骤（6）和步骤（7）显示了如何按列而不是按行进行过滤。需要使用冒号（:）来指示所有行的选择。冒号后面的逗号将分隔行和列的选择。但是，实际上还有一种更简单的方法来选择具有整数数据类型的列，即通过.select_dtypes 方法。

```
>>> movie.select_dtypes(int)
             num_voted_users  cast_total_facebook_likes
movie_title
Avatar              886204                       4834
Pirates o...        471220                      48350
Spectre             275868                      11700
The Dark ...       1144337                     106759
Star Wars...             8                        143
...                    ...                        ...
Signed Se...           629                       2283
The Follo...         73839                       1753
A Plague ...            38                          0
Shanghai ...          1255                       2386
My Date w...          4285                        163
```

步骤（8）和步骤（9）显示了如何同时进行这种行和列的选择。行是由布尔数组指定的，而列则是由列的列表指定的。在行和列选择之间需放置一个逗号。此外，步骤（9）还使用了列表推导式（list comprehension）遍历所有所需的列名称，以使用索引方法.get_loc 查找其整数位置。

第 8 章 对 齐 索 引

本章包含以下秘笈。
- 检查 Index 对象。
- 生成笛卡儿积。
- 了解索引暴增现象。
- 给不相等的索引填充值。
- 添加来自不同 DataFrames 中的列。
- 突出显示每列的最大值。
- 使用方法链复制.idxmax。
- 查找最常见的列的最大值。

8.1 介 绍

组合 Series 或 DataFrame 时，在进行任何计算之前，数据的每个维度会首先自动在每个轴上对齐。轴的这种静默且自动的对齐方式可能会使初学者感到困惑，但是它也为高级用户提供了灵活性。本章将深入探讨 Index 对象，然后演示利用其自动对齐功能的各种秘笈。

8.2 检查 Index 对象

如前文所述，每个 Series 和 DataFrame 的轴都有一个 Index 对象，用于标记值。有许多不同类型的 Index 对象，但是它们都有共同的行为。除 MultiIndex 以外的所有 Index 对象都是一维数据结构，结合了 Python 集合和 NumPy ndarray 的功能。

在本秘笈中，我们将检查 college 数据集的列索引并探索其许多功能。

8.2.1 实战操作

（1）读取 college 数据集，并创建一个保存列索引的变量 columns。

```
>>> import pandas as pd
>>> import numpy as np
>>> college = pd.read_csv("data/college.csv")
>>> columns = college.columns
>>> columns
Index(['INSTNM', 'CITY', 'STABBR', 'HBCU', 'MENONLY', 'WOMENONLY',
       'RELAFFIL', 'SATVRMID', 'SATMTMID', 'DISTANCEONLY', 'UGDS',
       'UGDS_WHITE', 'UGDS_BLACK', 'UGDS_HISP', 'UGDS_ASIAN',
       'UGDS_AIAN', 'UGDS_NHPI', 'UGDS_2MOR', 'UGDS_NRA', 'UGDS_UNKN',
       'PPTUG_EF', 'CURROPER', 'PCTPELL', 'PCTFLOAN', 'UG25ABV',
       'MD_EARN_WNE_P10', 'GRAD_DEBT_MDN_SUPP'],
      dtype='object')
```

（2）使用 .values 属性访问底层的 NumPy 数组。

```
>>> columns.values
array(['INSTNM', 'CITY', 'STABBR', 'HBCU', 'MENONLY', 'WOMENONLY',
       'RELAFFIL', 'SATVRMID', 'SATMTMID', 'DISTANCEONLY', 'UGDS',
       'UGDS_WHITE', 'UGDS_BLACK', 'UGDS_HISP', 'UGDS_ASIAN',
       'UGDS_AIAN', 'UGDS_NHPI', 'UGDS_2MOR', 'UGDS_NRA', 'UGDS_UNKN',
       'PPTUG_EF', 'CURROPER', 'PCTPELL', 'PCTFLOAN', 'UG25ABV',
       'MD_EARN_WNE_P10',
       'GRAD_DEBT_MDN_SUPP'], dtype=object)
```

（3）使用标量、列表或切片位置选择索引中的项目。

```
>>> columns[5]
'WOMENONLY'
>>> columns[[1, 8, 10]]
Index(['CITY', 'SATMTMID', 'UGDS'], dtype='object')
>>> columns[-7:-4]
Index(['PPTUG_EF', 'CURROPER', 'PCTPELL'], dtype='object')
```

（4）索引与 Series 和 DataFrame 共享许多相同的方法。

```
>>> columns.min(), columns.max(), columns.isnull().sum()
('CITY', 'WOMENONLY', 0)
```

（5）可以对 Index 对象使用基本的算术和比较运算符。

```
>>> columns + "_A"
Index([ 'INSTNM_A', 'CITY_A', 'STABBR_A', 'HBCU_A', 'MENONLY_A',
        'WOMENONLY_A', 'RELAFFIL_A', 'SATVRMID_A', 'SATMTMID_A',
        'DISTANCEONLY_A', 'UGDS_A', 'UGDS_WHITE_A', 'UGDS_BLACK_A',
        'UGDS_HISP_A', 'UGDS_ASIAN_A', 'UGDS_AIAN_A', 'UGDS_NHPI_A',
```

```
                    'UGDS_2MOR_A', 'UGDS_NRA_A', 'UGDS_UNKN_A', 'PPTUG_EF_A',
                    'CURROPER_A', 'PCTPELL_A', 'PCTFLOAN_A', 'UG25ABV_A',
                    'MD_EARN_WNE_P10_A', 'GRAD_DEBT_MDN_SUPP_A'], dtype='object')

>>> columns > "G"
array([ True, False,  True,  True,  True,  True,  True,  True,  True,
       False,  True,  True,  True,  True,  True,  True,  True,  True,
        True,  True, False,  True,  True,  True,  True,  True,  True])
```

（6）索引在创建之后，尝试更改索引值将会失败，因为索引是不可变的对象。

```
>>> columns[1] = "city"
Traceback (most recent call last):
    ...
TypeError: Index does not support mutable operations
```

8.2.2 原理解释

从许多 Index 对象操作中可以看到，它似乎与 Series 和 ndarray 有很多共同点。最重要的区别之一在于步骤（6）。索引是不可变的，一旦被创建后，就无法更改其值。

8.2.3 扩展知识

索引支持集合操作，即并集（union）、交集（intersection）、差集（difference）和对称差集（symmetric difference）。

```
>>> c1 = columns[:4]
>>> c1
Index(['INSTNM', 'CITY', 'STABBR', 'HBCU'], dtype='object')

>>> c2 = columns[2:6]
>>> c2
Index(['STABBR', 'HBCU', 'MENONLY', 'WOMENONLY'], dtype='object')

>>> c1.union(c2)           # 或 'c1 | c2'
Index(['CITY', 'HBCU', 'INSTNM', 'MENONLY', 'STABBR', 'WOMENONLY'],
dtype='object')

>>> c1.symmetric_difference(c2)      # 或 'c1 ^ c2'
Index(['CITY', 'INSTNM', 'MENONLY', 'WOMENONLY'], dtype='object')
```

索引具有许多与 Python 集合相同的操作。索引（通常）是使用哈希表实现的，当从

DataFrame 中选择行或列时，哈希表的访问速度非常快。因为值需要是可哈希的，所以 Index 对象的值必须是不可变的类型，如字符串、整数或元组，就像 Python 字典中的键一样。

索引支持重复值，并且如果在任何索引中碰巧有重复项，则哈希表将无法再用于其实现，并且对象访问会变得很慢。

8.3 生成笛卡儿积

每当 Series 或 DataFrame 与另一个 Series 或 DataFrame 一起操作时，每个对象的索引（行索引和列索引）都将在操作开始之前首先对齐。这种索引对齐发生在幕后，对于那些刚接触 Pandas 的人来说可能是非常令人惊讶的。除非索引相同，否则此对齐功能将始终在索引之间创建笛卡儿积。

笛卡儿积（cartesian product）是一个数学术语，通常出现在集合论中。两个集合之间的笛卡儿积是两个集合的对的所有组合。例如，标准纸牌中的 52 张纸牌代表 13 个等级（A，2，3，…，J，Q，K）和 4 个花色之间的笛卡儿积。

生成笛卡儿积并非总是预期的结果，但关键是我们必须了解发生的时间和方式，以避免意外的结果。在此秘笈中，将具有重叠但不相同的索引的两个 Series 相加在一起，产生了令人惊讶的结果。我们还将演示如果它们具有相同的索引会发生什么。

8.3.1 实战操作

请按照以下步骤创建笛卡儿积。

（1）构造两个具有不同索引但包含一些相同值的 Series。

```
>>> s1 = pd.Series(index=list("aaab"), data=np.arange(4))
>>> s1
a    0
a    1
a    2
b    3
dtype: int64

>>> s2 = pd.Series(index=list("cababb"), data=np.arange(6))
>>> s2
c    0
a    1
```

```
b    2
a    3
b    4
b    5
dtype: int64
```

（2）将两个 Series 加在一起以产生笛卡儿积。对于 s1 中的每个 a 索引值，我们将加上 s2 中的每个 a。

```
>>> s1 + s2
a    1.0
a    3.0
a    2.0
a    4.0
a    3.0
a    5.0
b    5.0
b    7.0
b    8.0
c    NaN
dtype: float64
```

8.3.2 原理解释

s1 中的每个 a 标签都将与 s2 中的每个 a 标签配对。该配对在结果 Series 中产生了 6 个 a 标签、3 个 b 标签和一个 c 标签。笛卡儿积在所有相同的索引标签之间发生。

由于带有标签 c 的元素在 Series s2 中是唯一的，因此 Pandas 默认其值为缺失值，因为在 s1 中没有与之对齐的标签。每当索引标签只存在于一个对象中时（即它是唯一的），Pandas 默认为缺失值。这样就会产生一个很糟糕的结果，即 Series 的数据类型被更改为 float，虽然每个 Series 中只有整数值。之所以发生类型更改，是因为 NumPy 的缺失值对象 np.nan 仅存在于浮点数中，而不存在于整数中。由于 Series 和 DataFrame 列必须具有相同的数值数据类型。因此，该列中的每个值都将被转换为浮点数。更改类型对于本示例中这个很小的数据集影响不大，但是对于较大的数据集来说，可能会显著影响内存使用量。

8.3.3 扩展知识

当索引是唯一的或同时包含相同的精确元素并且元素顺序相同时，则不会创建笛卡儿乘积。此外，当索引值是唯一的或一样的且具有相同的顺序时，也不会创建笛卡儿乘积，而是按索引的位置对齐。请注意，每个元素都按位置精确对齐，并且数据类型仍为整数。

```
>>> s1 = pd.Series(index=list("aaabb"), data=np.arange(5))
>>> s2 = pd.Series(index=list("aaabb"), data=np.arange(5))
>>> s1 + s2
a    0
a    2
a    4
b    6
b    8
dtype: int64
```

如果索引的元素相同，但是 Series 之间的顺序不同，则会出现笛卡儿积。下面在 s2 中更改索引的顺序，然后重新运行相同的操作。

```
>>> s1 = pd.Series(index=list("aaabb"), data=np.arange(5))
>>> s2 = pd.Series(index=list("bbaaa"), data=np.arange(5))
>>> s1 + s2
a    2
a    3
a    4
a    3
a    4
    ..
a    6
b    3
b    4
b    4
b    5
Length: 13, dtype: int64
```

请注意这一点，因为同一操作的 Pandas 会有两个截然不同的结果。发生这种情况的另一个实例是在 groupby 操作期间。如果对多列进行 groupby，并且其中一列的类型是 categorical，则将获得笛卡儿乘积，其中每个外部索引将具有每个内部索引值。

最后，我们将两个具有不同顺序索引值但没有重复值的 Series 加在一起。在相加时，我们没有得到笛卡儿积。

```
>>> s3 = pd.Series(index=list("ab"), data=np.arange(2))
>>> s4 = pd.Series(index=list("ba"), data=np.arange(2))
>>> s3 + s4
```

```
a    1
b    1
dtype: int64
```

在此秘笈中，每个 Series 具有不同数量的元素。一般来说，当操作维中不包含相同数量的元素时，Python 和其他语言中的类似数组的数据结构将不允许进行操作。Pandas 通过在完成操作之前首先对齐索引来实现此目的。

在第 7 章中，我们演示了可以设置一列为索引，然后通过它进行过滤。笔者比较喜欢的操作是将索引单独放在一边，然后在列上进行过滤。本节提供了另一方面的示例，即我们需要非常谨慎地使用索引。

8.4 了解索引暴增现象

在前面的秘笈中，我们介绍了一个很简单的将两个 Series 加在一起的示例，这两个 Series 包含了不相等的索引。其实这个秘笈更像一个"反秘笈"，因为它更多的意义是在演示不应执行的操作。在处理大量数据时，索引对齐的笛卡儿乘积会产生可笑的错误结果。

在此秘笈中，我们会将两个较大的 Series 加在一起，它们的索引只有几个唯一值，但是顺序不同，结果却是使索引中的值数量呈现爆炸性的增长。

8.4.1 实战操作

（1）读取 employee 数据并将索引设置为 RACE（种族）列。

```
>>> employee = pd.read_csv(
...     "data/employee.csv", index_col="RACE"
... )
>>> employee.head()
              UNIQUE_ID  POSITION_TITLE  ...  HIRE_DATE   JOB_DATE
RACE                                     ...
Hispanic/...         0   ASSISTAN...     ...  2006-06-12  2012-10-13
Hispanic/...         1   LIBRARY ...     ...  2000-07-19  2010-09-18
White                2   POLICE O...     ...  2015-02-03  2015-02-03
White                3   ENGINEER...     ...  1982-02-08  1991-05-25
White                4   ELECTRICIAN     ...  1989-06-19  1994-10-22
```

（2）选择 BASE_SALARY（基础薪水）列生成两个不同的 Series。现在来检查此操作是否创建了两个新对象。

```
>>> salary1 = employee["BASE_SALARY"]
>>> salary2 = employee["BASE_SALARY"]
>>> salary1 is salary2
True
```

（3）可以看到，salary1 和 salary2 变量引用的是同一对象。这意味着对其中一个的任何修改都会导致另一个的修改。为确保接收到数据的全新副本，可使用.copy 方法。

```
>>> salary2 = employee["BASE_SALARY"].copy()
>>> salary1 is salary2
False
```

（4）通过对其中一个 Series 进行排序来更改其索引顺序。

```
>>> salary1 = salary1.sort_index()
>>> salary1.head()
RACE
American Indian or Alaskan Native    78355.0
American Indian or Alaskan Native    26125.0
American Indian or Alaskan Native    98536.0
American Indian or Alaskan Native        NaN
American Indian or Alaskan Native    55461.0
Name: BASE_SALARY, dtype: float64

>>> salary2.head()
RACE
Hispanic/Latino    121862.0
Hispanic/Latino     26125.0
White               45279.0
White               63166.0
White               56347.0
Name: BASE_SALARY, dtype: float64
```

（5）可以看到，salary1 和 salary2 的索引顺序已经不一样了，下面将这些薪水 Series 加在一起。

```
>>> salary_add = salary1 + salary2

>>> salary_add.head()
RACE
American Indian or Alaskan Native    138702.0
American Indian or Alaskan Native    156710.0
American Indian or Alaskan Native    176891.0
American Indian or Alaskan Native    159594.0
```

```
American Indian or Alaskan Native   127734.0
Name: BASE_SALARY, dtype: float64
```

（6）操作成功完成。下面再创建一个 salary1 的 Series 并和它自身相加，然后输出每个 Series 的长度。可以看到，索引从 2 000 个值暴增到超过一百万个值。

```
>>> salary_add1 = salary1 + salary1
>>> len(salary1), len(salary2), len(salary_add), len(
...        salary_add1
... )
(2000, 2000, 1175424, 2000)
```

8.4.2 原理解释

在步骤（2）中，最初看似创建了两个唯一的对象，但实际上，它创建了一个由两个不同的变量名称引用的对象。表达式 employee['BASE_SALARY'] 从技术上讲是创建视图，而不是创建全新的副本。这通过 is 运算符得到了验证。

在 Pandas 中，视图不是新对象，而仅仅是对另一个对象的引用，通常是 DataFrame 的某些子集。此共享对象可能导致许多问题。

为了确保变量引用完全不同的对象，我们使用了 .copy 方法，然后使用 is 运算符验证它们是否为不同的对象。

步骤（4）使用 .sort_index 方法按种族对 Series 进行排序。请注意，该 Series 具有相同的索引条目，但它们现在的顺序与 salary1 的顺序不同。

步骤（5）将这些不同的 Series 相加在一起以得出总和。通过检查 head（前 5 个元素），仍然不清楚产生了什么。

步骤（6）将 salary1 和它自身相加，以显示两个不同 Series 的加法之间的比较。此秘笈中所有 Series 的长度都已经被输出，我们清楚地看到，salary_add 现在已经暴增到一百万个值以上。出现笛卡儿积是因为索引不是唯一的，而且顺序相同。此秘笈显示了索引不同时发生的更明显的示例。

8.4.3 扩展知识

可以通过做一些数学运算来验证 salary_add 的值的数量。当笛卡儿积在所有相同的索引值之间发生时，我们可以对它们的计数的平方求和，甚至索引中的缺失值也会与它们自身产生笛卡儿积。

```
>>> index_vc = salary1.index.value_counts(dropna=False)
```

```
>>> index_vc
Black or African American          700
White                              665
Hispanic/Latino                    480
Asian/Pacific Islander             107
NaN                                 35
American Indian or Alaskan Native   11
Others                               2
Name: RACE, dtype: int64

>>> index_vc.pow(2).sum()
1175424
```

8.5 给不相等的索引填充值

当使用加号运算符将两个 Series 加在一起并且一个索引的标签未出现在另一个索引中时，其结果值始终是缺失值。Pandas 具有 .add 方法，该方法提供了一个选项来填充缺失值。请注意，这些 Series 不包含重复的条目，因此无须担心笛卡儿积会让条目数量暴增。

在此秘笈中，我们使用 .add 方法和 fill_value 参数将棒球数据集中具有不等（但唯一）索引的多个 Series 加在一起，以确保结果中没有缺失值。

8.5.1 实战操作

（1）读取 3 个棒球数据集，并将 playerID 设置为索引。

```
>>> baseball_14 = pd.read_csv(
...     "data/baseball14.csv", index_col="playerID"
... )
>>> baseball_15 = pd.read_csv(
...     "data/baseball15.csv", index_col="playerID"
... )
>>> baseball_16 = pd.read_csv(
...     "data/baseball16.csv", index_col="playerID"
... )
>>> baseball_14.head()
           yearID  stint  teamID  lgID  ...  HBP   SH   SF  GIDP
playerID                                      ...
altuvjo01    2014      1     HOU    AL  ...  5.0  1.0  5.0  20.0
cartech02    2014      1     HOU    AL  ...  5.0  0.0  4.0  12.0
```

```
castrja01    2014        1       HOU     AL   ...   9.0   1.0   3.0   11.0
corpoca01    2014        1       HOU     AL   ...   3.0   1.0   2.0    3.0
dominma01    2014        1       HOU     AL   ...   5.0   2.0   7.0   23.0
```

（2）在索引上使用 .difference 方法，以发现哪些索引标签在 ballball_14 中而不在 ballball_15 中，反之亦然。

```
>>> baseball_14.index.difference(baseball_15.index)
Index(['corpoca01', 'dominma01', 'fowlede01', 'grossro01', 'guzmaje01',
       'hoeslj01', 'krausma01', 'preslal01', 'singljo02'],
      dtype='object', name='playerID')

>>> baseball_15.index.difference(baseball_14.index)
Index(['congeha01', 'correca01', 'gattiev01', 'gomezca01', 'lowrije01',
       'rasmuco01', 'tuckepr01', 'valbulu01'],
      dtype='object', name='playerID')
```

（3）每个索引都有很多独特的玩家。下面找出 3 年内每个玩家的总击球数。H 列包含的就是击球数。

```
>>> hits_14 = baseball_14["H"]
>>> hits_15 = baseball_15["H"]
>>> hits_16 = baseball_16["H"]
>>> hits_14.head()
playerID
altuvjo01    225
cartech02    115
castrja01    103
corpoca01     40
dominma01    121
Name: H, dtype: int64
```

（4）下面使用加号运算符将两个 Series 相加。

```
>>> (hits_14 + hits_15).head()
playerID
altuvjo01    425.0
cartech02    193.0
castrja01    174.0
congeha01      NaN
corpoca01      NaN
Name: H, dtype: float64
```

（5）可以看到，即使玩家 congeha01 和 corpoca01 具有 2015 年的击球数据（也就是说，在 hits_15 中包含了值），但由于他们在 hits_14 中没有值，因此他们最后的结果仍然是一个缺失值。可以将.add 方法与 fill_value 参数一起使用，以避免产生缺失值。

```
>>> hits_14.add(hits_15, fill_value=0).head()
playerID
altuvjo01    425.0
cartech02    193.0
castrja01    174.0
congeha01     46.0
corpoca01     40.0
Name: H, dtype: float64
```

（6）通过再次链接 add 方法添加 2016 年以来的击球数。

```
>>> hits_total = hits_14.add(hits_15, fill_value=0).add(
...        hits_16, fill_value=0
... )
>>> hits_total.head()
playerID
altuvjo01    641.0
bregmal01     53.0
cartech02    193.0
castrja01    243.0
congeha01     46.0
Name: H, dtype: float64
```

（7）检查结果中是否包含缺失值。

```
>>> hits_total.hasnans
False
```

8.5.2 原理解释

.add 方法的工作方式与加号运算符相似，但通过使用 fill_value 参数取代不匹配的索引，则可以提供更大的灵活性。在本示例中，将不匹配的索引值默认设置为 0 是有意义的，但是也可以使用其他数字。

有时每个 Series 都包含与缺失值相对应的索引标签。在此特定实例中，当将两个 Series 相加时，无论是否使用 fill_value 参数，索引标签仍将对应于缺失值。为了搞清楚这一点，请看以下示例，其中索引标签 a 对应于每个 Series 中的缺失值。

```
>>> s = pd.Series(
...        index=["a", "b", "c", "d"],
```

```
...         data=[np.nan, 3, np.nan, 1],
... )
>>> s
a    NaN
b    3.0
c    NaN
d    1.0
dtype: float64

>>> s1 = pd.Series(
...         index=["a", "b", "c"], data=[np.nan, 6, 10]
... )
>>> s1
a    NaN
b    6.0
c    10.0
dtype: float64

>>> s.add(s1, fill_value=5)
a    NaN
b    9.0
c    15.0
d    6.0
dtype: float64
```

8.5.3 扩展知识

此秘笈展示了如何将仅具有单个索引的 Series 添加在一起。也可以将 DataFrame 添加在一起。将两个 DataFrame 加在一起将在计算之前对齐索引和列，并为不匹配的索引插入缺失值。首先，从 2014 年棒球数据集中选择一些列。

```
>>> df_14 = baseball_14[["G", "AB", "R", "H"]]
>>> df_14.head()
            G    AB    R    H
playerID
altuvjo01   158  660   85   225
cartech02   145  507   68   115
castrja01   126  465   43   103
corpoca01    55  170   22    40
dominma01   157  564   51   121
```

还可以从 2015 年棒球数据集中分别选择一些相同和不同的列。

```
>>> df_15 = baseball_15[["AB", "R", "H", "HR"]]
>>> df_15.head()
           AB    R    H   HR
playerID
altuvjo01  638  86  200   15
cartech02  391  50   78   24
castrja01  337  38   71   11
congeha01  201  25   46   11
correca01  387  52  108   22
```

将两个 DataFrame 加在一起会在行或列标签无法对齐的地方创建缺失值。可以使用.style 属性并调用.highlight_null 方法，以查看缺失值的位置，如图 8-1 所示。

	AB	G	H	HR	R
playerID					
altuvjo01	1298	nan	425	nan	171
cartech02	898	nan	193	nan	118
castrja01	802	nan	174	nan	81
congeha01	nan	nan	nan	nan	nan
corpoca01	nan	nan	nan	nan	nan
correca01	nan	nan	nan	nan	nan
dominma01	nan	nan	nan	nan	nan
fowlede01	nan	nan	nan	nan	nan
gattiev01	nan	nan	nan	nan	nan
gomezca01	nan	nan	nan	nan	nan

`(df_14 + df_15).head(10).style.highlight_null('yellow')`

图 8-1　使用加号运算符时突出显示空值

只有在两个 DataFrame 中都显示 playerID 的行才可用。类似地，AB、H 和 R 列是两个 DataFrame 中唯一出现的列。即使在指定了 fill_value 参数的情况下使用.add 方法，仍然可能包含缺失值，这是因为在我们的输入数据中不存在行和列的某些组合。例如，playerID congeha01 与 G 列的交集就是如此，该玩家仅出现在 2015 年的数据集中而没有 G 列。因此，该值仍然缺失了，如图 8-2 所示。

```
(df_14
    .add(df_15, fill_value=0)
    .head(10)
    .style.highlight_null('yellow')
)
```

playerID	AB	G	H	HR	R
altuvjo01	1298	158	425	15	171
cartech02	898	145	193	24	118
castrja01	802	126	174	11	81
congeha01	201	nan	46	11	25
corpoca01	170	55	40	nan	22
correca01	387	nan	108	22	52
dominma01	564	157	121	nan	51
fowlede01	434	116	120	nan	61
gattiev01	566	nan	139	27	66
gomezca01	149	nan	36	4	19

图 8-2　使用.add 方法时突出显示空值

8.6　添加来自不同 DataFrames 中的列

所有 DataFrame 都可以向自己添加新列。但是，如前文所述，每当一个 DataFrame 添加来自另一个 DataFrame 或 Series 中的列时，索引将首先对齐，然后创建新列。

此秘笈将使用 employee 数据集添加一个新列，其中包含该员工所属部门的最高薪水。

8.6.1　实战操作

（1）导入 employee 数据，然后在新的 DataFrame 中选择 DEPARTMENT（部门）和 BASE_SALARY（基础薪水）列。

```
>>> employee = pd.read_csv("data/employee.csv")
>>> dept_sal = employee[["DEPARTMENT", "BASE_SALARY"]]
```

（2）按每个部门内的薪水对这个较小的 DataFrame 进行排序。

```
>>> dept_sal = dept_sal.sort_values(
...     ["DEPARTMENT", "BASE_SALARY"],
...     ascending=[True, False],
... )
```

（3）使用 .drop_duplicates 方法保留每个 DEPARTMENT 的第一行。

```
>>> max_dept_sal = dept_sal.drop_duplicates(
...     subset="DEPARTMENT"
... )
>>> max_dept_sal.head()
        DEPARTMENT  BASE_SALARY
              DEPARTMENT  BASE_SALARY
1494  Admn. & Regulatory Affairs    140416.0
149       City Controller's Office     64251.0
236              City Council    100000.0
647    Convention and Entertainment    38397.0
1500   Dept of Neighborhoods (DON)    89221.0
```

（4）将 DEPARTMENT 列放入每个 DataFrame 的索引中。

```
>>> max_dept_sal = max_dept_sal.set_index("DEPARTMENT")
>>> employee = employee.set_index("DEPARTMENT")
```

（5）现在索引包含匹配的值，我们可以向 employee DataFrame 中添加一个新列。

```
>>> employee = employee.assign(
...     MAX_DEPT_SALARY=max_dept_sal["BASE_SALARY"]
... )
>>> employee
                              UNIQUE_ID  ...  MAX_D/ALARY
DEPARTMENT                               ...
MunicipalCourts Department          0  ...    121862.0
Library                             1  ...    107763.0
Houston Police Department-HPD       2  ...    199596.0
Houston Fire Department (HFD)       3  ...    210588.0
General Services Department         4  ...     89194.0
...                               ...  ...         ...
Houston Police Department-HPD    1995  ...    199596.0
Houston Fire Department (HFD)    1996  ...    210588.0
Houston Police Department-HPD    1997  ...    199596.0
Houston Police Department-HPD    1998  ...    199596.0
Houston Fire Department (HFD)    1999  ...    210588.0
```

（6）我们可以使用 .query 查询方法来验证结果，以检查是否存在 BASE_SALARY

大于 MAX_DEPT_SALARY 的行。

```
>>> employee.query("BASE_SALARY > MAX_DEPT_SALARY")
Empty DataFrame
Columns: [UNIQUE_ID, POSITION_TITLE, BASE_SALARY, RACE, EMPLOYMENT_TYPE,
GENDER, EMPLOYMENT_STATUS, HIRE_DATE, JOB_DATE, MAX_DEPT_SALARY]
Index: []
```

（7）将代码重构为一个链。

```
>>> employee = pd.read_csv("data/employee.csv")
>>> max_dept_sal = (
...     employee
...     [["DEPARTMENT", "BASE_SALARY"]]
...     .sort_values(
...         ["DEPARTMENT", "BASE_SALARY"],
...         ascending=[True, False],
...     )
...     .drop_duplicates(subset="DEPARTMENT")
...     .set_index("DEPARTMENT")
... )

>>> (
...     employee
...     .set_index("DEPARTMENT")
...     .assign(
...         MAX_DEPT_SALARY=max_dept_sal["BASE_SALARY"]
...     )
... )
               UNIQUE_ID POSITION_TITLE ...   JOB_DATE MAX_DEPT_SALARY
DEPARTMENT                              ...
Municipal...          0       ASSISTAN... ... 2012-10-13       121862.0
Library               1       LIBRARY  ... ... 2010-09-18       107763.0
Houston P...          2       POLICE O... ... 2015-02-03       199596.0
Houston F...          3       ENGINEER... ... 1991-05-25       210588.0
General S...          4       ELECTRICIAN ... 1994-10-22        89194.0
...                 ...              ... ...        ...            ...
Houston P...       1995       POLICE O... ... 2015-06-09       199596.0
Houston F...       1996       COMMUNIC... ... 2013-10-06       210588.0
Houston P...       1997       POLICE O... ... 2015-10-13       199596.0
Houston P...       1998       POLICE O... ... 2011-07-02       199596.0
Houston F...       1999       FIRE FIG... ... 2010-07-12       210588.0
```

8.6.2 原理解释

步骤（2）和步骤（3）找到了每个部门的最高工资。为了使自动索引对齐能够正常工作，我们将每个 DataFrame 索引设置为 DEPARTMENT（部门）。

步骤（5）之所以能够正常工作，是因为左侧 DataFrame（employee）中的每行索引都与右侧 DataFrame（max_dept_sal）中的一个索引对齐（且仅与一个索引对齐）。如果 max_dept_sal 在其索引中具有任何部门重复项，那么我们将获得笛卡儿积。

例如，我们可以来试一试，看看在等式的右侧使用具有重复索引值的 DataFrame 时会发生什么。这里可以使用.sample DataFrame 方法随机选择 10 行而不进行替换。

```
>>> random_salary = dept_sal.sample(
...     n=10, random_state=42
... ).set_index("DEPARTMENT")
>>> random_salary
                                    BASE_SALARY
DEPARTMENT
Public Works & Engineering-PWE         34861.0
Houston Airport System (HAS)           29286.0
Houston Police Department-HPD          31907.0
Houston Police Department-HPD          66614.0
Houston Police Department-HPD          42000.0
Houston Police Department-HPD          43443.0
Houston Police Department-HPD          66614.0
Public Works & Engineering-PWE         52582.0
Finance                                93168.0
Houston Police Department-HPD          35318.0
```

现在可以看到索引中有了若干个重复的部门。当我们尝试创建新列时，会引发错误，提醒有重复。在 employee DataFrame 中，至少有一个索引标签与来自 random_salary 中的两个或多个索引标签结合在一起。

```
>>> employee["RANDOM_SALARY"] = random_salary["BASE_SALARY"]
Traceback (most recent call last):
...
ValueError: cannot reindex from a duplicate axis
```

8.6.3 扩展知识

在对齐期间，如果没有任何要对齐的 DataFrame 索引，则结果中将包含缺失值。现在可以创建一个发生这种情况的示例。

以下仅使用 max_dept_sal Series 的前 3 行创建新列。

```
>>> (
...     employee
...     .set_index("DEPARTMENT")
...     .assign(
...         MAX_SALARY2=max_dept_sal["BASE_SALARY"].head(3)
...     )
...     .MAX_SALARY2
...     .value_counts(dropna=False)
... )
NaN        1955
140416.0     29
100000.0     11
64251.0       5
Name: MAX_SALARY2, dtype: int64
```

操作成功完成,但只填充了 3 个部门的薪水。所有其他部门(它们未出现在 max_dept_sal Series 的前 3 行中)均产生了缺失值。

笔者更喜欢使用下面的代码,而不是步骤(7)中的代码。此代码将 .groupby 与 .transform 方法结合使用(下文将会对这两个方法展开详细讨论)。另外,这段代码更易于阅读和简短,并且不会因为重新分配索引而导致混乱。

```
>>> max_sal = (
...     employee
...     .groupby("DEPARTMENT")
...     .BASE_SALARY
...     .transform("max")
... )

>>> (employee.assign(MAX_DEPT_SALARY=max_sal))
   UNIQUE_ID  POSITION_TITLE  ...   JOB_DATE   MAX_DEPT_SALARY
0          0  ASSISTAN...     ...  2012-10-13         121862.0
1          1  LIBRARY ...     ...  2010-09-18         107763.0
2          2  POLICE O...     ...  2015-02-03         199596.0
3          3  ENGINEER...     ...  1991-05-25         210588.0
4          4  ELECTRICIAN     ...  1994-10-22          89194.0
...      ...  ...             ...  ...                     ...
1995    1995  POLICE O...     ...  2015-06-09         199596.0
1996    1996  COMMUNIC...     ...  2013-10-06         210588.0
1997    1997  POLICE O...     ...  2015-10-13         199596.0
```

```
1998      1998      POLICE O...  ...  2011-07-02      199596.0
1999      1999      FIRE FIG...  ...  2010-07-12      210588.0
```

这种方式之所以可行，是因为.transform 方法将保留原始索引。如果使用.groupby 方法创建了新索引，则可以使用.merge 方法合并数据。通过参数 how 指定合并方式为 left（左连接，即左侧 DataFrame 取全部数据），使用 left_on 参数指定 DEPARTMENT 作为连接关键字，设定 right_index 值为 True 来表示右侧 DataFrame 使用索引作为连接关键字。

```
>>> max_sal = (
...     employee
...     .groupby("DEPARTMENT")
...     .BASE_SALARY
...     .max()
... )

>>> (
...     employee.merge(
...         max_sal.rename("MAX_DEPT_SALARY"),
...         how="left",
...         left_on="DEPARTMENT",
...         right_index=True,
...     )
... )
      UNIQUE_ID  POSITION_TITLE  ...  JOB_DATE    MAX_DEPT_SALARY
0             0  ASSISTAN...     ...  2012-10-13         121862.0
1             1  LIBRARY ...     ...  2010-09-18         107763.0
2             2  POLICE O...     ...  2015-02-03         199596.0
3             3  ENGINEER...     ...  1991-05-25         210588.0
4             4  ELECTRICIAN     ...  1994-10-22          89194.0
...         ...  ...             ...  ...                     ...
1995       1995  POLICE O...     ...  2015-06-09         199596.0
1996       1996  COMMUNIC...     ...  2013-10-06         210588.0
1997       1997  POLICE O...     ...  2015-10-13         199596.0
1998       1998  POLICE O...     ...  2011-07-02         199596.0
1999       1999  FIRE FIG...     ...  2010-07-12         210588.0
```

8.7 突出显示每列的最大值

college 数据集有许多数字列，描述了有关每所学校的不同指标。许多人都对在特定指标上表现最好的学校感兴趣。

此秘笈将查找和揭示在每个数值列上具有最大值的学校，并设置 DataFrame 的样式以突出显示该信息。

8.7.1 实战操作

（1）读取 college 数据集，并以机构名称（INSTNM）列作为索引。

```
>>> college = pd.read_csv(
...     "data/college.csv", index_col="INSTNM"
... )
>>> college.dtypes
CITY                    object
STABBR                  object
HBCU                   float64
MENONLY                float64
WOMENONLY              float64
                        ...
PCTPELL                float64
PCTFLOAN               float64
UG25ABV                float64
MD_EARN_WNE_P10         object
GRAD_DEBT_MDN_SUPP      object
Length: 26, dtype: object
```

（2）除 CITY 和 STABBR 以外的所有其他列似乎都是数字。但是，经过仔细检查步骤（1）中的数据类型，意外发现 MD_EARN_WNE_P10 和 GRAD_DEBT_MDN_SUPP 列都是 object 类型，而不是数字。

为了更好地了解这些列中包含哪些类型的值，下面检查其中的一个样本。

```
>>> college.MD_EARN_WNE_P10.sample(10, random_state=42)
INSTNM
Career Point College                                  20700
Ner Israel Rabbinical College                      PrivacyS...
Reflections Academy of Beauty                           NaN
Capital Area Technical College                        26400
West Virginia University Institute of Technology      43400
Mid-State Technical College                           32000
Strayer University-Huntsville Campus                  49200
National Aviation Academy of Tampa Bay                45000
University of California-Santa Cruz                   43000
Lexington Theological Seminary                          NaN
Name: MD_EARN_WNE_P10, dtype: object
```

```
>>> college.GRAD_DEBT_MDN_SUPP.sample(10, random_state=42)
INSTNM
Career Point College                                    14977
Ner Israel Rabbinical College                           PrivacyS...
Reflections Academy of Beauty                           PrivacyS...
Capital Area Technical College                          PrivacyS...
West Virginia University Institute of Technology        23969
Mid-State Technical College                              8025
Strayer University-Huntsville Campus                    36173.5
National Aviation Academy of Tampa Bay                  22778
University of California-Santa Cruz                     19884
Lexington Theological Seminary                          PrivacyS...
Name: GRAD_DEBT_MDN_SUPP, dtype: object
```

（3）这些值是字符串，但我们希望它们是数字。在这种情况下，笔者喜欢使用.value_counts方法，看看是否能够找到任何迫使列为非数字的字符。

```
>>> college.MD_EARN_WNE_P10.value_counts()
PrivacySuppressed     822
38800                 151
21500                  97
49200                  78
27400                  46
                     ...
66700                   1
163900                  1
64400                   1
58700                   1
64100                   1
Name: MD_EARN_WNE_P10, Length: 598, dtype: int64

>>> set(college.MD_EARN_WNE_P10.apply(type))
{<class 'float'>, <class 'str'>}

>>> college.GRAD_DEBT_MDN_SUPP.value_counts()
PrivacySuppressed    1510
9500                  514
27000                 306
25827.5               136
25000                 124
                     ...
16078.5                 1
27763.5                 1
```

```
6382                  1
27625                 1
11300                 1
Name: GRAD_DEBT_MDN_SUPP, Length: 2038, dtype: int64
```

（4）罪魁祸首似乎是一些学校对这两列数据有隐私方面的关注。要强制将这些列转换为数字类型，可使用 Pandas 函数 to_numeric。如果使用 errors ="coerce"参数，那么它将把这些值转换为 NaN。

```
>>> cols = ["MD_EARN_WNE_P10", "GRAD_DEBT_MDN_SUPP"]
>>> for col in cols:
...     college[col] = pd.to_numeric(
...         college[col], errors="coerce"
...     )

>>> college.dtypes.loc[cols]
MD_EARN_WNE_P10       float64
GRAD_DEBT_MDN_SUPP    float64
dtype: object
```

（5）使用.select_dtypes 方法进行过滤，仅留下数字列。这将排除 STABBR 和 CITY 列，因为它们的最大值对于此问题没有意义。

```
>>> college_n = college.select_dtypes("number")
>>> college_n.head()
             HBCU   MENONLY  ...  MD_EARN_WNE_P10   GRAD_DEBT_MDN_SUPP
INSTNM                       ...
Alabama A...  1.0    0.0     ...      30300.0            33888.0
Universit...  0.0    0.0     ...      39700.0            21941.5
Amridge U...  0.0    0.0     ...      40100.0            23370.0
Universit...  0.0    0.0     ...      45500.0            24097.0
Alabama S...  1.0    0.0     ...      26600.0            33118.5
```

（6）有几列只有二进制（0 或 1）值，不会为最大值提供有用的信息。当查找这些列时，可以创建一个布尔系列，并使用.nunique 方法查找具有两个唯一值的所有列。

```
>>> binary_only = college_n.nunique() == 2
>>> binary_only.head()
HBCU          True
MENONLY       True
WOMENONLY     True
RELAFFIL      True
SATVRMID      False
dtype: bool
```

（7）使用布尔数组创建二进制列的列表。

```
>>> binary_cols = binary_only[binary_only].index
>>> binary_cols
Index(['HBCU', 'MENONLY', 'WOMENONLY', 'RELAFFIL', 'DISTANCEONLY',
'CURROPER'], dtype='object')
```

（8）由于我们正在寻找最大值，因此可以使用.drop方法删除二进制列。

```
>>> college_n2 = college_n.drop(columns=binary_cols)
>>> college_n2.head()
             SATVRMID  SATMTMID  ...  MD_EARN_WNE_P10  GRAD_DEBT_MDN_SUPP
INSTNM                            ...
Alabama A...    424.0     420.0  ...          30300.0             33888.0
Universit...    570.0     565.0  ...          39700.0             21941.5
Amridge U...      NaN       NaN  ...          40100.0             23370.0
Universit...    595.0     590.0  ...          45500.0             24097.0
Alabama S...    425.0     430.0  ...          26600.0             33118.5
```

（9）现在可以使用.idxmax方法查找每列最大值的索引标签。

```
>>> max_cols = college_n2.idxmax()
>>> max_cols
SATVRMID                       California Institute of Technology
SATMTMID                       California Institute of Technology
UGDS                                   University of Phoenix-Arizona
UGDS_WHITE                   Mr Leon's School of Hair Design-Moscow
UGDS_BLACK                       Velvatex College of Beauty Culture
                                          ...
PCTPELL                                    MTI Business College Inc
PCTFLOAN                                    ABC Beauty College Inc
UG25ABV                               Dongguk University-Los Angeles
MD_EARN_WNE_P10                       Medical College of Wisconsin
GRAD_DEBT_MDN_SUPP      Southwest University of Visual Arts-Tucson
Length: 18, dtype: object
```

（10）在max_cols Series上调用.unique方法。这将返回college_n2中具有最大值的索引值的ndarray。

```
>>> unique_max_cols = max_cols.unique()
>>> unique_max_cols[:5]
array([ 'California Institute of Technology',
        'University of Phoenix-Arizona',
        "Mr Leon's School of Hair Design-Moscow",
        'Velvatex College of Beauty Culture',
```

```
'Thunderbird School of Global Management'], dtype=object)
```

（11）使用 max_cols 的值选择仅具有最大值的学校的那些行，然后使用.style 属性突出显示这些值。

```
college_n2.loc[unique_max_cols].style.highlight_max()
```

其输出结果如图 8-3 所示。

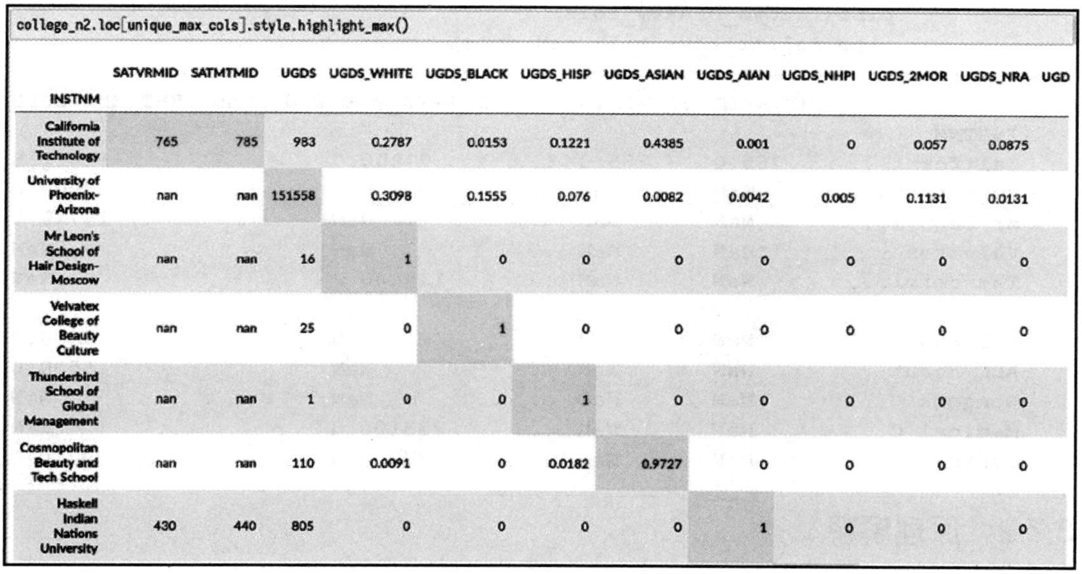

```
college_n2.loc[unique_max_cols].style.highlight_max()
```

INSTNM	SATVRMID	SATMTMID	UGDS	UGDS_WHITE	UGDS_BLACK	UGDS_HISP	UGDS_ASIAN	UGDS_AIAN	UGDS_NHPI	UGDS_2MOR	UGDS_NRA	UGD
California Institute of Technology	765	785	983	0.2787	0.0153	0.1221	0.4385	0.001	0	0.057	0.0875	
University of Phoenix-Arizona	nan	nan	151558	0.3098	0.1555	0.076	0.0082	0.0042	0.005	0.1131	0.0131	
Mr Leon's School of Hair Design-Moscow	nan	nan	16	1	0	0	0	0	0	0	0	
Velvatex College of Beauty Culture	nan	nan	25	0	1	0	0	0	0	0	0	
Thunderbird School of Global Management	nan	nan	1	0	0	1	0	0	0	0	0	
Cosmopolitan Beauty and Tech School	nan	nan	110	0.0091	0	0.0182	0.9727	0	0	0	0	
Haskell Indian Nations University	430	440	805	0	0	0	0	1	0	0	0	

图 8-3　突出显示每列的最大值

（12）重构代码以使其更易于阅读。

```
>>> def remove_binary_cols(df):
...     binary_only = df.nunique() == 2
...     cols = binary_only[binary_only].index.tolist()
...     return df.drop(columns=cols)

>>> def select_rows_with_max_cols(df):
...     max_cols = df.idxmax()
...     unique = max_cols.unique()
...     return df.loc[unique]

>>> (
...     college
...     .assign(
```

```
...             MD_EARN_WNE_P10=pd.to_numeric(
...                 college.MD_EARN_WNE_P10, errors="coerce"
...             ),
...             GRAD_DEBT_MDN_SUPP=pd.to_numeric(
...                 college.GRAD_DEBT_MDN_SUPP, errors="coerce"
...             ),
...         )
...         .select_dtypes("number")
...         .pipe(remove_binary_cols)
...         .pipe(select_rows_with_max_cols)
... )
             SATVRMID  SATMTMID  ...  MD_EARN_WNE_P10  GRAD_DEBT_MDN_SUPP
INSTNM                           ...
Californi...    765.0    785.0  ...         77800.0             11812.5
Universit...      NaN      NaN  ...             NaN             33000.0
Mr Leon's...      NaN      NaN  ...             NaN             15710.0
Velvatex ...      NaN      NaN  ...             NaN                 NaN
Thunderbi...      NaN      NaN  ...        118900.0                 NaN
       ...       ...      ...  ...             ...                 ...
MTI Busin...      NaN      NaN  ...         23000.0              9500.0
ABC Beaut...      NaN      NaN  ...             NaN             16500.0
Dongguk U...      NaN      NaN  ...             NaN                 NaN
Medical C...      NaN      NaN  ...        233100.0                 NaN
Southwest...      NaN      NaN  ...         27200.0             49750.0
```

8.7.2 原理解释

.idxmax 方法是一种很有用的方法，尤其是当索引的标记非常有意义时。本示例比较出乎意料的是，MD_EARN_WNE_P10 和 GRAD_DEBT_MDN_SUPP 都是 object 数据类型。加载 CSV 文件时，如果列中至少包含一个字符串，那么 Pandas 会将该列作为 object 类型列出（即使该列可能同时包含数字和字符串类型）。

通过在步骤（2）中检查特定的列值，我们能够发现这些列中包含字符串。

在步骤（3）中，使用了 .value_counts 方法显示有问题的字符。我们发现问题的来源是 PrivacySuppressed 值。

Pandas 可以使用 to_numeric 函数将仅包含数字字符的所有字符串强制转换为数字数据类型。步骤（4）即执行了此操作。要覆盖在 to_numeric 函数遇到无法转换的字符串时引发错误的默认行为，必须将 coerce 传递给 errors 参数。这将强制所有非数字字符串变为缺失值（即 np.nan）。

还有几列包含的是无用或无意义的最大值。步骤（5）～步骤（8）执行了对它们的删除操作。.select_dtypes 方法对于包含许多列的宽 DataFrames 可能是有益的。

在步骤（9）中，.idxmax 方法遍历了所有列以找到每列的最大值的索引。它将结果作为 Series 输出。可以看到，SAT 数学（SATMTMID 列）和词汇（SATVRMID 列）成绩最高的学校是 California Institute of Technology（加州理工学院），而 Dongguk University Los Angeles（东国大学洛杉矶分校）的 25 岁以上学生人数（UG25ABV 列）最多。

尽管.idxmax 方法提供的信息很方便，但不会产生相应的最大值。为此，我们在步骤（10）中从 max_cols Series 的值中收集了所有唯一的学校名称。

接下来，在步骤（11）中，我们对.loc 进行索引，以基于索引标签选择行，该索引标签在步骤（1）中加载 CSV 时被设置为学校名称。此过滤器仅适用于具有最大值的学校。DataFrame 具有.style 属性，该属性本身具有一些方法来更改 DataFrame 的外观。突出显示最大值可使结果更加清晰。

最后，我们重构了代码以使其成为一个非常整洁的管道。

8.7.3　扩展知识

默认情况下，.highlight_max 方法将突出显示每列的最大值。也可以使用 axis 参数来突出显示每行的最大值。这里，我们只选择 college 数据集中的种族百分比列，并突出显示每所学校百分比最高的种族。

```
>>> college = pd.read_csv(
...     "data/college.csv", index_col="INSTNM"
... )
>>> college_ugds = college.filter(like="UGDS_").head()
```

其输出结果如图 8-4 所示。

```
college_ugds.style.highlight_max(axis='columns')
```

INSTNM	UGDS_WHITE	UGDS_BLACK	UGDS_HISP	UGDS_ASIAN	UGDS_AIAN	UGDS_NHPI	UGDS_2MOR	UGDS_NRA	UGDS_UNKN
Alabama A & M University	0.0333	0.9353	0.0055	0.0019	0.0024	0.0019	0	0.0059	0.0138
University of Alabama at Birmingham	0.5922	0.26	0.0283	0.0518	0.0022	0.0007	0.0368	0.0179	0.01
Amridge University	0.299	0.4192	0.0069	0.0034	0	0	0	0	0.2715
University of Alabama in Huntsville	0.6988	0.1255	0.0382	0.0376	0.0143	0.0002	0.0172	0.0332	0.035
Alabama State University	0.0158	0.9208	0.0121	0.0019	0.001	0.0006	0.0098	0.0243	0.0137

图 8-4　显示每行的最大值

8.8 使用方法链复制.idxmax

在学习 Pandas 时，一个很好的做法是尝试自行实现内置 DataFrame 方法。这种复制类型可以使你对通常不会遇到的其他 Pandas 方法有更深入的了解。如果要求你仅使用本书到目前为止介绍的方法复制.idxmax，那么这是一项很有挑战性的任务。

此秘笈将各种基础方法链接在一起，以最终找到包含最大列值的所有行索引值。

8.8.1 实战操作

（1）加载 college 数据集并执行与 8.7 节"突出显示每列的最大值"秘笈相同的操作，以仅获取感兴趣的数字列。

```
>>> def remove_binary_cols(df):
...     binary_only = df.nunique() == 2
...     cols = binary_only[binary_only].index.tolist()
...     return df.drop(columns=cols)

>>> college_n = (
...     college
...     .assign(
...         MD_EARN_WNE_P10=pd.to_numeric(
...             college.MD_EARN_WNE_P10, errors="coerce"
...         ),
...         GRAD_DEBT_MDN_SUPP=pd.to_numeric(
...             college.GRAD_DEBT_MDN_SUPP, errors="coerce"
...         ),
...     )
...     .select_dtypes("number")
...     .pipe(remove_binary_cols)
... )
```

（2）使用.max 方法查找每列的最大值。

```
>>> college_n.max().head()
SATVRMID             765.0
SATMTMID             785.0
UGDS              151558.0
UGDS_WHITE             1.0
UGDS_BLACK             1.0
```

```
dtype: float64
```

（3）使用.eq DataFrame 方法针对列的.max 方法测试每个值。默认情况下，.eq 方法将列 DataFrame 的列与传递的 Series 索引的标签对齐。

```
>>> college_n.eq(college_n.max()).head()
              SATVRMID  SATMTMID  ...  MD_EARN_WNE_P10  GRAD_DEBT_MDN_SUPP
INSTNM                            ...
Alabama A...     False     False  ...            False               False
Universit...     False     False  ...            False               False
Amridge U...     False     False  ...            False               False
Universit...     False     False  ...            False               False
Alabama S...     False     False  ...            False               False
```

（4）此 DataFrame 中所有至少具有一个 True 值的行必须包含一列最大值。下面使用.any 方法查找至少具有一个 True 值的所有此类行。

```
>>> has_row_max = (
...    college_n
...    .eq(college_n.max())
...    .any(axis="columns")
... )
>>> has_row_max.head()
INSTNM
Alabama A & M University                False
University of Alabama at Birmingham     False
Amridge University                      False
University of Alabama in Huntsville     False
Alabama State University                False
dtype: bool
```

（5）只有 18 列，这意味着 has_row_max 中最多只能有 18 个 True 值。下面找出具体有多少个 True 值。

```
>>> college_n.shape
(7535, 18)
>>> has_row_max.sum()
401
```

（6）这有点出乎意料，但是事实证明，有些列的许多行都等于最大值。这对于许多最大值为 1 的百分比列来说很常见。.idxmax 方法返回的是第一次出现的最大值。下面稍作备份，删除.any 方法，然后看看步骤（3）的输出。现在可以运行.cumsum 方法，而不是累计所有 True 值。

```
>>> college_n.eq(college_n.max()).cumsum()
                SATVRMID  SATMTMID  ...  MD_EARN_WNE_P10  GRAD_DEBT_MDN_SUPP
INSTNM                                ...
Alabama A...          0         0   ...                0                   0
Universit...          0         0   ...                0                   0
Amridge U...          0         0   ...                0                   0
Universit...          0         0   ...                0                   0
Alabama S...          0         0   ...                0                   0
...                 ...       ...   ...              ...                 ...
SAE Insti...          1         1   ...                1                   2
Rasmussen...          1         1   ...                1                   2
National...           1         1   ...                1                   2
Bay Area...           1         1   ...                1                   2
Excel Lea...          1         1   ...                1                   2
```

（7）有些列有一个唯一的最大值，如 SATVRMID（SAT 词汇成绩）和 SATMTMID（SAT 数学成绩），而另一些列（如 UGDS_WHITE）则有许多个最大值，这是因为 109 所学校的本科生中 100%为白人，所以它们在 UGDS_WHITE（白人本科生）列中的值均为 1。如果再一次链接.cumsum 方法，则值 1 在每列中只会出现一次，并且该值将是最大值的第一次出现。

```
>>> (college_n.eq(college_n.max()).cumsum().cumsum())
                SATVRMID  SATMTMID  ...  MD_EARN_WNE_P10  GRAD_DEBT_MDN_SUPP
INSTNM                                ...
Alabama A...          0         0   ...                0                   0
Universit...          0         0   ...                0                   0
Amridge U...          0         0   ...                0                   0
Universit...          0         0   ...                0                   0
Alabama S...          0         0   ...                0                   0
...                 ...       ...   ...              ...                 ...
SAE Insti...       7305      7305   ...             3445               10266
Rasmussen...       7306      7306   ...             3446               10268
National...        7307      7307   ...             3447               10270
Bay Area...        7308      7308   ...             3448               10272
Excel Lea...       7309      7309   ...             3449               10274
```

（8）现在可以使用.eq 方法测试每个值是否等于 1，然后使用.any 方法查找至少具有一个 True 值的行。

```
>>> has_row_max2 = (
...     college_n.eq(college_n.max())
...     .cumsum()
```

```
...         .cumsum()
...         .eq(1)
...         .any(axis="columns")
... )

>>> has_row_max2.head()
INSTNM
Alabama A & M University                  False
University of Alabama at Birmingham       False
Amridge University                        False
University of Alabama in Huntsville       False
Alabama State University                  False
dtype: bool
```

（9）检查 has_row_max2 的 True 值是否不超过列数。

```
>>> has_row_max2.sum()
16
```

（10）现在我们需要 has_row_max2 为 True 的所有机构。可以在 Series 本身上使用布尔索引来找到它们。

```
>>> idxmax_cols = has_row_max2[has_row_max2].index
>>> idxmax_cols
Index([ 'Thunderbird School of Global Management',
        'Southwest University of Visual Arts-Tucson','ABC Beauty College Inc',
        'Velvatex College of Beauty Culture',
        'California Institute of Technology',
        'Le Cordon Bleu College of Culinary Arts-San Francisco',
        'MTI Business College Inc', 'Dongguk University-Los Angeles',
        'Mr Leon's School of Hair Design-Moscow',
        'Haskell Indian Nations University', 'LIU Brentwood',
        'Medical College of Wisconsin', 'Palau Community College',
        'California University of Management and Sciences',
        'Cosmopolitan Beauty and Tech School', 'University of Phoenix-Arizona'],
      dtype='object', name='INSTNM')
```

（11）这 16 家机构都是至少其中一列第一次出现最大值的索引。可以检查它们是否与.idxmax 方法找到的相同。

```
>>> set(college_n.idxmax().unique()) == set(idxmax_cols)
True
```

（12）重构为 idx_max 函数。

```
>>> def idx_max(df):
...     has_row_max = (
...         df
...         .eq(df.max())
...         .cumsum()
...         .cumsum()
...         .eq(1)
...         .any(axis="columns")
...     )
...     return has_row_max[has_row_max].index

>>> idx_max(college_n)
Index([ 'Thunderbird School of Global Management',
        'Southwest University of Visual Arts-Tucson','ABC Beauty College Inc',
        'Velvatex College of Beauty Culture',
        'California Institute of Technology',
        'Le Cordon Bleu College of Culinary Arts-San Francisco',
        'MTI Business College Inc', 'Dongguk University-Los Angeles',
        'Mr Leon's School of Hair Design-Moscow',
        'Haskell Indian Nations University', 'LIU Brentwood',
        'Medical College of Wisconsin', 'Palau Community College',
        'California University of Management and Sciences',
        'Cosmopolitan Beauty and Tech School', 'University of Phoenix-Arizona'],
       dtype='object', name='INSTNM')
```

8.8.2 原理解释

步骤（1）通过将两列转换为数字并消除二进制列来复制 8.7 节"突出显示每列的最大值"秘笈的工作。

在步骤（2）中找到了每列的最大值。在这里需要谨慎一些，因为 Pandas 会默默地丢弃无法产生最大值的列。如果发生这种情况，则步骤（3）仍将完成，但它将为每列提供 False 值而没有可用的最大值。

步骤（4）使用 .any 方法扫描每行，以搜索至少一个 True 值。至少具有一个 True 值的任何行都将包含某个列的最大值。

在步骤（5）中对所得的布尔 Series 进行了求和，以确定有多少行包含最大值。

出乎意料的是，包含最大值的行数远多于列数。步骤（6）深入说明了为什么会发生这种情况。我们对步骤（3）的输出结果进行了累加求和，然后检测了与每列的最大值相等的行总数。

许多大学的本科生只有一个种族（白人），因此其白人本科生的比例为100%。到目前为止，这是具有最大值的多个行的最大贡献者。可以看到，虽然SAT成绩列和大学本科生种族百分比列都只有一个最大值，但是白人百分比列却有许多行都包含了最大值。

我们的目标是找到具有最大值的第一行，因此可以再次取累加总和，以便每列都只有一行等于1。

步骤（8）将代码格式化为每行具有一个方法，并像在步骤（4）中一样运行.any方法。如果步骤（8）成功，那么我们的True值应该不超过列数。步骤（9）断言这是真的。

为了验证采用这种方式找到的列是否与通过.idxmax方法找到的列相同，我们对has_row_max2本身使用了布尔选择。列将以不同的顺序排列，因此我们将列名称的顺序转换为集合，这些集合本质上将以无序方式比较相等性。

8.8.3 扩展知识

可以用一行比较长的代码将索引操作符与匿名函数链接起来，从而完成此秘笈。这个小技巧使你无须执行步骤（10）。在此秘笈中，我们还可以确定.idxmax方法与手动实现之间的时间差。

```
>>> def idx_max(df):
...     has_row_max = (
...         df
...         .eq(df.max())
...         .cumsum()
...         .cumsum()
...         .eq(1)
...         .any(axis="columns")
...         [lambda df_: df_]
...         .index
...     )
...     return has_row_max

>>> %timeit college_n.idxmax().values
1.12 ms ± 28.4 µs per loop (mean ± std. dev. of 7 runs, 1000 loops each)
>>> %timeit idx_max(college_n)
5.35 ms ± 55.2 µs per loop (mean ± std. dev. of 7 runs, 100 loops each)
```

非常遗憾的是，我们的手动实现的工作速度是Pandas内置.idxmax方法的5倍，但是不管其性能如何下降，许多创造性的实用解决方案都使用累计方法（如 .cumsum）和 Boolean Series 来沿着轴查找特征或特定模式。

8.9 查找最常见的列的最大值

  college 数据集包含超过 7500 所大学的 8 个不同种族的本科生人口百分比数据。找到每所学校本科生人数最多的种族，然后为整个数据集找到该结果的分布将是一件很有意义的事情。我们将能够回答这样一个问题："有百分之几的机构，其拥有的白人学生要比其他种族更多？"

  在此秘笈中，我们使用 .idxmax 方法查找每所学校的本科生百分比最高的种族，然后查找这些最大值的分布情况。

8.9.1 实战操作

（1）读取 college 数据集，然后仅选择包含大学种族百分比信息的列。

```
>>> college = pd.read_csv(
...     "data/college.csv", index_col="INSTNM"
... )
>>> college_ugds = college.filter(like="UGDS_")
>>> college_ugds.head()
              UGDS_WHITE  UGDS_BLACK  ...  UGDS_NRA  UGDS_UNKN
INSTNM                                ...
Alabama A...      0.0333      0.9353  ...    0.0059     0.0138
Universit...      0.5922      0.2600  ...    0.0179     0.0100
Amridge U...      0.2990      0.4192  ...    0.0000     0.2715
Universit...      0.6988      0.1255  ...    0.0332     0.0350
Alabama S...      0.0158      0.9208  ...    0.0243     0.0137
```

（2）对列应用 .idxmax 方法以获得每行中包含最高种族百分比的大学名称。

```
>>> highest_percentage_race = college_ugds.idxmax(
...     axis="columns"
... )
>>> highest_percentage_race.head()
INSTNM
Alabama A & M University
University of Alabama at Birmingham
Amridge University
```

```
University of Alabama in Huntsville
Alabama State University
dtype: object
```

（3）使用.value_counts方法返回最大值出现次数的分布情况。添加 normalize = True 参数进行归一化，使其总和为1。

```
>>> highest_percentage_race.value_counts(normalize=True)
UGDS_WHITE      0.670352
UGDS_BLACK      0.151586
UGDS_HISP       0.129473
UGDS_UNKN       0.023422
UGDS_ASIAN      0.012074
UGDS_AIAN       0.006110
UGDS_NRA        0.004073
UGDS_NHPI       0.001746
UGDS_2MOR       0.001164
dtype: float64
```

8.9.2 原理解释

此秘笈的关键是要找到所有代表相同信息单元的列，然后对它们进行相互比较。

一般不会对列进行比较。例如，将 SAT 词汇成绩与大学生人数进行比较是没有意义的。

对于以相同方式构造的数据，可以将.idxmax 方法应用于数据的每行中，以找到具有最大值的列。我们需要使用 axis 参数更改其默认行为。

步骤（3）完成此操作并返回一个 Series，现在可以对其应用.value_counts 方法以返回分布情况。本示例将 True 值传递给了 normalize 参数，这是因为我们仅对分布（相对频率）感兴趣，而对原始计数不感兴趣。

8.9.3 扩展知识

我们可能还想要探索更多并回答这个问题：对于那些黑人学生多于其他种族的学校，其第二高种族百分比的分布是什么样的？

```
>>> (
...     college_ugds
...     [highest_percentage_race == "UGDS_BLACK"]
...     .drop(columns="UGDS_BLACK")
```

```
...         .idxmax(axis="columns")
...         .value_counts(normalize=True)
... )
UGDS_WHITE    0.661228
UGDS_HISP     0.230326
UGDS_UNKN     0.071977
UGDS_NRA      0.018234
UGDS_ASIAN    0.009597
UGDS_2MOR     0.006718
UGDS_AIAN     0.000960
UGDS_NHPI     0.000960
dtype: float64
```

在应用与此秘笈相同的方法之前，我们需要删除 UGDS_BLACK 列。从上述结果可以看到，在黑人人口较多的学校中，似乎也会具有较高的西班牙裔人口百分比（UGDS_HISP）。

第 9 章 分组以进行聚合、过滤和转换

本章包含以下秘笈。
- 定义聚合。
- 使用多个列和函数进行分组和聚合。
- 分组后删除多重索引。
- 使用自定义聚合函数进行分组。
- 使用*args 和**kwargs 自定义聚合函数。
- 检查 groupby 对象。
- 筛选少数族裔占多数的州。
- 通过减肥赌注做出改变。
- 计算每个州的 SAT 加权平均成绩。
- 按连续变量分组。
- 计算城市之间的航班总数。
- 寻找最长的准点航班连续记录。

9.1 介 绍

数据分析过程中最基本的任务之一是，在对每个组执行计算之前将数据分成独立的组。这种方法已经存在了很长一段时间，但最近被称为拆分-应用-合并（split-apply-combine）。本章介绍功能强大的.groupby 方法，该方法允许以任何可以想象的方式对数据进行分组，并在返回单个数据集之前将任何类型的函数独立地应用于每个组中。

在开始使用秘笈之前，我们需要先了解一些术语。所有基本的 groupby 操作都具有分组列，并且这些列中值的每个唯一组合都代表数据的独立分组。其语法如下。

```
df.groupby(['list', 'of', 'grouping', 'columns'])
df.groupby('single_column')      # 按单个列分组时
```

调用.groupby 方法的结果是一个 groupby 对象。正是这个 groupby 对象将成为驱动本章所有计算的引擎。创建此 groupby 对象时，Pandas 做的工作很少，仅验证了分组是可

能的。数据分析人员必须在此 groupby 对象上链接方法以充分发挥其功能。

.groupby 方法最常见的用途是执行聚合（aggregation）。什么是聚合？当许多输入的序列被汇总或组合为单个值输出时，发生的就是聚合。例如，汇总一列的所有值或找到其最大值就是应用于数据序列的聚合。聚合将采用一个序列并将其归约为单个值。

除了前面定义的分组列之外，大多数聚合还有两个其他组件，即聚合列（aggregation column）和聚合函数（aggregation function）。聚合列是其值将被聚合的列。聚合函数则被定义发生什么类型的聚合。聚合函数包括 sum（求和）、min（最小值）、max（最大值）、mean（平均值）、count（计数）、variance（方差）和 std（标准差）等。

9.2 定义聚合

在本秘笈中，我们将检查 flights（航班）数据集并执行最简单的聚合，仅涉及一个分组列、一个聚合列和一个聚合函数。我们将找到每家航空公司的平均到达延误时间。Pandas 将使用不同的语法来创建聚合，此秘笈将展示这一点。

9.2.1 实战操作

（1）读取 flights 数据集。

```
>>> import pandas as pd
>>> import numpy as np
>>> flights = pd.read_csv('data/flights.csv')
>>> flights.head()
0  1  1  4  ...   65.0  0  0
1  1  1  4  ...  -13.0  0  0
2  1  1  4  ...   35.0  0  0
3  1  1  4  ...   -7.0  0  0
4  1  1  4  ...   39.0  0  0
```

（2）定义分组列（AIRLINE）、聚合列（ARR_DELAY）和聚合函数（mean）。将分组列放在.groupby 方法中，然后调用.agg 方法，采用字典形式，将聚合列和聚合函数配对。如果传入一个字典，那么它将返回一个 DataFrame 实例。

```
>>> (flights
...     .groupby('AIRLINE')
...     .agg({'ARR_DELAY':'mean'})
... )
```

```
              ARR_DELAY
AIRLINE
AA            5.542661
AS           -0.833333
B6            8.692593
DL            0.339691
EV            7.034580
...                ...
OO            7.593463
UA            7.765755
US            1.681105
VX            5.348884
WN            6.397353
```

或者,你也可以将聚合列放在索引操作符中,然后将聚合函数作为字符串传递给.agg 方法。这将返回一个 Series。

```
>>> (flights
...    .groupby('AIRLINE')
...    ['ARR_DELAY']
...    .agg('mean')
... )
AIRLINE
AA     5.542661
AS    -0.833333
B6     8.692593
DL     0.339691
EV     7.034580
         ...
OO     7.593463
UA     7.765755
US     1.681105
VX     5.348884
WN     6.397353
Name: ARR_DELAY, Length: 14, dtype: float64
```

(3)步骤(2)中使用的字符串名称是一种方便措施,Pandas 提供了可引用特定聚合函数的便利。我们可以将任何聚合函数直接传递给.agg 方法,如 NumPy 的 mean 函数。其输出与步骤(2)相同。

```
>>> (flights
...    .groupby('AIRLINE')
...    ['ARR_DELAY']
```

```
...        .agg(np.mean)
...    )
AIRLINE
AA    5.542661
AS   -0.833333
B6    8.692593
DL    0.339691
EV    7.034580
         ...
OO    7.593463
UA    7.765755
US    1.681105
VX    5.348884
WN    6.397353
Name: ARR_DELAY, Length: 14, dtype: float64
```

（4）在本示例中，可以完全跳过.agg方法，而直接使用文本方法中的代码。此输出结果也与步骤（3）相同。

```
>>> (flights
...    .groupby('AIRLINE')
...    ['ARR_DELAY']
...    .mean()
...    )
AIRLINE
AA    5.542661
AS   -0.833333
B6    8.692593
DL    0.339691
EV    7.034580
         ...
OO    7.593463
UA    7.765755
US    1.681105
VX    5.348884
WN    6.397353
Name: ARR_DELAY, Length: 14, dtype: float64
```

9.2.2 原理解释

.groupby方法的语法不像其他方法那么简单。我们可以通过将.groupby方法的结果存储为它自己的变量来拦截步骤（2）中的方法链。

```
>>> grouped = flights.groupby('AIRLINE')
>>> type(grouped)
<class 'pandas.core.groupby.generic.DataFrameGroupBy'>
```

在上面的语句中，首先使用其自己独特的属性和方法产生一个全新的中间对象。在此阶段不进行任何计算。Pandas 仅验证分组列。该 groupby 对象具有 .agg 方法，可以执行聚合。使用此方法的方式之一是向其传递一个字典，该字典将聚合列映射到聚合函数中，如步骤（2）所示。如果向 .agg 方法中传入一个字典，则结果将是一个 DataFrame。

Pandas 库通常具有不止一种方式来执行相同的操作。步骤（3）即显示了另一种执行 groupby 分组的方式。它不是标识字典中的聚合列，而是将其放在索引操作符中，就好像你是从 DataFrame 中选择它作为列一样。然后，将函数字符串名称作为标量传递给 .agg 方法。在这种情况下，其结果是一个 Series。

你可以将任何聚合函数传递给 .agg 方法。为了简单起见，Pandas 允许你使用字符串名称，但是你也可以像在步骤（4）中一样显式调用聚合函数。NumPy 提供了许多可以聚合值的函数。

当仅应用单个聚合函数时，通常可以直接将其作为对 groupby 对象本身执行的方法进行调用，而无须 .agg 方法。

值得一提的是，虽然并非所有聚合函数都有等效的方法，但确实大多数聚合函数都具有等效的方法。

9.2.3 扩展知识

如果对 .agg 方法不使用聚合函数，则 Pandas 会引发异常。例如，下面看看将平方根函数应用于每个组中会发生什么。

```
>>> (flights
...    .groupby('AIRLINE')
...    ['ARR_DELAY']
...    .agg(np.sqrt)
... )
Traceback (most recent call last):
...
ValueError: function does not reduce
```

9.3 使用多个列和函数进行分组和聚合

可以使用多个列进行分组和聚合。其语法与使用单个列进行分组和聚合的语法略有不同。与任何类型的分组操作一样，使用多个列分组有助于识别 3 个组件，即分组列、聚合列和聚合函数。

在本秘笈中，我们通过回答以下查询来展示 .groupby 方法的灵活性。

- 查找每个工作日每家航空公司的已取消航班的数量。
- 查找每个工作日每家航空公司的已取消和改航航班的数量和百分比。
- 对于每个始发地和目的地，查找航班总数、已取消航班的数量和百分比，以及飞航时间的平均值和方差。

9.3.1 实战操作

（1）读取 flights 数据集，然后通过定义分组列（AIRLINE、WEEKDAY）、汇总列（CANCELLED）和汇总函数（sum）来回答第一个查询。

```
>>> (flights
...     .groupby(['AIRLINE', 'WEEKDAY'])
...     ['CANCELLED']
...     .agg('sum')
... )
AIRLINE  WEEKDAY
AA       1          41
         2           9
         3          16
         4          20
         5          18
                    ..
WN       3          18
         4          10
         5           7
         6          10
         7           7
Name: CANCELLED, Length: 98, dtype: int64
```

（2）通过为每对分组和聚合列使用一个列表来回答第二个查询，并为聚合函数使用一个列表。

```
>>> (flights
...     .groupby(['AIRLINE', 'WEEKDAY'])
...     [['CANCELLED', 'DIVERTED']]
...     .agg(['sum', 'mean'])
... )
```

第 9 章 分组以进行聚合、过滤和转换

		CANCELLED		DIVERTED	
		sum	mean	sum	mean
AIRLINE	WEEKDAY				
AA	1	41	0.032106	6	0.004699
	2	9	0.007341	2	0.001631
	3	16	0.011949	2	0.001494
	4	20	0.015004	5	0.003751
	5	18	0.014151	1	0.000786
...		...	...	...	...
WN	3	18	0.014118	2	0.001569
	4	10	0.007911	4	0.003165
	5	7	0.005828	0	0.000000
	6	10	0.010132	3	0.003040
	7	7	0.006066	3	0.002600

（3）使用 .agg 方法中的字典回答第三个查询，以便将特定的聚合列映射到特定的聚合函数中。

```
>>> (flights
...     .groupby(['ORG_AIR', 'DEST_AIR'])
...     .agg({'CANCELLED':['sum', 'mean', 'size'],
...           'AIR_TIME':['mean', 'var']})
... )
```

		CANCELLED		...	AIR_TIME	
		sum	mean	...	mean	var
ORG_AIR	DEST_AIR			...		
ATL	ABE	0	0.000000	...	96.387097	45.778495
	ABQ	0	0.000000	...	170.500000	87.866667
	ABY	0	0.000000	...	28.578947	6.590643
	ACY	0	0.000000	...	91.333333	11.466667
	AEX	0	0.000000	...	78.725000	47.332692
...		...	...	...	...	...
SFO	SNA	4	0.032787	...	64.059322	11.338331
	STL	0	0.000000	...	198.900000	101.042105
	SUN	0	0.000000	...	78.000000	25.777778
	TUS	0	0.000000	...	100.200000	35.221053
	XNA	0	0.000000	...	173.500000	0.500000

（4）在 Pandas 0.25 中，有一个命名的聚合对象，该对象可以创建非分层列。我们将使用这些列重复上述查询。

```
>>> (flights
...     .groupby(['ORG_AIR', 'DEST_AIR'])
```

```
...        .agg(sum_cancelled=pd.NamedAgg(column='CANCELLED',
aggfunc='sum'),
...               mean_cancelled=pd.NamedAgg(column='CANCELLED',
aggfunc='mean'),
...               size_cancelled=pd.NamedAgg(column='CANCELLED',
aggfunc='size'),
...               mean_air_time=pd.NamedAgg(column='AIR_TIME',
aggfunc='mean'),
...               var_air_time=pd.NamedAgg(column='AIR_TIME',
aggfunc='var'))
... )
                 sum_cancelled  mean_cancelled  ...  mean_air_time
ORG_AIR DEST_AIR                                 ...
ATL     ABE                  0        0.000000  ...      96.387097
        ABQ                  0        0.000000  ...     170.500000
        ABY                  0        0.000000  ...      28.578947
        ACY                  0        0.000000  ...      91.333333
        AEX                  0        0.000000  ...      78.725000
...                        ...             ...  ...            ...
SFO     SNA                  4        0.032787  ...      64.059322
        STL                  0        0.000000  ...     198.900000
        SUN                  0        0.000000  ...      78.000000
        TUS                  0        0.000000  ...     100.200000
        XNA                  0        0.000000  ...     173.500000
```

9.3.2 原理解释

为了像步骤（1）一样按多列分组，我们将字符串名称列表传递给 .groupby 方法。AIRLINE 和 WEEKDAY 的每个独特组合都构成了自己的组。在每个组中，可以计算已取消的航班总数，然后将其返回为一个 Series。

步骤（2）按 AIRLINE 和 WEEKDAY 分组，但这一次聚合了两列。此步骤应用了两个聚合函数中的每一个，使用了字符串 sum 和 mean 将这两个聚合函数中的每一个应用于每列，从而每组返回 4 个列。

步骤（3）更进一步，使用了字典将特定的聚合列映射到不同的聚合函数中。

请注意，size 聚合函数返回的是每个组的总行数。这与 count 聚合函数不一样，后者返回的是每个组中非缺失值的数量。

步骤（4）显示了用于创建扁平列（即所谓的聚合）的新语法。

9.3.3 扩展知识

要展平（Flat）步骤（3）中的列，可以使用 .to_flat_index 方法（自 Pandas 0.24 版本起可用）。

```
>>> res = (flights
...    .groupby(['ORG_AIR', 'DEST_AIR'])
...    .agg({'CANCELLED':['sum', 'mean', 'size'],
...          'AIR_TIME':['mean', 'var']})
... )
>>> res.columns = ['_'.join(x) for x in
...     res.columns.to_flat_index()]
>>> res
```

		CANCELLED_sum	CANCELLED_mean	...	AIR_TIME_mean
ORG_AIR	DEST_AIR			...	
ATL	ABE	0	0.000000	...	96.387097
	ABQ	0	0.000000	...	170.500000
	ABY	0	0.000000	...	28.578947
	ACY	0	0.000000	...	91.333333
	AEX	0	0.000000	...	78.725000
...	...	...	...	...	...
SFO	SNA	4	0.032787	...	64.059322
	STL	0	0.000000	...	198.900000
	SUN	0	0.000000	...	78.000000
	TUS	0	0.000000	...	100.200000
	XNA	0	0.000000	...	173.500000

这种方式有点麻烦，我们希望通过链式操作来展平列。遗憾的是，.reindex 方法不支持展平。相反，我们将必须利用 .pipe 方法。

```
>>> def flatten_cols(df):
...     df.columns = ['_'.join(x) for x in
...         df.columns.to_flat_index()]
...     return df

>>> res = (flights
...    .groupby(['ORG_AIR', 'DEST_AIR'])
...    .agg({'CANCELLED':['sum', 'mean', 'size'],
...          'AIR_TIME':['mean', 'var']})
...    .pipe(flatten_cols)
... )

>>> res
```

		CANCELLED_sum	CANCELLED_mean	...	AIR_TIME_mean
ORG_AIR	DEST_AIR			...	
ATL	ABE	0	0.000000	...	96.387097
	ABQ	0	0.000000	...	170.500000
	ABY	0	0.000000	...	28.578947
	ACY	0	0.000000	...	91.333333
	AEX	0	0.000000	...	78.725000
...		...	...	...	...
SFO	SNA	4	0.032787	...	64.059322
	STL	0	0.000000	...	198.900000
	SUN	0	0.000000	...	78.000000
	TUS	0	0.000000	...	100.200000
	XNA	0	0.000000	...	173.500000

请注意，在使用多个列进行分组时，Pandas 会创建一个层次结构索引或多重索引。在前面的示例中，它返回了 1130 行。但是，如果我们分组的一列是分类的（也就是说，它的数据类型是 category，而不是 object），则 Pandas 将为每个级别创建所有组合的笛卡儿积。在这种情况下，它将返回 2710 行。当然，如果你的 category 列具有更高的基数，则可以获得更多的值。

```
>>> res = (flights
...     .assign(ORG_AIR=flights.ORG_AIR.astype('category'))
...     .groupby(['ORG_AIR', 'DEST_AIR'])
...     .agg({'CANCELLED':['sum', 'mean', 'size'],
...           'AIR_TIME':['mean', 'var']})
... )
>>> res
```

		CANCELLED		...	AIR_TIME	
		sum	mean	...	mean	var
ORG_AIR	DEST_AIR			...		
ATL	ABE	0.0	0.0	...	96.387097	45.778495
	ABI	NaN	NaN	...	NaN	NaN
	ABQ	0.0	0.0	...	170.500000	87.866667
	ABR	NaN	NaN	...	NaN	NaN
	ABY	0.0	0.0	...	28.578947	6.590643
...		...	...	...	...	...
SFO	TYS	NaN	NaN	...	NaN	NaN
	VLD	NaN	NaN	...	NaN	NaN
	VPS	NaN	NaN	...	NaN	NaN
	XNA	0.0	0.0	...	173.500000	0.500000
	YUM	NaN	NaN	...	NaN	NaN

要纠正这种组合暴增的情况，可使用 observed=True 参数。这使得分类列的分组将像

字符串类型列的分组一样工作，仅显示观察到的值，而不是产生笛卡儿积。

```
>>> res = (flights
...     .assign(ORG_AIR=flights.ORG_AIR.astype('category'))
...     .groupby(['ORG_AIR', 'DEST_AIR'], observed=True)
...     .agg({'CANCELLED':['sum', 'mean', 'size'],
...           'AIR_TIME':['mean', 'var']})
... )
>>> res
```

		CANCELLED		...	AIR_TIME	
		sum	mean	...	mean	var
ORG_AIR	DEST_AIR			...		
LAX	ABQ	1	0.018182	...	89.259259	29.403215
	ANC	0	0.000000	...	307.428571	78.952381
	ASE	1	0.038462	...	102.920000	102.243333
	ATL	0	0.000000	...	224.201149	127.155837
	AUS	0	0.000000	...	150.537500	57.897310
...		...	...	...	...	...
MSP	TTN	1	0.125000	...	124.428571	57.952381
	TUL	0	0.000000	...	91.611111	63.075163
	TUS	0	0.000000	...	176.000000	32.000000
	TVC	0	0.000000	...	56.600000	10.300000
	XNA	0	0.000000	...	90.642857	115.939560

9.4 分组后删除多重索引

不可避免地，当使用 groupby 时，你将创建一个多重索引（multiIndex）。在索引和列中都可能发生多重索引。具有多重索引的 DataFrame 将更加难以导航定位，并且有时列名称也容易令人感到困惑。

在此秘笈中，我们将使用 .groupby 方法执行聚合，以创建包含行和列多重索引的 DataFrame。然后，我们对索引进行操作，使其具有单个级别，并且列名称具有描述性。

9.4.1 实战操作

（1）读取 flights 数据集，编写一条语句以查找飞行的总里程和平均里程，以及每家航空公司在每个工作日的最大和最小抵港延迟时间。

```
>>> flights = pd.read_csv('data/flights.csv')
>>> airline_info = (flights
```

```
...         .groupby(['AIRLINE', 'WEEKDAY'])
...         .agg({'DIST':['sum', 'mean'],
...              'ARR_DELAY':['min', 'max']})
...         .astype(int)
... )
>>> airline_info
                    DIST            ARR_DELAY
                    sum    mean     min    max
AIRLINE WEEKDAY
AA      1        1455386   1139     -60    551
        2        1358256   1107     -52    725
        3        1496665   1117     -45    473
        4        1452394   1089     -46    349
        5        1427749   1122     -41    732
...                 ...    ...      ...    ...
WN      3         997213   782      -38    262
        4        1024854   810      -52    284
        5         981036   816      -44    244
        6         823946   834      -41    290
        7         945679   819      -45    261
```

（2）行和列均由具有两个级别的多重索引标记。下面将这两个级别压缩到一个级别上。为了解决这些列，可以使用多重索引方法 .to_flat_index。我们可以先显示每个级别的输出，然后将两个级别连接起来，再将其设置为新列的值。

```
>>> airline_info.columns.get_level_values(0)
Index(['DIST', 'DIST', 'ARR_DELAY', 'ARR_DELAY'], dtype='object')
>>> airline_info.columns.get_level_values(1)
Index(['sum', 'mean', 'min', 'max'], dtype='object')

>>> airline_info.columns.to_flat_index()
Index([ ('DIST', 'sum'), ('DIST', 'mean'), ('ARR_DELAY', 'min'),
       ('ARR_DELAY', 'max')],
      dtype='object')

>>> airline_info.columns = ['_'.join(x) for x in
...      airline_info.columns.to_flat_index()]

>>> airline_info
                 DIST_sum  DIST_mean  ARR_DELAY_min  ARR_DELAY_max
AIRLINE WEEKDAY
AA      1        1455386    1139         -60            551
        2        1358256    1107         -52            725
```

第 9 章 分组以进行聚合、过滤和转换

```
           3    1496665    1117    -45    473
           4    1452394    1089    -46    349
           5    1427749    1122    -41    732
...             ...         ...     ...    ...
WN         3     997213     782    -38    262
           4    1024854     810    -52    284
           5     981036     816    -44    244
           6     823946     834    -41    290
           7     945679     819    -45    261
```

（3）去除多重索引行的一种快速方法是使用 .reset_index 方法。

```
>>> airline_info.reset_index()
    AIRLINE  WEEKDAY  ...  ARR_DELAY_min  ARR_DELAY_max
0        AA        1  ...            -60            551
1        AA        2  ...            -52            725
2        AA        3  ...            -45            473
3        AA        4  ...            -46            349
4        AA        5  ...            -41            732
..      ...      ...  ...            ...            ...
93       WN        3  ...            -38            262
94       WN        4  ...            -52            284
95       WN        5  ...            -44            244
96       WN        6  ...            -41            290
97       WN        7  ...            -45            261
```

（4）重构代码以使其更易于阅读。使用 Pandas 0.25 版本的功能可以自动展平列。

```
>>> (flights
...    .groupby(['AIRLINE', 'WEEKDAY'])
...    .agg(dist_sum=pd.NamedAgg(column='DIST', aggfunc='sum'),
...         dist_mean=pd.NamedAgg(column='DIST', aggfunc='mean'),
...         arr_delay_min=pd.NamedAgg(column='ARR_DELAY', aggfunc='min'),
...         arr_delay_max=pd.NamedAgg(column='ARR_DELAY', aggfunc='max'))
...    .astype(int)
...    .reset_index()
... )
   AIRLINE  WEEKDAY  ...  ARR_DELAY_min  ARR_DELAY_max
0       AA        1  ...            -60            551
1       AA        2  ...            -52            725
2       AA        3  ...            -45            473
3       AA        4  ...            -46            349
```

4	AA	5	...	-41	732
...	...	...	...	...	...
93	WN	3	...	-38	262
94	WN	4	...	-52	284
95	WN	5	...	-44	244
96	WN	6	...	-41	290
97	WN	7	...	-45	261

9.4.2 原理解释

当使用 .agg 方法对多个列执行聚合时，Pandas 将创建一个具有两个级别的索引对象。聚合列变为顶层，聚合函数变为底层。Pandas 显示的多重索引级别与单级别的列不同。而在 Jupyter 或 Python Shell 中，除了最里面的级别，重复的索引值都不会显示。

你可以检查步骤（1）中的 DataFrame 以进行验证。例如，DIST 列仅显示一次，但它引用了前两列。

步骤（2）通过首先使用多重索引方法 .get_level_values 检索每个级别的基础值来定义新列。此方法接收标识索引级别的整数。它们从外部（顶层/左侧）开始，从零开始编号。我们将结合使用最近添加的索引方法 .to_flat_index 与列表推导式（list comprehension），可为每列创建字符串。这些新值将被分配给列属性。

在步骤（3）中，使用了 .reset_index 方法将两个索引级别都推入列中。这很简单，其实应该有一个类似的方法来压缩列名称。

在步骤（4）中，使用了 NamedAgg 类（这是 Pandas 0.25 版本中的新增功能）创建平面聚合列。

9.4.3 扩展知识

默认情况下，在 groupby 操作结束时，Pandas 会将所有分组列都放入索引中。数据分析人员可以通过将 .groupby 方法中的 as_index 参数设置为 False，以避免此行为。你也可以在分组之后链接 .reset_index 方法以获得相同的效果，步骤（3）就是这样做的。

来看以下示例。本示例查找的是每家航空公司每个航班飞行的平均距离，并就使用了 as_index=False 的参数设置。

```
>>> (flights
...     .groupby(['AIRLINE'], as_index=False)
...     ['DIST']
...     .agg('mean')
...     .round(0)
```

```
...   )
      AIRLINE     DIST
0          AA   1114.0
1          AS   1066.0
2          B6   1772.0
3          DL    866.0
4          EV    460.0
..        ...      ...
9          OO    511.0
10         UA   1231.0
11         US   1181.0
12         VX   1240.0
13         WN    810.0
```

看看上述结果中航空公司的顺序。默认情况下，Pandas 对分组列进行排序。sort 参数存在于.groupby 方法中，并默认为 True。你可以将 sort 参数设置为 False，以使分组列的顺序与在数据集中遇到分组列的顺序相同。通过不对数据进行排序，性能会有很小的提高。

9.5 使用自定义聚合函数进行分组

Pandas 提供了许多与 groupby 对象一起使用的聚合函数。在某些时候，你可能需要编写自定义的用户定义函数（user-defined function，UDF），该函数在 Pandas 或 NumPy 中均不存在。

在此秘笈中，我们将使用 college 数据集计算每个州的本科生总体的均值和标准差。然后，使用此信息从每个州的任何单一总体值的均值中找到最大标准差数。

9.5.1 实战操作

（1）读取 college 数据集，按州（STABBR 列）找到本科生总体（UGDS 列）的 mean（均值）和 std（标准差）。

```
>>> college = pd.read_csv('data/college.csv')
>>> (college
...     .groupby('STABBR')
...     ['UGDS']
...     .agg(['mean', 'std'])
...     .round(0)
... )
```

```
            mean      std
STABBR
AK        2493.0    4052.0
AL        2790.0    4658.0
AR        1644.0    3143.0
AS        1276.0       NaN
AZ        4130.0   14894.0
...          ...       ...
VT        1513.0    2194.0
WA        2271.0    4124.0
WI        2655.0    4615.0
WV        1758.0    5957.0
WY        2244.0    2745.0
```

（2）此输出不是我们想要的。我们不是在寻找整个组的均值和标准差，而是寻找任何一家机构的均值的最大标准差数值。要计算此值，需要从每家机构的本科生人数中减去各州的本科生平均人数，然后除以标准差。这样就可以使每组的本科生总体标准化，得到一个 std_score。然后，可以利用 std_score 分数的绝对值的最大值来找到离均值最远的那个。Pandas 没有提供执行此操作的函数。因此，我们需要创建一个自定义函数 max_deviation。

```
>>> def max_deviation(s):
...     std_score = (s - s.mean()) / s.std()
...     return std_score.abs().max()
```

（3）定义 max_deviation 函数后，将该函数直接传递给.agg 方法以完成聚合。

```
>>> (college
...   .groupby('STABBR')
...   ['UGDS']
...   .agg(max_deviation)
...   .round(1)
... )
STABBR
AK     2.6
AL     5.8
AR     6.3
AS     NaN
AZ     9.9
       ...
VT     3.8
WA     6.6
WI     5.8
```

```
WV    7.2
WY    2.8
Name: UGDS, Length: 59, dtype: float64
```

9.5.2 原理解释

没有一个预定义的 Pandas 函数来计算远离平均值的最大标准偏差数,因此我们需要编写自己的函数。请注意,此自定义函数 max_deviation 接收单个参数 s。

在步骤(3)中,你会注意到函数名称被放置在 .agg 方法内,而不是直接被调用。这样,参数 s 就没有显式传递给 max_deviation。相反,Pandas 将 UGDS 列作为一个 Series 隐式传递给了 max_deviation。

每组调用一次 max_deviation 函数。由于 s 是一个 Series,因此所有正常的 Series 方法均可用。在所谓的标准化(standardization)过程中,它会先从组的每个值中减去该特定组的平均值,然后除以标准差。

由于我们对均值的绝对偏差感兴趣,因此可从所有标准化分数中获取绝对值并返回最大值。.agg 方法要求从函数中返回标量,否则将引发异常。

Pandas 默认使用样本标准偏差,这对于只有一个值的任何组都是未定义的。例如,州缩写 AS(美属萨摩亚)将返回缺失值,因为它在数据集中只有一家机构。

9.5.3 扩展知识

可以将上述自定义函数应用于多个聚合列。只需将更多的列名称添加到索引操作符中即可。max_deviation 函数仅与数字列协同工作。

```
>>> (college
...     .groupby('STABBR')
...     [['UGDS', 'SATVRMID', 'SATMTMID']]
...     .agg(max_deviation)
...     .round(1)
... )
        UGDS   SATVRMID   SATMTMID
STABBR
AK       2.6        NaN        NaN
AL       5.8        1.6        1.8
AR       6.3        2.2        2.3
AS       NaN        NaN        NaN
AZ       9.9        1.9        1.4
...      ...        ...        ...
```

```
VT      3.8     1.9     1.9
WA      6.6     2.2     2.0
WI      5.8     2.4     2.2
WV      7.2     1.7     2.1
WY      2.8     NaN     NaN
```

还可以将自定义聚合函数与预构建函数一起使用。以下是按州（STABBR 列）和宗教信仰（RELAFFIL 列）进行的分组。

```
>>> (college
...     .groupby(['STABBR', 'RELAFFIL'])
...     [['UGDS', 'SATVRMID', 'SATMTMID']]
...     .agg([max_deviation, 'mean', 'std'])
...     .round(1)
... )
                        UGDS                 ...   SATMTMID
                max_deviation    mean        ...       mean     std
STABBR  RELAFFIL                              ...
AK      0                 2.1  3508.9        ...        NaN     NaN
        1                 1.1   123.3        ...      503.0     NaN
AL      0                 5.2  3248.8        ...      515.8    56.7
        1                 2.4   979.7        ...      485.6    61.4
AR      0                 5.8  1793.7        ...      503.6    39.0
...                       ...     ...        ...        ...     ...
WI      0                 5.3  2879.1        ...      591.2    85.7
        1                 3.4  1716.2        ...      526.6    42.5
WV      0                 6.9  1873.9        ...      480.0    27.7
        1                 1.3   716.4        ...      484.8    17.7
WY      0                 2.8  2244.4        ...      540.0     NaN
```

可以看到，Pandas 使用函数名称作为返回列的名称。可以使用.rename 方法直接更改列名称，也可以修改该函数的属性.__name__。

```
>>> max_deviation.__name__
'max_deviation'
>>> max_deviation.__name__ = 'Max Deviation'
>>> (college
...     .groupby(['STABBR', 'RELAFFIL'])
...     [['UGDS', 'SATVRMID', 'SATMTMID']]
...     .agg([max_deviation, 'mean', 'std'])
...     .round(1)
... )
                        UGDS                 ...   SATMTMID
```

		Max Deviation	mean	...	mean	std
STABBR	RELAFFIL			...		
AK	0	2.1	3508.9	...	NaN	NaN
	1	1.1	123.3	...	503.0	NaN
AL	0	5.2	3248.8	...	515.8	56.7
	1	2.4	979.7	...	485.6	61.4
AR	0	5.8	1793.7	...	503.6	39.0
...		...	...	...	...	...
WI	0	5.3	2879.1	...	591.2	85.7
	1	3.4	1716.2	...	526.6	42.5
WV	0	6.9	1873.9	...	480.0	27.7
	1	1.3	716.4	...	484.8	17.7
WY	0	2.8	2244.4	...	540.0	NaN

9.6 使用*args和**kwargs自定义聚合函数

当编写自定义聚合函数时，Pandas会隐式地将每个聚合列作为一个Series传递给它，一次一列。有时候，需要传递更多的参数给函数，而不仅仅是Series本身。为此，你需要了解Python将任意数量的参数传递给函数的能力。

.agg的签名是agg(func, *args, **kwargs)。func参数是一个归约函数、归约方法的字符串名称、归约函数列表，或将列映射到函数或函数列表的字典。此外，如前文所述，你可以使用关键字参数来创建命名的聚合。

如果你的归约函数需要使用其他参数，则可以利用*args和**kwargs参数，将参数传递给归约函数。可以使用*args将任意数量的位置参数传递给自定义的聚合函数。类似地，**kwargs允许传递任意数量的关键字参数。

在此秘笈中，将为college数据集构建一个自定义函数，该函数可按州和宗教信仰找到大学本科生总体在这两个值之间的学校的百分比。

9.6.1 实战操作

（1）定义一个函数，该函数返回本科生总体为1000～3000的学校的百分比。

```
>>> def pct_between_1_3k(s):
...     return (s
...             .between(1_000, 3_000)
...             .mean()
...             * 100
...     )
```

(2) 计算按州和宗教信仰划分的百分比分组。

```
>>> (college
...     .groupby(['STABBR', 'RELAFFIL'])
...     ['UGDS']
...     .agg(pct_between_1_3k)
...     .round(1)
... )
STABBR  RELAFFIL
AK      0           14.3
        1            0.0
AL      0           23.6
AR      0           27.9
                    ...
WI      0           13.8
        1           36.0
WV      0           24.6
        1           37.5
WY      0           54.5
Name: UGDS, Length: 112, dtype: float64
```

(3) 步骤（1）中定义的函数能正常工作，但不能给用户提供选择上下限的灵活性。因此可以创建一个新函数，该函数允许用户参数化这些上下限范围。

```
>>> def pct_between(s, low, high):
...     return s.between(low, high).mean() * 100
```

(4) 将步骤（3）中创建的新函数连同上下限一起传递给.agg 方法。

```
>>> (college
...     .groupby(['STABBR', 'RELAFFIL'])
...     ['UGDS']
...     .agg(pct_between, 1_000, 10_000)
...     .round(1)
... )
STABBR  RELAFFIL
AK      0           42.9
        1            0.0
AL      0           45.8
        1           37.5
AR      0           39.7
                    ...
```

```
WI         0         31.0
           1         44.0
WV         0         29.2
           1         37.5
WY         0         72.7
Name: UGDS, Length: 112, dtype: float64
```

9.6.2 原理解释

步骤（1）创建了一个不接收任何额外参数的函数。上下限被硬编码到函数中，这样做的灵活性较差。

步骤（2）显示了此聚合的结果。

步骤（3）创建了一个更为灵活的函数，在该函数中，我们动态地设置了上下限。

步骤（4）是*args 和**kwargs 的魔力发挥作用的地方。

在此特定示例中，我们将两个非关键字参数 1_000 和 10_000 传递给.agg 方法。Pandas 分别将这两个参数传递给 pct_between 的 low 和 high 参数。

在步骤（4）中，有若干种方法可以达到相同的结果。可以显式地使用关键字参数来产生相同的结果。

```
(college
    .groupby(['STABBR', 'RELAFFIL'])
    ['UGDS']
    .agg(pct_between, high=10_000, low=1_000)
    .round(1)
)
```

9.6.3 扩展知识

如果我们要调用多个聚合函数，并且其中一些需要参数，则可以利用 Python 的闭包功能创建一个新函数，该函数的参数在其调用环境中被封闭。

```
>>> def between_n_m(n, m):
...     def wrapper(ser):
...         return pct_between(ser, n, m)
...     wrapper.__name__ = f'between_{n}_{m}'
...     return wrapper

>>> (college
...     .groupby(['STABBR', 'RELAFFIL'])
```

```
...         ['UGDS']
...         .agg([between_n_m(1_000, 10_000), 'max', 'mean'])
...         .round(1)
... )
                 between_1000_10000      max     mean
STABBR  RELAFFIL
AK      0                      42.9  12865.0   3508.9
        1                       0.0    275.0    123.3
AL      0                      45.8  29851.0   3248.8
        1                      37.5   3033.0    979.7
AR      0                      39.7  21405.0   1793.7
...                             ...      ...      ...
WI      0                      31.0  29302.0   2879.1
        1                      44.0   8212.0   1716.2
WV      0                      29.2  44924.0   1873.9
        1                      37.5   1375.0    716.4
WY      0                      72.7   9910.0   2244.4
```

9.7 检查 groupby 对象

在 DataFrame 上使用 .groupby 方法的直接结果是 groupby 对象。一般来说，可以在此对象上链接操作以进行聚合或转换，而无须将中间值存储在变量中。

在此秘笈中，我们将检查 groupby 对象以检查各个组。

9.7.1 实战操作

（1）将 college 数据集中的州（STABBR）和宗教信仰（RELAFFIL）列进行分组，将结果保存到变量中并确认其类型。

```
>>> college = pd.read_csv('data/college.csv')
>>> grouped = college.groupby(['STABBR', 'RELAFFIL'])
>>> type(grouped)
<class 'pandas.core.groupby.generic.DataFrameGroupBy'>
```

（2）使用 dir 函数查找 groupby 对象的属性。

```
>>> print([attr for attr in dir(grouped) if not
...        attr.startswith('_')])
['CITY', 'CURROPER', 'DISTANCEONLY', 'GRAD_DEBT_MDN_SUPP', 'HBCU',
'INSTNM', 'MD_EARN_WNE_P10', 'MENONLY', 'PCTFLOAN', 'PCTPELL',
```

第9章 分组以进行聚合、过滤和转换

```
 'PPTUG_EF', 'RELAFFIL', 'SATMTMID', 'SATVRMID' , 'STABBR', 'UG25ABV',
 'UGDS', 'UGDS_2MOR', 'UGDS_AIAN', 'UGDS_ASIAN', 'UGDS_BLACK',
 'UGDS _HISP', 'UGDS_NHPI', 'UGDS_NRA', 'UGDS_UNKN', 'UGDS_WHITE',
 'WOMENONLY', 'agg', 'aggregate ', 'all', 'any', 'apply', 'backfill',
 'bfill', 'boxplot', 'corr', 'corrwith', 'count', 'cov', 'cumcount',
 'cummax', 'cummin', 'cumprod', 'cumsum', 'describe', 'diff', 'dtypes',
 'expanding', 'ffill', 'fillna', 'filter', 'first', 'get_group', 'groups',
 'head', 'hist', 'id xmax', 'idxmin', 'indices', 'last', 'mad', 'max',
 'mean', 'median', 'min', 'ndim', 'ngroup ', 'ngroups', 'nth', 'nunique',
 'ohlc', 'pad', 'pct_change', 'pipe', 'plot', 'prod', 'quan tile', 'rank',
 'resample', 'rolling', 'sem', 'shift', 'size', 'skew', 'std', 'sum',
 'tail', 'take', 'transform', 'tshift', 'var']
```

（3）使用.ngroups 属性查找组数。

```
>>> grouped.ngroups
112
```

（4）要查找每个组的唯一标识标签，可查看.groups 属性，该属性包含映射到该组的所有相应索引标签的每个唯一组的字典。因为我们按两列分组，所以每个键都有一个元组，一个值用于 STABBR 列，另一个值用于 RELAFFIL 列。

```
>>> groups = list(grouped.groups)
>>> groups[:6]
[('AK', 0), ('AK', 1), ('AL', 0), ('AL', 1), ('AR', 0), ('AR', 1)]
```

（5）要使用 .get_group 方法检索单个组，可以给它传递一个确切的组标签的元组。例如，要获得佛罗里达州（州名称缩写为 FL）的所有宗教附属学校，可执行以下操作。

```
>>> grouped.get_group(('FL', 1))
        INSTNM          CITY   ... MD_EARN_WNE_P10  GRAD_DEBT_MDN_SUPP
712   The Bapt...   Graceville ...      30800              20052
713   Barry Un...       Miami  ...      44100              28250
714   Gooding  ... Panama City ...        NaN           PrivacyS...
715   Bethune-...     Daytona  ...      29400              36250
724   Johnson  ...   Kissimmee ...      26300              20199
...       ...           ...    ...        ...                ...
7486  Strayer  ... Coral Sp... ...      49200             36173.5
7487  Strayer  ... Fort Lau... ...      49200             36173.5
7488  Strayer  ...    Miramar  ...      49200             36173.5
7489  Strayer  ...      Miami  ...      49200             36173.5
7490  Strayer  ...      Miami  ...      49200             36173.5
```

（6）你可能想看看每个单独的小组。这是可能的，因为 groupby 对象是可迭代的。

如果在 Jupyter 中，则可以利用 display 函数在单个单元格中显示每个组（否则，Jupyter 将仅显示该单元格的最后一条语句的结果）。

```
from IPython.display import display
for name, group in grouped:
    print(name)
    display(group.head(3))
```

其输出结果如图 9-1 所示。

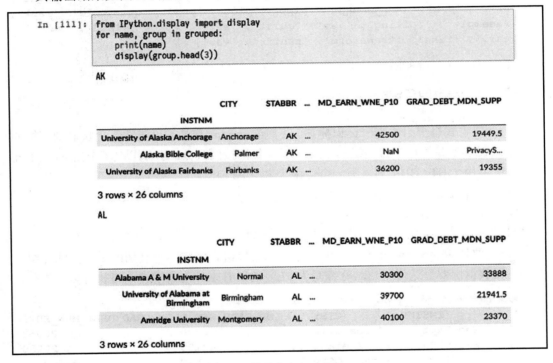

图 9-1　显示多个 DataFrame

当然，我们通常还希望查看单个组中的一些示例数据，以确定要对这些组应用什么函数。如果我们从分组的列中知道值的名称，则可以使用步骤（6）。一般来说，我们并不知道这些名称，但也不需要查看所有组。以下是对代码的一些调试，这些调试操作通常足以了解组的外观。

```
>>> for name, group in grouped:
...     print(name)
...     print(group)
```

```
...         break
('AK', 0)
           INSTNM           CITY   ...   MD_EARN_WNE_P10   GRAD_DEBT_MDN_SUPP
60      Universi...     Anchorage  ...             42500              19449.5
62      Universi...     Fairbanks  ...             36200                19355
63      Universi...        Juneau  ...             37400                16875
65      AVTEC-Al...        Seward  ...             33500            PrivacyS...
66      Charter ...     Anchorage  ...             39200                13875
67      Alaska C...     Anchorage  ...             28700                 8994
5171    Ilisagvi...        Barrow  ...             24900            PrivacyS...
```

（7）还可以在 groupby 对象上调用 .head 方法，以在单个 DataFrame 中将每个组的第一行放在一起。

```
>>> grouped.head(2)
           INSTNM           CITY   ...   MD_EARN_WNE_P10   GRAD_DEBT_MDN_SUPP
0       Alabama  ...        Normal  ...             30300                33888
1      Universi...    Birmingham  ...             39700              21941.5
2       Amridge ...    Montgomery  ...             40100                23370
10     Birmingh...    Birmingham  ...             44200                27000
43     Prince I...      Elmhurst  ...       PrivacyS...                20992
...         ...           ...     ...             ...                  ...
5289    Pacific ...      Mangilao  ...       PrivacyS...          PrivacyS...
6439    Touro Un...     Henderson  ...               NaN          PrivacyS...
7352    Marinell...     Henderson  ...             21200               9796.5
7404    Universi...     St. Croix  ...             31800                15150
7419    Computer...    Las Cruces  ...             21300                14250
```

9.7.2 原理解释

步骤（1）创建了 groupby 对象。

步骤（2）显示了所有公共属性和方法，以显示对象的功能。每个组由元组唯一标识，该元组包含分组列中值的唯一组合。Pandas 允许使用步骤（5）中显示的 .get_group 方法选择特定的组作为 DataFrame。

很少需要遍历整个小组。实际上，你应该避免这样做，因为这样可能会很慢。但是，有时你别无选择。遍历 groupby 对象时，将为你提供一个元组，其中包含组名和 DataFrame，并将分组列移入索引中。

在步骤（6）中，此元组在 for 循环中被解压缩到 name 和 group 变量中。

遍历组时，你可以做的一件事就是直接在 notebook 中显示每个组的一些行。为此，如果使用的是 Jupyter，则可以使用 IPython.display 模块中的 print 函数或 display 函数。

9.7.3 扩展知识

在步骤（2）中，还有几个很实用的方法没有被加入列表中，如 .nth 方法。当提供的是整数列表时，.nth 方法可以从每个组中选择那些特定的行。例如，以下操作将从每个组中选择第一行和最后一行。

```
>>> grouped.nth([1, -1])
                        INSTNM          CITY    ...    MD_EARN_WNE_P10
                                                ...
STABBR  RELAFFIL
AK      0               Universi...     Fairbanks  ...    36200
        0               Ilisagvi...     Barrow     ...    24900
        1               Alaska P...     Anchorage  ...    47000
        1               Alaska C...     Soldotna   ...    NaN
AL      0               Universi...     Birmingham ...    39700
...                     ...             ...        ...
WV      0               BridgeVa...     South C... ...    NaN
        1               Appalach...     Mount Hope ...    28700
        1               West Vir...     Nutter Fort ...   16700
WY      0               Central  ...    Riverton   ...    25200
        0               CollegeA...     Cheyenne   ...    25600
```

9.8 筛选少数族裔占多数的州

前文我们已经介绍过使用布尔数组过滤/筛选行。以类似的方式，当使用 .groupby 方法时，我们也可以过滤出组。groupby 对象的 .filter 方法接收一个函数，该函数必须返回 True 或 False 来指示是否保留组。

调用 .groupby 方法之后应用的此 .filter 方法与 2.3 节"使用方法选择列"中介绍的 DataFrame .filter 方法完全不同。

要注意的一件事是，当应用 .filter 方法时，结果不使用分组列作为索引，而是保留原始索引。DataFrame .filter 方法过滤列，而不是值。

在此秘笈中，我们将使用 college 数据集查找非白人大学生比白人大学生多的所有州。这是来自美国的数据集，其中白人占多数，因此，我们实际上寻找的是少数族裔占多数的州。

9.8.1 实战操作

（1）读取 college 数据集，按州分组，并显示分组的总数。这应该等于从 .nunique Series 方法检索的唯一州的数量。

```
>>> college = pd.read_csv('data/college.csv', index_col='INSTNM')
>>> grouped = college.groupby('STABBR')
>>> grouped.ngroups
59
>>> college['STABBR'].nunique()  # 验证数字是否相同
59
```

（2）分组变量具有 .filter 方法，该方法接收一个自定义函数，该函数确定是否保留分组。自定义函数接收当前组的 DataFrame，并且需要返回布尔值。下面定义一个函数，该函数将计算少数族裔学生的总百分比，如果该百分比大于用户定义的阈值，则返回 True。

```
>>> def check_minority(df, threshold):
...     minority_pct = 1 - df['UGDS_WHITE']
...     total_minority = (df['UGDS'] * minority_pct).sum()
...     total_ugds = df['UGDS'].sum()
...     total_minority_pct = total_minority / total_ugds
...     return total_minority_pct > threshold
```

（3）使用 .filter 方法，将 check_minority 函数传递给它，并加上一个 50%的阈值，以查找所有少数族裔占大多数的州。

```
>>> college_filtered = grouped.filter(check_minority, threshold=.5)
>>> college_filtered
                   CITY    STABBR ... MD_EARN_WNE_P10  GRAD_DEBT_MDN_SUPP
INSTNM
Everest C...     Phoenix    AZ   ...      28600              9500
Collins C...     Phoenix    AZ   ...      25700             47000
Empire Be...     Phoenix    AZ   ...      17800              9588
Empire Be...     Tucson     AZ   ...      18200              9833
Thunderbi...     Glendale   AZ   ...     118900          PrivacyS...
    ...            ...      ... ...         ...               ...
WestMed C...     Merced     CA   ...       NaN             15623.5
Vantage C...     El Paso    TX   ...       NaN              9500
SAE Insti...     Emeryville CA   ...       NaN              9500
Bay Area  ...    San Jose   CA   ...       NaN          PrivacyS...
Excel Lea...     San Antonio TX  ...       NaN             12125
```

（4）仅查看输出结果可能看不出发生了什么。该 DataFrame 以亚利桑那州（缩写为 AZ）而非阿拉斯加州（缩写为 AK）开头，因此可以从视觉上确认某些东西已被更改。将这个过滤后的 DataFrame 的形状与原始 DataFrame 进行比较。查看结果可以发现，大约 60% 的行都已被过滤，只有 20 个州是少数族裔占大多数的。

```
>>> college.shape
(7535, 26)
>>> college_filtered.shape
(3028, 26)
>>> college_filtered['STABBR'].nunique()
20
```

9.8.2　原理解释

此秘笈以州为单位查看所有大学机构的本科生总体。目标是保留州（州作为一个整体）中少数族裔占大多数的所有行。

这要求我们按州对数据进行分组，这在步骤（1）中已经执行。在输出结果中可以看到，有 59 个独立的组。

分组变量的.filter 方法可以将所有行保留在一个组中或将其过滤掉，并且不会更改列数。.filter 方法可通过此秘笈中的用户定义函数 check_minority 执行此过滤操作。check_minority 函数可接收每个组的一个 DataFrame，并且需要返回一个布尔值。

在 check_minority 函数内部，首先计算每所大学机构的非白人学生的百分比和总数，然后是所有学生的总数。最后，根据给定的阈值检查整个州的非白人学生的百分比，看看是否超过 50%，这会产生一个布尔值。

最终结果是一个 DataFrame，该 DataFrame 的列与原始列相同（并且具有相同的索引，而不是分组索引），但是已经过滤了来自州的未超过阈值的行。由于过滤后的 DataFrame 的标题可能与原始标题相同，因此需要进行一些检查以确保操作成功完成。我们可以通过检查行数和唯一州来验证这一点。

9.8.3　扩展知识

函数 check_minority 是很灵活的，该函数接收一个参数来降低或提高少数族裔占比的阈值。下面检查其他两个阈值的唯一州的形状和数量。

```
>>> college_filtered_20 = grouped.filter(check_minority, threshold=.2)
>>> college_filtered_20.shape
(7461, 26)
```

```
>>> college_filtered_20['STABBR'].nunique()
57
>>> college_filtered_70 = grouped.filter(check_minority, threshold=.7)
>>> college_filtered_70.shape
(957, 26)
>>> college_filtered_70['STABBR'].nunique()
10
```

9.9 通过减肥赌注做出改变

增加减肥动机的一种方法是与他人打赌。本秘笈中的方案将跟踪 4 个月内两个人的体重减轻情况，并确定获胜者。

在此秘笈中，我们将使用来自两个人的模拟数据来跟踪他们在 4 个月内体重减轻的百分比。在每个月末，将根据当月体重减轻的百分比来宣布获胜者，体重减轻比例高者胜。为了跟踪体重减轻，我们按月和人对数据进行分组，然后调用 .transform 方法以查找每月相对于月初的体重减轻百分比。

在此秘笈中将使用 .transform 方法。此方法返回一个新对象，该对象将保留原始 DataFrame 的索引，但允许对数据组进行计算。

9.9.1 实战操作

（1）读取原始的 weight_loss 数据集，并检查来自 Amy 和 Bob 两个人的第一个月的数据。每月总共有 4 个称量数据。

```
>>> weight_loss = pd.read_csv('data/weight_loss.csv')
>>> weight_loss.query('Month == "Jan"')
   Name  Month   Week    Weight
0   Bob   Jan   Week 1    291
1   Amy   Jan   Week 1    197
2   Bob   Jan   Week 2    288
3   Amy   Jan   Week 2    189
4   Bob   Jan   Week 3    283
5   Amy   Jan   Week 3    189
6   Bob   Jan   Week 4    283
7   Amy   Jan   Week 4    190
```

（2）要确定每个月的赢家，只需要比较每个月的第一周到最后一周的体重减轻。但是，如果想要每周更新一次，则可以计算从当前周到每月第一周的体重减轻。接下来我们就可以创建这样一个函数（即 percent_loss），以提供每周的体重减轻更新的功能。这将需要一个 Series 并返回相同大小的 Series。

```
>>> def percent_loss(s):
...     return ((s - s.iloc[0]) / s.iloc[0]) * 100
```

（3）现在为一月份 Bob 的数据测试该函数。

```
>>> (weight_loss
...     .query('Name=="Bob" and Month=="Jan"')
...     ['Weight']
...     .pipe(percent_loss)
... )
0    0.000000
2   -1.030928
4   -2.749141
6   -2.749141
Name: Weight, dtype: float64
```

（4）可以看到，第一周之后，Bob 的体重减轻了约 1%。他在第二周继续减肥，但在最后一周没有任何进展。我们可以将此函数应用于人和月的每个单一组合，以获得每月相对于该月第一周的体重减轻百分比。为此，我们需要按 Name 和 Month 对数据进行分组，然后使用 .transform 方法应用此自定义函数。我们传递给 .transform 的函数需要维护传递给它的组的索引，因此可以在此处使用 percent_loss。

```
>>> (weight_loss
...     .groupby(['Name', 'Month'])
...     ['Weight']
...     .transform(percent_loss)
... )
0     0.000000
1     0.000000
2    -1.030928
3    -4.060914
4    -2.749141
       ...
27   -3.529412
28   -3.065134
29   -3.529412
30   -4.214559
31   -5.294118
Name: Weight, Length: 32, dtype: float64
```

（5）.transform 方法采用了一个函数，该函数将返回一个对象，这个对象与传递给它的对象有相同的索引（和相同的行数）。因为它具有相同的索引，所以可以将其插入为列。.transform 方法可用于汇总组中的信息，然后将其添加回原始 DataFrame。我们还将过滤 Bob 的两个月数据。

```
>>> (weight_loss
...     .assign(percent_loss=(weight_loss
...         .groupby(['Name', 'Month'])
...         ['Weight']
...         .transform(percent_loss)
...         .round(1)))
...     .query('Name=="Bob" and Month in ["Jan", "Feb"]')
... )
    Name  Month   Week    Weight  percent_loss
0   Bob   Jan     Week 1  291      0.0
2   Bob   Jan     Week 2  288     -1.0
4   Bob   Jan     Week 3  283     -2.7
6   Bob   Jan     Week 4  283     -2.7
8   Bob   Feb     Week 1  283      0.0
10  Bob   Feb     Week 2  275     -2.8
12  Bob   Feb     Week 3  268     -5.3
14  Bob   Feb     Week 4  268     -5.3
```

（6）请注意，在新的月份之后，体重减轻的百分比会重置。使用此新的 percent_loss 列，可以手动确定获胜者，但是这里也不妨看看是否可以找到一种自动执行此操作的方法。由于唯一重要的一周是最后一周，所以我们选择第 4 周。

```
>>> (weight_loss
...     .assign(percent_loss=(weight_loss
...         .groupby(['Name', 'Month'])
...         ['Weight']
...         .transform(percent_loss)
...         .round(1)))
...     .query('Week == "Week 4"')
... )
    Name  Month   Week    Weight  percent_loss
6   Bob   Jan     Week 4  283     -2.7
7   Amy   Jan     Week 4  190     -3.6
14  Bob   Feb     Week 4  268     -5.3
15  Amy   Feb     Week 4  173     -8.9
22  Bob   Mar     Week 4  261     -2.6
23  Amy   Mar     Week 4  170     -1.7
```

```
30    Bob       Apr    Week 4    250             -4.2
31    Amy       Apr    Week 4    161             -5.3
```

（7）这缩小了周数，但仍然无法自动找出每个月的赢家。可以考虑使用 .pivot 方法重塑此数据，以便 Bob 和 Amy 两个人每月的减肥百分比并排。

```
>>> (weight_loss
...     .assign(percent_loss=(weight_loss
...         .groupby(['Name', 'Month'])
...         ['Weight']
...         .transform(percent_loss)
...         .round(1)))
...     .query('Week == "Week 4"')
...     .pivot(index='Month', columns='Name',
...            values='percent_loss')
... )
Name   Amy   Bob
Month
Apr    -5.3  -4.2
Feb    -8.9  -5.3
Jan    -3.6  -2.7
Mar    -1.7  -2.6
```

（8）此输出使每个月的获胜者一目了然，但是我们仍然可以更进一步。NumPy 具有向量化的 if then else 函数，该函数被称为 where，它可以将 Series 或布尔数组映射到其他值上。下面创建一个 winner 列，用来表示获胜者。

```
>>> (weight_loss
...     .assign(percent_loss=(weight_loss
...         .groupby(['Name', 'Month'])
...         ['Weight']
...         .transform(percent_loss)
...         .round(1)))
...     .query('Week == "Week 4"')
...     .pivot(index='Month', columns='Name',
...            values='percent_loss')
...     .assign(winner=lambda df_:
...             np.where(df_.Amy < df_.Bob, 'Amy', 'Bob'))
... )
Name   Amy   Bob   winner
Month
Apr    -5.3  -4.2    Amy
Feb    -8.9  -5.3    Amy
```

```
Jan     -3.6    -2.7        Amy
Mar     -1.7    -2.6        Bob
```

在 Jupyter 中，可以使用 .style 属性突出显示每个月的获胜百分比。

```
(weight_loss
    .assign(percent_loss=(weight_loss
        .groupby(['Name', 'Month'])
        ['Weight']
        .transform(percent_loss)
        .round(1)))
    .query('Week == "Week 4"')
    .pivot(index='Month', columns='Name',
           values='percent_loss')
    .assign(winner=lambda df_:
            np.where(df_.Amy < df_.Bob, 'Amy', 'Bob'))
    .style.highlight_min(axis=1)
)
```

其输出结果如图 9-2 所示。

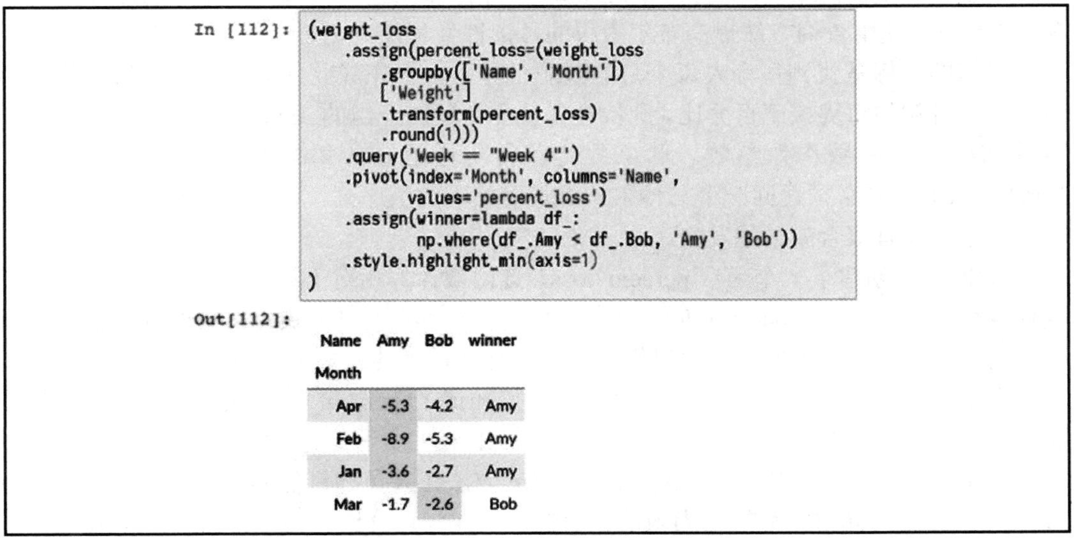

```
In [112]: (weight_loss
            .assign(percent_loss=(weight_loss
                .groupby(['Name', 'Month'])
                ['Weight']
                .transform(percent_loss)
                .round(1)))
            .query('Week == "Week 4"')
            .pivot(index='Month', columns='Name',
                   values='percent_loss')
            .assign(winner=lambda df_:
                    np.where(df_.Amy < df_.Bob, 'Amy', 'Bob'))
            .style.highlight_min(axis=1)
          )
Out[112]:
Name   Amy   Bob   winner
Month
Apr    -5.3  -4.2  Amy
Feb    -8.9  -5.3  Amy
Jan    -3.6  -2.7  Amy
Mar    -1.7  -2.6  Bob
```

图 9-2 突出显示获胜的减肥百分比

（9）使用 .value_counts 方法返回各人的最终得分，以获胜的月数为单位。

```
>>> (weight_loss
...     .assign(percent_loss=(weight_loss
```

```
...            .groupby(['Name', 'Month'])
...            ['Weight']
...            .transform(percent_loss)
...            .round(1)))
...         .query('Week == "Week 4"')
...         .pivot(index='Month', columns='Name',
...                values='percent_loss')
...         .assign(winner=lambda df_:
...                 np.where(df_.Amy < df_.Bob, 'Amy', 'Bob'))
...         .winner
...         .value_counts()
... )
Amy    3
Bob    1
Name: winner, dtype: int64
```

9.9.2 原理解释

在整个秘笈中，.query 方法用于过滤数据，而不是使用布尔数组。有关更多信息，请参阅 7.8 节"使用查询方法提高布尔索引的可读性"秘笈。

我们的目标是找到每个人每个月的减肥百分比。要完成此任务，一种方法是计算相对于每个月初的每周减肥百分比。此特定任务非常适合分组的 .transform 方法。.transform 方法需要一个函数作为参数。此函数将获得传递给它的每个组（可以是 Series 或 DataFrame）。它必须返回一个与传入的组长度相同的值序列，否则将引发异常。在此处理过程中，没有聚合或过滤发生。

步骤（2）创建了一个函数 percent_loss，该函数计算相对于第一个值的减肥（也可能是体重增加）百分比。percent_loss 函数使用传递给其自身 Series，将其他周的值减去第一周的值，然后将获得的结果除以第一个值。这样得到的就是体重减轻百分比。

在步骤（3）中，使用某人在某月内的数值测试了 percent_loss 函数，证明此函数是正常有效的。

在步骤（4）中，使用.groupby 和.transform 方法在人员和月份的每个组合上运行此函数。将 Weight（体重）列转换为当前一周的体重减轻百分比。

在步骤（6）中，为每个人输出了第一个月的数据。Pandas 将新数据作为 Series 返回。该 Series 本身单拿出来没有什么用处，只有将该 Series 作为新列附加到原始 DataFrame 中才更有意义。在步骤（5）中已经完成了此操作。

为了确定获胜者，每个月仅需要第 4 周的数据。在这里其实就可以停下来，手动确定获胜者，但 Pandas 提供了自动化的功能，所以我们还可以更进一步。步骤（7）中的.pivot 函数通过将一列的唯一值旋转为新的列名称来重塑我们的数据集。index 参数可用于指定不想旋转的列。传递给 values 参数的列将平铺在 index 和 columns 参数中列的每个唯一组合上。

请注意，.pivot 方法仅在 index 和 columns 参数中的列的每个唯一组合仅出现一次的情况下才有效。如果唯一的组合不止一个，则会引发异常。在这种情况下，可以考虑使用.pivot_table 或.groupby 方法。

以下是将.groupyby 与.unstack 方法结合使用以模拟数据透视（pivot）功能的示例。

```
>>> (weight_loss
...    .assign(percent_loss=(weight_loss
...        .groupby(['Name', 'Month'])
...        ['Weight']
...        .transform(percent_loss)
...        .round(1)))
...    .query('Week == "Week 4"')
...    .groupby(['Month', 'Name'])
...    ['percent_loss']
...    .first()
...    .unstack()
... )
Name    Amy   Bob
Month
Apr    -5.3  -4.2
Feb    -8.9  -5.3
Jan    -3.6  -2.7
Mar    -1.7  -2.6
```

在经过数据透视处理之后，即可使用 NumPy where 函数，该函数的第一个参数是一个条件判断语句，通过该语句可以产生一个布尔值 Series。True 值被映射到 Amy 身上，False 值被映射到 Bob 身上。最后突出显示每个月的获胜者，并使用 .value_counts 方法统计最终得分。

9.9.3 扩展知识

现在来研究步骤（7）中的 DataFrame 输出。你是否注意到月份是按字母顺序而不是按时间顺序排列的？遗憾的是，至少在本示例中，Pandas 以字母顺序对月份进行了排序。可以通过将 Month 的数据类型更改为分类变量来解决此问题。分类变量将每列的所有值映射为一个整数。我们可以选择此映射为月份的正常时间顺序。这样，Pandas 就可以在 .pivot 方法期间使用此底层整数映射来按时间顺序排列月份。

```
>>> (weight_loss
...     .assign(percent_loss=(weight_loss
...         .groupby(['Name', 'Month'])
...         ['Weight']
...         .transform(percent_loss)
...         .round(1)),
...         Month=pd.Categorical(weight_loss.Month,
...             categories=['Jan', 'Feb', 'Mar', 'Apr'],
...             ordered=True))
...     .query('Week == "Week 4"')
...     .pivot(index='Month', columns='Name',
...         values='percent_loss')
... )
Name    Amy   Bob
Month
Jan    -3.6  -2.7
Feb    -8.9  -5.3
Mar    -1.7  -2.6
Apr    -5.3  -4.2
```

要将 Month 转换为有序的分类列，可使用 Categorical 构造函数。将原始列作为 Series 传递，并将所有分类的唯一序列按所需顺序传递给 Categories 参数。一般来说，要将 object 数据类型的列按字母顺序以外的其他方式排序，都可以先将其转换为分类。

9.10 计算每个州的 SAT 加权平均成绩

groupby 对象具有 4 个方法，这些方法接收一个（或多个）函数对每个组执行计算。这 4 个方法分别为.agg、.filter、.transform 和.apply。这 4 个方法的前 3 个方法中的每个方法都有一个非常特定的输出，函数必须返回该输出。例如，.agg 方法必须返回一个标量值，.filter 方法必须返回一个布尔值，而.transform 方法则必须返回与传递的组具有相同长度的 Series 或 DataFrame。这 3 个方法在前面的秘笈中都已经分别详细介绍过。

但是，.apply 方法不一样，该方法既可能返回一个标量值，也可能返回一个 Series 甚至是任何形状的 DataFrame，因此该方法是非常灵活的。另外，.apply 方法对于每个组（在 DataFrame 上）仅调用一次，而 .transform 和 .agg 方法则是对于每个聚合列（在 Series 上）仅调用一次。

当同时对多个列进行操作时，.apply 方法可以返回单个对象，这一功能使得此秘笈中的计算成为可能。

在此秘笈中，我们将从 college 数据集中计算每个州的 SAT 数学和词汇分数的加权平均值。分数的加权方式是基于每所学校的本科生人数。

9.10.1 实战操作

（1）读取 college 数据集，然后分别在 UGDS（本科生人数）、SATMTMID（SAT 数学成绩）或 SATVRMID（SAT 词汇成绩）列中删除所有包含缺失值的行。这些列不应该包含缺失值。

```
>>> college = pd.read_csv('data/college.csv')
>>> subset = ['UGDS', 'SATMTMID', 'SATVRMID']
>>> college2 = college.dropna(subset=subset)
>>> college.shape
(7535, 27)
>>> college2.shape
(1184, 27)
```

（2）可以看到，绝大多数大学机构都在这 3 个必填列上包含了缺失值数据，但这不妨碍本示例继续进行（因为仍然有 1184 所大学在这 3 个必填列上都包含了数据）。接下来，创建一个用户定义的函数 weighted_math_average 来计算 SAT 数学分数的加权平均值。

```
>>> def weighted_math_average(df):
...     weighted_math = df['UGDS'] * df['SATMTMID']
...     return int(weighted_math.sum() / df['UGDS'].sum())
```

（3）按州分组，然后将 weighted_math_average 函数传递给 .apply 方法。因为每个组都有多个列，并且我们希望将其归约为一个值，所以这里需要使用 .apply 方法。weighted_math_average 函数对于每个组仅调用一次（不是针对该组的各个列）。

```
>>> college2.groupby('STABBR').apply(weighted_math_average)
STABBR
AK    503
AL    536
AR    529
AZ    569
CA    564
     ...
VT    566
WA    555
```

```
WI    593
WV    500
WY    540
Length: 53, dtype: int64
```

（4）可以看到，我们成功返回了每个组的标量值。这里也可以做试验，看看将相同的函数 weighted_math_average 传递给 .agg 方法（使用每列调用该函数）的结果是什么。

```
>>> (college2
...     .groupby('STABBR')
...     .agg(weighted_math_average)
... )
Traceback (most recent call last):
    ...
KeyError: 'UGDS'
```

（5）weighted_math_average 函数将应用于 DataFrame 中的每个非聚合列。如果尝试将列限制为仅 SATMTMID，则会收到错误消息，因为你无权访问 UGDS。因此，完成对多个列执行操作的最佳方法就是使用 .apply 方法。

```
>>> (college2
...     .groupby('STABBR')
...     ['SATMTMID']
...     .agg(weighted_math_average)
... )
Traceback (most recent call last):
    ...
KeyError: 'UGDS'
```

（6）.apply 方法的一个很不错的功能是，你可以通过返回一个 Series 来创建多个新列。此返回的 Series 的索引将是新的列名称。下面修改自定义函数以计算两个 SAT 分数的加权平均值（w_math_avg 和 w_verbal_avg）和算术平均值（math_avg 和 verbal_avg），以及每个组中大学机构数目的计数（count）。可以在 Series 中返回这 5 个值。

```
>>> def weighted_average(df):
...     weight_m = df['UGDS'] * df['SATMTMID']
...     weight_v = df['UGDS'] * df['SATVRMID']
...     wm_avg = weight_m.sum() / df['UGDS'].sum()
...     wv_avg = weight_v.sum() / df['UGDS'].sum()
...     data = {'w_math_avg': wm_avg,
...             'w_verbal_avg': wv_avg,
...             'math_avg': df['SATMTMID'].mean(),
...             'verbal_avg': df['SATVRMID'].mean(),
```

```
...                'count': len(df)
...            }
...        return pd.Series(data)
>>> (college2
...     .groupby('STABBR')
...     .apply(weighted_average)
...     .astype(int)
... )
        w_math_avg  w_verbal_avg  math_avg  verbal_avg  count
STABBR
AK             503           555       503         555      1
AL             536           533       504         508     21
AR             529           504       515         491     16
AZ             569           557       536         538      6
CA             564           539       562         549     72
...            ...           ...       ...         ...    ...
VT             566           564       526         527      8
WA             555           541       551         548     18
WI             593           556       545         516     14
WV             500           487       481         473     17
WY             540           535       540         535      1
```

9.10.2 原理解释

为了使此秘笈正确完成，我们需要筛选在 UGDS、SATMTMID 和 SATVRMID 3 列中没有缺失值的大学机构。默认情况下，.dropna 方法将删除包含一个或多个缺失值的行。这里必须使用 subset 参数来限制其查看的列，因为我们仅需要 .dropna 方法考虑 UGDS、SATMTMID 和 SATVRMID 列的缺失值问题。

如果不删除缺失值，那么它将导致加权平均值的计算失真。接下来，你可以看到 AK 的加权分数约为 5 分和 6 分，这显然没有意义。

```
>>> (college
...     .groupby('STABBR')
...     .apply(weighted_average)
... )
        w_math_avg  w_verbal_avg   math_avg   verbal_avg  count
STABBR
AK        5.548091      6.121651  503.000000  555.000000   10.0
AL      261.895658    260.550109  504.285714  508.476190   96.0
AR      301.054792    287.264872  515.937500  491.875000   86.0
AS        0.000000      0.000000         NaN         NaN    1.0
```

AZ	61.815821	60.511712	536.666667	538.333333	133.0
...	...	...	...	...	...
VT	389.967094	388.696848	526.875000	527.500000	27.0
WA	274.885878	267.880280	551.222222	548.333333	123.0
WI	153.803086	144.160115	545.071429	516.857143	112.0
WV	224.697582	218.843452	481.705882	473.411765	73.0
WY	216.761180	214.754132	540.000000	535.000000	11.0

在步骤（2）中，我们定义了一个仅计算 SATMTMID 列的加权平均值的函数 weighted_math_average。加权平均值与算术平均值不同，因为每个值都需要乘以一个权重，然后将这个数量相加并除以权重之和。在本示例中，我们的权重就是在校本科生人数。

在步骤（3）中，我们将 weighted_math_average 函数传递给了 .apply 方法。函数 weighted_math_average 获得传递的每个组所有原始列的 DataFrame。.apply 方法返回的是单个标量值，即 SATMTMID 的加权平均值。

在此阶段，你可能认为也可以使用 .agg 方法进行此计算。但是，在步骤（4）和步骤（5）中的试验证明，用 .agg 方法直接替换 .apply 方法不能正常工作，因为 .agg 方法返回的是其每个聚合列的值。

步骤（6）显示了 .apply 方法的多功能性。我们建立了一个新函数 weighted_average，该函数可以计算两个 SAT 列的加权平均值和算术平均值，以及每组的行数。若要使用 .apply 方法创建多个列，则必须返回一个 Series。其索引值用作结果 DataFrame 中的列名。使用.apply 方法可以返回任意多个值。

请注意，由于我们使用的 Python 版本大于 3.5，因此可以在 weighted_average 中使用普通字典来创建 Series。这是因为从 Python 3.6 开始，默认对字典进行了排序。

9.10.3 扩展知识

在此秘笈中，我们为每个组返回一行作为一个 Series。通过返回 DataFrame，可以为每个组返回任意数量的行和列。

除了查找算术平均值（arithmetic mean）和加权平均值（weighted mean）之外，还可以查找两个 SAT 列的几何平均值（geometric mean）和谐波平均值（harmonic mean），然后将结果作为 DataFrame 返回，其中数据行是均值类型的名称，列是 SAT 类型。

为了减轻编码负担，可以使用 NumPy 函数平均值来计算加权平均值，并使用 SciPy 函数 gmean 和 hmean 分别表示几何平均值和谐波平均值。

```
>>> from scipy.stats import gmean, hmean
>>> def calculate_means(df):
```

第 9 章 分组以进行聚合、过滤和转换

```
...     df_means = pd.DataFrame(index=['Arithmetic', 'Weighted',
...                                    'Geometric', 'Harmonic'])
...     cols = ['SATMTMID', 'SATVRMID']
...     for col in cols:
...         arithmetic = df[col].mean()
...         weighted = np.average(df[col], weights=df['UGDS'])
...         geometric = gmean(df[col])
...         harmonic = hmean(df[col])
...         df_means[col] = [arithmetic, weighted,
...                          geometric, harmonic]
...     df_means['count'] = len(df)
...     return df_means.astype(int)
>>> (college2
...     .groupby('STABBR')
...     .apply(calculate_means)
... )
```

		SATMTMID	SATVRMID	count
STABBR				
AK	Arithmetic	503	555	1
	Weighted	503	555	1
	Geometric	503	555	1
	Harmonic	503	555	1
AL	Arithmetic	504	508	21
...		...	...	...
WV	Harmonic	480	472	17
WY	Arithmetic	540	535	1
	Weighted	540	535	1
	Geometric	540	534	1
	Harmonic	540	535	1

9.11 按连续变量分组

在 Pandas 中进行分组时，通常使用具有离散重复值的列。如果没有重复的值，则分组将毫无意义，因为每个组只有一行。连续数字列通常具有很少的重复值，并且通常不用于形成组。但是，如果我们可以将具有连续值的列转换为离散列，方法是将每个值放置在一个箱子中，通过四舍五入或使用其他映射，则将它们分组是有意义的。有关对数字列进行分箱处理的操作，也可以参考 5.4 节"分类数据"秘笈。

在此秘笈中，我们探索 flights（航班）数据集以发现不同旅行距离的航空公司分布。例如，这使我们能够找到 500 英里～1000 英里飞行最多的航空公司。为此，我们可以使

用 Pandas cut 函数来离散每个航班飞行的距离。

9.11.1 实战操作

(1) 读取 flights 数据集。

```
>>> flights = pd.read_csv('data/flights.csv')
>>> flights
      MONTH  DAY  WEEKDAY  ...  ARR_DELAY  DIVERTED  CANCELLED
0         1    1        4  ...       65.0         0          0
1         1    1        4  ...      -13.0         0          0
2         1    1        4  ...       35.0         0          0
3         1    1        4  ...       -7.0         0          0
4         1    1        4  ...       39.0         0          0
...     ...  ...      ...  ...        ...       ...        ...
58487    12   31        4  ...      -19.0         0          0
58488    12   31        4  ...        4.0         0          0
58489    12   31        4  ...       -5.0         0          0
58490    12   31        4  ...       34.0         0          0
58491    12   31        4  ...       -1.0         0          0
```

(2) 如果要查找一定航班距离范围内的航空公司分布,则需要将 DIST 列的值放入离散的箱子 (bin) 中。让我们使用 Pandas cut 函数将数据分成 5 个 bin。

```
>>> bins = [-np.inf, 200, 500, 1000, 2000, np.inf]
>>> cuts = pd.cut(flights['DIST'], bins=bins)
>>> cuts
0          (500.0, 1000.0]
1         (1000.0, 2000.0]
2          (500.0, 1000.0]
3         (1000.0, 2000.0]
4         (1000.0, 2000.0]
                ...
58487     (1000.0, 2000.0]
58488        (200.0, 500.0]
58489        (200.0, 500.0]
58490      (500.0, 1000.0]
58491      (500.0, 1000.0]
Name: DIST, Length: 58492, dtype: category
Categories (5, interval[float64]): [(-inf, 200.0] < (200.0, 500.0]
< (500.0, 1000.0] < (1000.0, 2000.0] < (2000.0, inf]]
```

(3) 创建一个有序的分类 Series。为了理解到底发生了什么,现在可以统计每个类别的计数。

```
>>> cuts.value_counts()
(500.0, 1000.0]     20659
(200.0, 500.0]      15874
(1000.0, 2000.0]    14186
(2000.0, inf]        4054
(-inf, 200.0]        3719
Name: DIST, dtype: int64
```

（4）这个 cuts Series 现在可以用于形成组。Pandas 允许将许多类型传递给 .groupby 方法。在本示例中，可以将 cuts Series 传递给 .groupby 方法，然后在 AIRLINE 列上调用 .value_counts 方法以查找每个距离组的分布。可以看到，天西航空公司（SkyWest，代码为 OO）在少于 200 英里的航班中占据了约 33%的比例，而在 200 英里～500 英里的航班中，其占比约为 16%。

```
>>> (flights
...     .groupby(cuts)
...     ['AIRLINE']
...     .value_counts(normalize=True)
...     .round(3)
... )
DIST            AIRLINE
(-inf, 200.0]   OO         0.326
                EV         0.289
                MQ         0.211
                DL         0.086
                AA         0.052
                            ...
(2000.0, inf]   WN         0.046
                HA         0.028
                NK         0.019
                AS         0.012
                F9         0.004
Name: AIRLINE, Length: 57, dtype: float64
```

9.11.2 原理解释

在步骤（2）中，.cut 函数将 DIST 列的每个值放入 5 个箱子中的一个。

箱子由定义边缘的 6 个数字序列创建。请注意，你始终需要比箱子数多一个边缘。可以将 bins 参数传递给一个整数，这将自动创建该数目的等宽箱子。NumPy 中提供了负无穷大和正无穷大值，并确保将所有值都放在箱子中。如果值在箱子的边缘之外，则这些值将成为缺失值并且不会被放置在箱子中。

cuts 变量现在是一个包含 5 个有序分类的 Series。该变量具有所有常规的 Series 方法，在步骤（3）中，使用了.value_counts 方法来了解其分布。

.groupby 方法允许传递任何对象进行分组。这意味着数据分析人员可以从与当前 DataFrame 完全无关的内容中形成组。在本示例中，我们对 cuts 变量中的值进行了分组。对于每个分组，我们使用了.value_counts 方法（将 normalize 参数设置为 True），以找到每个航空公司的航班的百分比。

从这个结果可以得出一些有趣的见解。从全部结果来看，天西航空公司在 200 英里以下的航班中是领先的，但该公司没有超过 2000 英里的航班。相比之下，美国航空公司（american airline，代码为 AA）在 200 英里以下的航班中排名第五，但到目前为止，该公司拥有 1000 英里～2000 英里的航班最多。

9.11.3 扩展知识

当按 cuts 变量分组时，可以找到更多结果。例如，我们可以为每个距离分组找到第 25%、50%和 75%的分位数 Airtime。由于 Airtime 以分钟为单位，因此可以除以 60 得到小时。这将返回具有多重索引的 Series。

```
>>> (flights
...     .groupby(cuts)
...     ['AIR_TIME']
...     .quantile(q=[.25, .5, .75])
...     .div(60)
...     .round(2)
... )
DIST
(-inf, 200.0]     0.25    0.43
                  0.50    0.50
                  0.75    0.57
(200.0, 500.0]    0.25    0.77
                  0.50    0.92
                           ...
(1000.0, 2000.0]  0.50    2.93
                  0.75    3.40
(2000.0, inf]     0.25    4.30
                  0.50    4.70
                  0.75    5.03
```

```
Name: AIR_TIME, Length: 15, dtype: float64
```

使用 cut 函数时，可以通过此信息创建具有提示性的字符串标签。这些标签替换了索引中的区间符号。在这里还可以链接.unstack 方法，再由该方法将内部索引级别转换为列名称，具体如下。

```
>>> labels=['Under an Hour', '1 Hour', '1-2 Hours',
...         '2-4 Hours', '4+ Hours']
>>> cuts2 = pd.cut(flights['DIST'], bins=bins, labels=labels)
>>> (flights
...    .groupby(cuts2)
...    ['AIRLINE']
...    .value_counts(normalize=True)
...    .round(3)
...    .unstack()
... )
AIRLINE            AA     AS     B6  ...     US     VX     WN
DIST                                 ...
Under an Hour   0.052    NaN    NaN  ...    NaN    NaN  0.009
1 Hour          0.071  0.001  0.007  ...  0.016  0.028  0.194
1-2 Hours       0.144  0.023  0.003  ...  0.025  0.004  0.138
2-4 Hours       0.264  0.016  0.003  ...  0.040  0.012  0.160
4+ Hours        0.212  0.012  0.080  ...  0.065  0.074  0.046
```

9.12 计算城市之间的航班总数

在 flights 航班数据集中，包含了始发地和目的地机场的数据。这意味着可以轻松计算出始发地和目的地之间的航班数量。例如，计算从休斯敦出发并降落在亚特兰大的航班数量非常简单，比较困难的是计算两座城市之间的航班总数。

在此秘笈中，我们将计算两座城市之间的航班总数，而不管始发地或目的地是哪一个。为此，我们需要按字母顺序对始发地和目的地机场进行排序，以使机场的每种组合始终以相同的顺序出现。然后，可以使用这种新的列安排方式形成组，并进行计数。

9.12.1 实战操作

（1）读取 flights 数据集，并找到每个始发地机场（ORG_AIR 列）与目的地机场（DEST_AIR 列）之间的航班总数。

```
>>> flights = pd.read_csv('data/flights.csv')
```

```
>>> flights_ct = flights.groupby(['ORG_AIR', 'DEST_AIR']).size()
>>> flights_ct
ORG_AIR  DEST_AIR
ATL      ABE         31
         ABQ         16
         ABY         19
         ACY          6
         AEX         40
                    ...
SFO      SNA        122
         STL         20
         SUN         10
         TUS         20
         XNA          2
Length: 1130, dtype: int64
```

（2）选择在休斯敦（IAH）和亚特兰大（ATL）之间的双向航班总数。

```
>>> flights_ct.loc[[('ATL', 'IAH'), ('IAH', 'ATL')]]
ORG_AIR  DEST_AIR
ATL      IAH         121
IAH      ATL         148
dtype: int64
```

（3）可以简单地将这两个数字求和，以得出城市之间的总航班数，但是有一种更有效、更自动化的解决方案可以适用于所有航班。下面按字母顺序对每一行的起点和终点列进行排序。可以使用 axis ='columns' 来做到这一点。

```
>>> f_part3 = (flights
...     [['ORG_AIR', 'DEST_AIR']]
...     .apply(lambda ser:
...          ser.sort_values().reset_index(drop=True),
...          axis='columns')
... )
>>> f_part3
       DEST_AIR  ORG_AIR
0           SLC      LAX
1           IAD      DEN
2           VPS      DFW
3           DCA      DFW
4           MCI      LAX
...         ...      ...
58487       DFW      SFO
```

```
58488          SFO          LAS
58489          SBA          SFO
58490          ATL          MSP
58491          BOI          SFO
```

（4）现在已经对每行中的起点和终点值进行了排序，因此列名称是不正确的。下面将其重命名为更通用的名称，然后再次找到所有城市之间的航班总数。

```
>>> rename_dict = {0:'AIR1', 1:'AIR2'}
>>> (flights
...     [['ORG_AIR', 'DEST_AIR']]
...     .apply(lambda ser:
...             ser.sort_values().reset_index(drop=True),
...             axis='columns')
...     .rename(columns=rename_dict)
...     .groupby(['AIR1', 'AIR2'])
...     .size()
... )
AIR1  AIR2
ATL   ABE      31
      ABQ      16
      ABY      19
      ACY       6
      AEX      40
               ..
SFO   SNA     122
      STL      20
      SUN      10
      TUS      20
      XNA       2
Length: 1130, dtype: int64
```

（5）现在选择亚特兰大（ATL）和休斯敦（IAH）之间的所有航班，并验证这些航班是否与步骤（2）中的两个值之和匹配。

```
>>> (flights
...     [['ORG_AIR', 'DEST_AIR']]
...     .apply(lambda ser:
...             ser.sort_values().reset_index(drop=True),
...             axis='columns')
...     .rename(columns=rename_dict)
...     .groupby(['AIR1', 'AIR2'])
...     .size()
```

```
...        .loc[('ATL', 'IAH')]
... )
269
```

（6）在步骤（2）中，从亚特兰大到休斯敦的航班数为 121，从休斯敦到亚特兰大的航班数为 148，双向合计正是 269。值得一提的是，如果尝试先选择休斯敦（IAH），后选择亚特兰大（ATL）以计算它们之间的航班数，则会出现错误。

```
>>> (flights
...     [['ORG_AIR', 'DEST_AIR']]
...     .apply(lambda ser:
...            ser.sort_values().reset_index(drop=True),
...            axis='columns')
...     .rename(columns=rename_dict)
...     .groupby(['AIR1', 'AIR2'])
...     .size()
...     .loc[('IAH', 'ATL')]
... )
Traceback (most recent call last)
    ...
KeyError: 'ATL'
```

9.12.2 原理解释

在步骤（1）中，我们按始发地和目的地机场列形成分组，然后将 .size 方法应用于 groupby 对象，该对象返回每个组的总行数。注意，我们可以将字符串 size 传递给 .agg 方法以实现相同的结果。在步骤（2）中，选择了亚特兰大和休斯敦之间每个方向的航班总数。结果是具有两个级别的多重索引的 Series。从多重索引中选择行的一种方法是将 .loc 索引操作符传递给精确级别值的元组。

在本示例中，可以选择两个不同的行，即('ATL', 'IAH')和('IAH', 'ATL')。我们使用了元组列表来正确执行此操作。

步骤（3）是此秘笈中最重要的步骤。我们希望亚特兰大和休斯敦之间的所有航班只有一个标签，到目前为止，有两个标签。如果按字母顺序对始发地机场和目的地机场的每种组合进行排序，那么这将为机场之间的航班使用一个标签。为此，可以在 DataFrame 上使用 .apply 方法。这与 groupby .apply 方法不同。在步骤（3）中没有形成任何组。

我们必须向 DataFrame .apply 方法中传递一个函数。在本示例中，它是一个 lambda 函数，它对每一行进行排序。默认情况下，lambda 函数被传递给每一列。我们可以通过使用 axis ='columns'（或 axis = 1）来更改计算方向。lambda 函数会将每一行数据隐式地作为 Series 传递给它。它返回带有已排序机场代码的 Series。我们必须调用.reset_index 方法，以便在应用 lambda 函数后列不会重新对齐。

.apply 方法使用 lambda 函数遍历所有行。完成此操作后，将为每一行对两列中的值进行排序。列名称现在已无意义。下一步对列名称进行了重命名，然后执行与步骤（2）中相同的分组和汇总。这次，亚特兰大和休斯敦之间的所有航班都属于同一标签。

9.12.3 扩展知识

步骤（3）~步骤（6）是成本非常高的操作，需要几秒钟才能完成。这里只有大约 60000 行，因此该解决方案无法很好地扩展到更大的数据。

使用 axis ='columns'（或 axis = 1）调用 .apply 方法是所有 Pandas 中性能最低的操作之一。在内部，Pandas 在每行上循环，并且不提供 NumPy 的任何速度提升。如果可能的话，请避免将 .apply 方法与 axis = 1 一起使用。

使用 NumPy sort 函数，可以大大提高速度。因此，我们可以使用 sort 函数并分析其输出。默认情况下，sort 函数将对每一行进行排序。

```
>>> data_sorted = np.sort(flights[['ORG_AIR', 'DEST_AIR']])
>>> data_sorted[:10]
array([ ['LAX', 'SLC'],
        ['DEN', 'IAD'],
        ['DFW', 'VPS'],
        ['DCA', 'DFW'],
        ['LAX', 'MCI'],
        ['IAH', 'SAN'],
        ['DFW', 'MSY'],
        ['PHX', 'SFO'],
        ['ORD', 'STL'],
        ['IAH', 'SJC']], dtype=object)
```

这里返回的是二维 NumPy 数组。NumPy 不执行分组操作，因此下面使用 DataFrame 构造函数创建一个新的 DataFrame 并检查它是否等于步骤（3）中的 DataFrame。

```
>>> flights_sort2 = pd.DataFrame(data_sorted, columns=['AIR1', 'AIR2'])
>>> flights_sort2.equals(f_part3.rename(columns={0:'AIR1', 1:'AIR2'}))
True
```

可以看到，DataFrames 是相同的，因此可以将步骤（3）替换为上述更快的排序例程。现在不妨对这两种不同的排序方法进行计时，以了解它们之间的区别。

```
>>> %%timeit
```

```
>>> flights_sort = (flights
...     [['ORG_AIR', 'DEST_AIR']]
...     .apply(lambda ser:
...            ser.sort_values().reset_index(drop=True),
...            axis='columns')
... )
1min 5s ± 2.67 s per loop (mean ± std. dev. of 7 runs, 1 loop each)
```

```
>>> %%timeit
>>> data_sorted = np.sort(flights[['ORG_AIR', 'DEST_AIR']])
>>> flights_sort2 = pd.DataFrame(data_sorted,
...     columns=['AIR1', 'AIR2'])
14.6 ms ± 173 µs per loop (mean ± std. dev. of 7 runs, 100 loops each)
```

在此示例中，NumPy 解决方案比通过 Pandas 使用 .apply 方法的速度快了 4452 倍。

9.13 寻找最长的准点航班连续记录

对于航空公司而言，最重要的指标之一是其准时飞行连续记录。美国联邦航空管理局（Federal Aviation Administration）认为，航班延误的标准是在比原定抵达时间至少晚 15min 后才抵达。Pandas 提供了计算每家航空公司准点航班总数和百分比的方法。尽管这些基本摘要统计数据是一个重要的指标，但还有其他一些有趣的计算，如查找每家航空公司在其始发机场的连续按时飞行的连续记录。

在此秘笈中，我们将查找每个始发机场的每家航空公司的最长连续航班准点率。这就要求列中的每个值都必须知道紧随其后的值。在将此方法应用于每个组之前，可以巧妙地使用 .diff 和 .cumsum 方法查找记录。

📝 注意：

我们在本秘笈中开发的 max_streak 函数公开了 Pandas 1.0 和 1.0.1 的回归。此 bug 应该在 Pandas 1.0.2 中被修复。有关详细信息可访问 https://github.com/pandas-dev/pandas/issues/31802。

9.13.1 实战操作

（1）在开始使用 flights 数据集之前，不妨使用一个小样本的 Series 来练习计算它们的连续记录。

```
>>> s = pd.Series([0, 1, 1, 0, 1, 1, 1, 0])
>>> s
0    0
1    1
2    1
3    0
4    1
5    1
6    1
7    0
dtype: int64
```

（2）我们对一个连续记录的最终表示方式将是与原始序列长度相同的 Series，每个记录从 1 开始计数。首先，可以使用.cumsum 方法。

```
>>> s1 = s.cumsum()
>>> s1
0    0
1    1
2    2
3    2
4    3
5    4
6    5
7    5
dtype: int64
```

（3）现在，我们已经累计了该 Series 中的所有计数（5）。接下来可以将此 Series 乘以原始 Series。

```
>>> s.mul(s1)
0    0
1    1
2    2
3    0
4    3
5    4
6    5
7    0
dtype: int64
```

（4）这样，在原始值为 1 的地方，我们已经有了一个非 0 值。这个结果非常接近我们的期望。接下来，我们需要在原始值为 1 的地方，重新开始每个连续记录的计数，而不是让累计总和不变。在本示例中，可以链接.diff 方法，该方法将从当前值中减去前一个值。

```
>>> s.mul(s1).diff()
0    NaN
1    1.0
2    1.0
3   -2.0
4    3.0
5    1.0
6    1.0
7   -5.0
dtype: float64
```

（5）负值表示记录结束。我们需要将负值向下传播到 Series 上，并使用它们减去步骤（2）中多余的累计量。为此，需要使用 .where 方法使所有非负值都变成缺失值。

```
>>> (s
...    .mul(s.cumsum())
...    .diff()
...    .where(lambda x: x < 0)
... )
0    NaN
1    NaN
2    NaN
3   -2.0
4    NaN
5    NaN
6    NaN
7   -5.0
dtype: float64
```

（6）现在可以使用 .ffill 方法向下传播这些负值。

```
>>> (s
...    .mul(s.cumsum())
...    .diff()
...    .where(lambda x: x < 0)
...    .ffill()
... )
0    NaN
1    NaN
2    NaN
3   -2.0
4   -2.0
```

```
5   -2.0
6   -2.0
7   -5.0
dtype: float64
```

（7）可以将此 Series 加回到累计总数上，以清除多余的累计。

```
>>> (s
...     .mul(s.cumsum())
...     .diff()
...     .where(lambda x: x < 0)
...     .ffill()
...     .add(s.cumsum(), fill_value=0)
... )
0   0.0
1   1.0
2   2.0
3   0.0
4   1.0
5   2.0
6   3.0
7   0.0
dtype: float64
```

（8）现在我们已经有了可以正常工作的记录查找器，可用于找到每家航空公司和始发地机场最长的准点航班记录。下面读取 flights 数据集并创建一列以表示准时到达。

```
>>> flights = pd.read_csv('data/flights.csv')
>>> (flights
...     .assign(ON_TIME=flights['ARR_DELAY'].lt(15).astype(int))
...     [['AIRLINE', 'ORG_AIR', 'ON_TIME']]
... )
       AIRLINE ORG_AIR  ON_TIME
0           WN     LAX        0
1           UA     DEN        1
2           MQ     DFW        0
3           AA     DFW        1
4           WN     LAX        0
...        ...     ...      ...
58487       AA     SFO        1
58488       F9     LAS        1
58489       OO     SFO        1
58490       WN     MSP        0
58491       OO     SFO        1
```

（9）使用前 7 个步骤中的逻辑来定义一个函数 max_streak，该函数将返回给定 Series 的最大记录数。

```
>>> def max_streak(s):
...     s1 = s.cumsum()
...     return (s
...         .mul(s1)
...         .diff()
...         .where(lambda x: x < 0)
...         .ffill()
...         .add(s1, fill_value=0)
...         .max()
...     )
```

（10）查找每家航空公司和始发机场的最大准点到达记录，以及航班总数和准点到达率（百分比）。首先，对一年中的日期和预定的出发时间进行排序。

```
>>> (flights
...     .assign(ON_TIME=flights['ARR_DELAY'].lt(15).astype(int))
...     .sort_values(['MONTH', 'DAY', 'SCHED_DEP'])
...     .groupby(['AIRLINE', 'ORG_AIR'])
...     ['ON_TIME']
...     .agg(['mean', 'size', max_streak])
...     .round(2)
... )
                    mean    size    max_streak
AIRLINE ORG_AIR
AA      ATL         0.82    233         15
        DEN         0.74    219         17
        DFW         0.78    4006        64
        IAH         0.80    196         24
        LAS         0.79    374         29
...                 ...     ...         ...
WN      LAS         0.77    2031        39
        LAX         0.70    1135        23
        MSP         0.84    237         32
        PHX         0.77    1724        33
        SFO         0.76    445         17
```

9.13.2 原理解释

在 Pandas 中查找数据中的连续记录并不是一项简单的操作，对此需要向前或向后看的方法（如.diff 或.shift），还需要记住当前状态的方法（如.cumsum）。前 7 个步骤的最终结果是与原始 Series 长度相同的 Series，该 Series 将跟踪所有连续的序列。在这些步骤中，我们使用.mul 和.add 方法分别代替等效的运算符*和+。这样可以使从左到右的计算过程更加简洁。当然，也可以将它们替换为实际的运算符。

理想情况下，我们希望告诉 Pandas 在每个连续记录开始时都应用.cumsum 方法，并在每个连续记录结束后重置自身。要将此信息传达给 Pandas，需要采取许多步骤。

步骤（2）将整个 Series 中所有的 1 累计起来。其余步骤将慢慢清除所有多余的累计。为了识别这种多余的累计，我们需要找到每个连续记录的结尾，并从下一个连续记录的开头减去该值。

要找到每个连续记录的结尾，可以通过在步骤（3）中将累计和（cumulative sum）乘以原始的 0 和 1 的 Series 来巧妙地使所有值都不是连续 0 的一部分，第一个 0 之后是非 0，标记连续记录的结尾。这种处理方式很好，但是同样，我们还需要消除多余的累计。知道连续记录结束的地方并不能使我们达到目的。

在步骤（4）中，我们使用了.diff 方法来查找此多余累计。.diff 方法可以获取当前值与位于距离其一定行数的任何值之间的差。默认情况下，返回当前值与前一个值之间的差。

在步骤（4）中，只有负值才有意义。那些是连续结束后的 1。这些值需要向下传播，直到后续的连续记录结束。为了消除所有我们不关心的值（将它们变成缺失值），可以使用.where 方法（与 NumPy where 函数不同），该方法采用与调用 Series 大小相同的布尔数组。默认情况下，所有 True 值保持不变，而 False 值则称为缺失值。.where 方法允许采用函数作为其第一个参数，将调用 Series 用作条件的一部分。在本示例中使用的是一个匿名函数，该函数获得隐式传递的 Series，并检查每个值是否小于 0。步骤（5）的结果是一个 Series，其中仅保留负值，其余的值则被改为缺失值。

步骤（6）中的.ffill 方法将缺失值替换为沿着 Series 下降的最后一个非缺失值。由于前 3 个值不跟随非缺失值，因此它们仍然为缺失值。最终我们将获得一个消除多余累计的 Series。我们将累计 Series 添加到步骤（6）的结果中，以使连续记录全部从 0 开始。

.add 方法允许我们用 fill_value 参数替换缺失值。这样就完成了在数据集中查找连续记录的过程。

当执行这样的复杂逻辑时，最好使用一个较小的数据集，这样你就可以清晰地知道最终的输出结果是什么。本示例的前 7 个步骤就是这样做的。

从步骤（8）开始，在分组时构建这种连续记录逻辑将是一件非常困难的任务。

在步骤（8）中，我们创建了 ON_TIME 列。值得注意的一项是，被取消的航班将在 ARR_DELAY 列中包含缺失值，这样它就不会通过布尔条件，因此 ON_TIME 列的值为 0。也就是说，被取消的航班也被视为延迟航班。

步骤（9）将前 7 个步骤转变为一个函数 max_streak，并链接了 .max 方法以返回最长的准点航班连续记录。由于 max_streak 函数将返回单个值，因此该函数实际上是一个聚合函数，在步骤（10）中，该函数将被传递给 .agg 方法。为确保查看连续航班，我们使用了 .sort_values 方法按日期和计划出发时间进行排序。

9.13.3 扩展知识

既然已经找到了准点到达的最长连续记录，那么就可以很容易地找到相反的特征，即延迟到达的最长连续记录。下列函数可为传递给它的每个组返回两行。第一行是连续记录的起点，最后一行是连续记录的终点。每一行包含连续记录开始和结束的月份和日期，以及连续记录的总长度。

```
>>> def max_delay_streak(df):
...     df = df.reset_index(drop=True)
...     late = 1 - df['ON_TIME']
...     late_sum = late.cumsum()
...     streak = (late
...         .mul(late_sum)
...         .diff()
...         .where(lambda x: x < 0)
...         .ffill()
...         .add(late_sum, fill_value=0)
...     )
...     last_idx = streak.idxmax()
...     first_idx = last_idx - streak.max() + 1
...     res = (df
...         .loc[[first_idx, last_idx], ['MONTH', 'DAY']]
...         .assign(streak=streak.max())
...     )
...     res.index = ['first', 'last']
...     return res

>>> (flights
...     .assign(ON_TIME=flights['ARR_DELAY'].lt(15).astype(int))
...     .sort_values(['MONTH', 'DAY', 'SCHED_DEP'])
...     .groupby(['AIRLINE', 'ORG_AIR'])
...     .apply(max_delay_streak)
...     .sort_values('streak', ascending=False)
... )
```

			MONTH	DAY	streak
AIRLINE	ORG_AIR				
AA	DFW	first	2.0	26.0	38.0
		last	3.0	1.0	38.0
MQ	ORD	last	1.0	12.0	28.0
		first	1.0	6.0	28.0
	DFW	last	2.0	26.0	25.0
...			...	...	...
US	LAS	last	1.0	7.0	1.0
AS	ATL	first	5.0	4.0	1.0
OO	LAS	first	2.0	8.0	1.0
EV	PHX	last	8.0	1.0	0.0
		first	NaN	NaN	0.0

当我们使用.apply groupby 方法时，每个组的 DataFrame 都将被传递给 max_delay_streak 函数。在此函数内部，将删除 DataFrame 的索引并用 RangeIndex 代替，以便能够轻松找到连续记录的第一行和最后一行。ON_TIME 列将被反转，然后使用相同的逻辑来查找被延迟航班的连续记录。该连续记录的第一行和最后一行的索引将被存储为变量。然后这些索引用于选择连续记录结束时的月份和日期。我们使用 DataFrame 返回结果。最后，标记并命名索引以使最终结果更清晰。

最终结果将显示延迟航班的最长连续记录以及第一个和最后一个日期。

最后，我们还可以进行一些调查研究，看看是否可以找出导致这些延迟的原因。天气恶劣是航班延误或取消的常见原因。先来看第一行，美国航空（AA）从达拉斯沃思堡（DFW）机场开始连续 38 班延误航班，其记录从 2015 年 2 月 26 日—3 月 1 日。查看 2015 年 2 月 27 日的历史气象数据，降雪量为 2in（这是当天的记录）。可见正是由于 DFW 的这个重大天气事件，给整座城市的交通出行带来了障碍。值得一提的是，第三个最长的连续记录也和 DFW 有关，但这是降雪前几天的事，并且是针对另一家航空公司的。

第 10 章 将数据重组为规整形式

本章包含以下秘笈。
- 使用 stack 将变量值规整为列名称。
- 使用 melt 将变量值规整为列名称。
- 同时堆叠多组变量。
- 反转已堆叠的数据。
- 在 groupby 聚合之后取消堆叠。
- 使用 groupby 聚合复制 .pivot_table 方法的功能。
- 重命名轴的级别以方便数据的重塑。
- 对多个变量存储为列名称的情况进行规整。
- 对多个变量存储为单个列的情况进行规整。
- 对多个值存储在同一单元格中的情况进行规整。
- 对变量存储在列名称和值中的情况进行规整。

10.1 介 绍

前面章节中使用的所有数据集都没有做太多整理，也没有做任何工作来更改其结构。我们都是在将数据集导入之后立即开始以原始形状对其进行处理，其实在对数据集开始进行更详细的分析之前，许多从外部获得的数据集都需要进行大量的重组整理。在某些情况下，整个项目可能只关心格式化数据，以便他人可以轻松处理。

有许多术语可用于描述数据重组的过程和结果，数据科学家最常用的术语是规整数据（tidy data）。规整数据是 Hadley Wickham 创造的一个术语，用于描述一种使分析变得容易进行的数据形式。本章将讨论 Hadley 提出的许多想法以及如何用 Pandas 对其进行实现。要了解有关规整数据的更多信息，请阅读 Hadley 的论文。其网址如下。

http://vita.had.co.nz/papers/tidy-data.pdf

表 10-1 是不规整数据的示例。

表 10-1 不规整数据

Name	Category	Value
Jill	Bank	2300
Jill	Color	Red
John	Bank	1100
Jill	Age	40
John	Color	Purple

表 10-2 是规整数据示例。

表 10-2 规整数据

Name	Age	Bank	Color
Jill	40	2300	Red
John	38		Purple

什么是规整数据？Hadley 提出了确定数据集是否规整的 3 个指导原则。
- 每个变量组成一列。
- 每个观察结果组成一行。
- 每种观测单位组成一个表格。

任何不符合这些准则的数据集都被认为是混乱的。一旦开始将数据重组为规整格式，此定义就会变得更有意义，但就目前而言，我们需要知道什么是变量（variable）、观测值（observation）和观测单位（observational unit）。

在使用"变量"这个术语时需要注意，这里的"变量"不是指 Python 变量，而是指一个数据片段。对此比较好的理解方式就是考虑变量名和变量值之间的区别。变量名称是标签，如性别、种族、薪水和职位；变量值是那些每次观察结果都可能发生改变的事物，如男性、女性或白人、黑人等。

单个观测值是单个观测单位的所有变量值的集合。

为了更好地理解观测单位可能是什么，这里不妨以零售商店为例。该商店包含了有关每笔交易、员工、客户、商品和商店本身的数据。这些数据中的每个都可以被视为一个观察单位，并且需要自己的表格。将员工信息（如已工作的时间）与客户信息（如花费的金额）组合在同一张表中，将破坏上述规整原则。

解决混乱数据（messy data）的第一步是对其进行识别，混乱的存在有无限可能性。Hadley 明确提到了 5 种最常见的混乱数据类型。
- 列名称是值，而不是变量名。

- ❑ 多个变量被存储在列名称中。
- ❑ 变量同时被存储在行和列中。
- ❑ 多种类型的观测单位被存储在同一表格中。
- ❑ 一个观察单位被存储在多张表中。

重要的是要理解，规整数据通常不涉及更改数据集的值、填充缺失值或进行任何类型的分析。规整数据操作包括更改数据的形状或结构以符合上述规整原则。规整数据类似于将所有工具都整理好放在工具箱中，而不是到处乱扔，散布在整个房屋中。

在日常生活中，将工具整理好放置在工具箱中，用的时候就可以顺手抬来而不是像无头苍蝇一样到处寻找。显然，正确放置工具可以帮助我们轻松完成所有其他任务。同样的道理，数据格式经过正确规整后，再做进一步分析就变得容易得多。

发现混乱的数据后，即可使用 Pandas 库重新规整数据。Pandas 提供的主要规整工具是 DataFrame 方法.stack、.melt、.unstack 和.pivot。更复杂的规整工作还包括拆分文本，这需要.str 访问器。其他辅助方法，如.rename、.rename_axis、.reset_index 和.set_index 将有助于对规整数据进行最终处理。

10.2 使用 stack 将变量值规整为列名称

为了帮助理解规整数据和混乱数据之间的差异，不妨来看一个表格示例，此表格可能是也可能不是规整表格。

```
>>> import pandas as pd
>>> import numpy as np
>>> state_fruit = pd.read_csv('data/state_fruit.csv', index_col=0)
>>> state_fruit
         Apple  Orange  Banana
Texas       12      10      40
Arizona      9       7      12
Florida      0      14     190
```

上述表似乎没有什么混乱之处，并且该信息很容易使用。但是，按照上述规整原则来评判的话，该表并不是规整表格。该表的每个列名称都是变量的值。实际上，在该 DataFrame 中甚至都没有变量名。将混乱的数据集转换为规整数据的第一步就是识别所有变量。在上面的特定数据集中，其实包含了两个变量，一个是 state（州），另一个是 fruit（水果）。现在的问题是，对于这些数字数据，不知道它们的上下文含义。可以考虑将此变量标记为 weight（重量）或其他有意义的名称。

这个特定的混乱数据集包含了变量值作为列名称，因此需要将这些列名称转换为列值。在本秘笈中，可使用 stack 方法将该 DataFrame 重组为规整形式。

10.2.1 实战操作

（1）首先请注意，州的名称位于 DataFrame 的索引中。这些州名称已经被正确地垂直放置，所以不需要重组。问题是列名。.stack 方法可采用所有列名称，并将这些列名称旋转到索引中。一般来说，当调用 .stack 方法时，数据会变得更高。

（2）可以看到，在这种情况下，结果将从 DataFrame 折叠为一个 Series。

```
>>> state_fruit.stack()
Texas    Apple    12
         Orange   10
         Banana   40
Arizona  Apple    9
         Orange   7
         Banana   12
Florida  Apple    0
         Orange   14
         Banana   190
dtype: int64
```

（3）请注意，现在我们有了一个带有多重索引的 Series。也就是说，现在索引中有两个级别。原始索引已被推到左侧，以便为水果列名称腾出空间。使用这一个命令之后，我们就基本上有了规整数据。每个变量（包括州、水果和重量）都是垂直的。可以使用 .reset_index 方法将该结果转换为一个 DataFrame。

```
>>> (state_fruit
...     .stack()
...     .reset_index()
... )
    level_0  level_1   0
0   Texas    Apple     12
1   Texas    Orange    10
2   Texas    Banana    40
3   Arizona  Apple     9
4   Arizona  Orange    7
5   Arizona  Banana    12
6   Florida  Apple     0
7   Florida  Orange    14
8   Florida  Banana    190
```

（4）现在的结构是正确的，但是列名称没有体现出应有的意义。因此可以使用适当的标识符替换这些列名称。

```
>>> (state_fruit
...     .stack()
...     .reset_index()
...     .rename(columns={'level_0':'state',
...          'level_1': 'fruit', 0: 'weight'})
... )
     state    fruit  weight
0    Texas    Apple      12
1    Texas   Orange      10
2    Texas   Banana      40
3  Arizona    Apple       9
4  Arizona   Orange       7
5  Arizona   Banana      12
6  Florida    Apple       0
7  Florida   Orange      14
8  Florida   Banana     190
```

（5）除了使用 .rename 方法，还可以使用鲜为人知的 Series 方法 .rename_axis 设置索引级别的名称，然后使用 .reset_index 方法。

```
>>> (state_fruit
...     .stack()
...     .rename_axis(['state', 'fruit'])
... )
state    fruit
Texas    Apple     12
         Orange    10
         Banana    40
Arizona  Apple      9
         Orange     7
         Banana    12
Florida  Apple      0
         Orange    14
         Banana   190
dtype: int64
```

（6）在步骤（5）的基础上，可以将 .reset_index 方法与 name 参数链接起来，以重现步骤（4）的输出结果。

```
>>> (state_fruit
```

```
...         .stack()
...         .rename_axis(['state', 'fruit'])
...         .reset_index(name='weight')
... )
     state     fruit  weight
0    Texas     Apple      12
1    Texas    Orange      10
2    Texas    Banana      40
3  Arizona     Apple       9
4  Arizona    Orange       7
5  Arizona    Banana      12
6  Florida     Apple       0
7  Florida    Orange      14
8  Florida    Banana     190
```

10.2.2 原理解释

.stack 方法的功能非常强大，需要花费一些时间才能完全理解和欣赏。默认情况下，.stack 方法采用（层次结构列中的最内层）列名并对其进行转置，因此这些列名成为新的最内层索引。请注意，每个旧列名称仍标记其原始值，并且与每个州配对。在 3×3 的 DataFrame 中有 9 个原始值，这些值被转换到具有相同数量值的单个 Series 中。相应地，原始的第一行数据成为结果 Series 中的前 3 个值，以此类推。

在步骤（2）中重置索引后，Pandas 将 DataFrame 列默认设置为 level_0、level_1 和 0（两个字符串和一个整数）。这是因为调用此方法的 Series 具有两个未正式命名的索引级别。Pandas 还通过从外部 0 开始的整数来引用索引。

步骤（3）显示了使用.rename 方法重命名列的直观方法。

或者，也可以通过链接 .rename_axis 方法来设置列名称。.rename_axis 方法可以使用值列表作为索引级别名称。

重置索引时，Pandas 可使用这些索引级别名称作为新的列名称。此外，.reset_index 方法还具有一个 name 参数，该 name 参数对应于 Series 值的新列名称。

所有 Series 都有一个 name 属性，可以使用 .rename 方法对该属性进行分配或更改。当使用 .reset_index 方法时，name 属性将成为列名。

10.2.3 扩展知识

使用 .stack 方法的关键之一是将所有你不希望转换的列都放在索引中。最初使用索引中的州读取此秘笈中的数据集。下面查看如果不将州读取为索引，将会发生什么。

```
>>> state_fruit2 = pd.read_csv('data/state_fruit2.csv')
>>> state_fruit2
     State  Apple  Orange  Banana
0    Texas     12      10      40
1   Arizona     9       7      12
2   Florida     0      14     190
```

由于州名称不在索引中，因此在此 DataFrame 上使用 .stack 方法会将所有值整形为一个很长的值的 Series。

```
>>> state_fruit2.stack()
0  State      Texas
   Apple         12
   Orange        10
   Banana        40
1  State      Arizona
               ...
   Banana        12
2  State      Florida
   Apple          0
   Orange        14
   Banana       190
Length: 12, dtype: object
```

上述命令将重塑所有列，这次包括 state（州），而这根本不是我们需要的。因此，为了正确地重塑数据，需要首先使用 .set_index 方法将所有不需要重塑的列放入索引中，然后使用 .stack 方法。下面的代码给出与步骤（1）类似的结果。

```
>>> state_fruit2.set_index('State').stack()
State
Texas    Apple      12
         Orange     10
         Banana     40
Arizona  Apple       9
         Orange      7
         Banana     12
Florida  Apple       0
         Orange     14
         Banana    190
dtype: int64
```

10.3 使用 melt 将变量值规整为列名称

像大多数大型 Python 库一样，Pandas 具有许多完成同一任务的不同方法，区别通常是代码的可读性和性能。DataFrame 具有一个名为 .melt 的方法，该方法与前面的秘笈中介绍的 .stack 方法类似，但具有更大的灵活性。stack 的英文原意为堆叠，melt 的意思是融化，从它们的英文意思区别中也可以稍稍窥见这两种方法的不同。

在本秘笈中，我们将使用 .melt 方法规整一个以变量值作为列名称的 DataFrame。

10.3.1 实战操作

（1）读取 state_fruit2.csv 数据集。

```
>>> state_fruit2 = pd.read_csv('data/state_fruit2.csv')
>>> state_fruit2
     State  Apple  Orange  Banana
0    Texas     12      10      40
1  Arizona      9       7      12
2  Florida      0      14     190
```

（2）使用 .melt 方法，注意将适当的列传递给 id_vars 和 value_vars 参数。

```
>>> state_fruit2.melt(id_vars=['State'],
...      value_vars=['Apple', 'Orange', 'Banana'])
     State  variable  value
0    Texas     Apple     12
1  Arizona     Apple      9
2  Florida     Apple      0
3    Texas    Orange     10
4  Arizona    Orange      7
5  Florida    Orange     14
6    Texas    Banana     40
7  Arizona    Banana     12
8  Florida    Banana    190
```

（3）步骤（2）创建了规整数据。默认情况下，.melt 方法将转换后的列名称称为 variable，并将相应的值称为 value。为方便起见，.melt 方法还有两个附加参数，即 var_name 和 value_name，使用这两个参数可以重命名这两列。

```
>>> state_fruit2.melt( id_vars=['State'],
```

```
...                   value_vars=['Apple', 'Orange', 'Banana'],
...                   var_name='Fruit',
...                   value_name='Weight')
     State    Fruit  Weight
0    Texas    Apple      12
1   Arizona   Apple       9
2   Florida   Apple       0
3    Texas   Orange      10
4   Arizona  Orange       7
5   Florida  Orange      14
6    Texas   Banana      40
7   Arizona  Banana      12
8   Florida  Banana     190
```

10.3.2 原理解释

.melt 方法可以重塑 DataFrame。该方法最多包含 5 个参数，其中，以下两个参数对于了解如何正确重塑数据至关重要。

- id_vars：这是要保留为列而不要整形的列名列表。
- value_vars：这是要重塑为单列的列名称的列表。

id_vars 或标识变量将保留在同一列中，但对传递给 value_vars 的每一列重复此操作。.melt 方法的一个关键方面是它会忽略索引中的值，并且会默默地删除索引，将其替换为默认的 RangeIndex。这意味着，如果你确实要保留索引中的值，则需要在使用.melt 方法之前先重置索引。

10.3.3 扩展知识

.melt 方法的所有参数都是可选的，并且如果你希望所有值都位于单个列中，而这些值的旧列标签位于另一个列中，则可以使用默认参数调用.melt。

```
>>>state_fruit2.melt()
   variable    value
0    State     Texas
1    State    Arizona
2    State    Florida
3    Apple       12
4    Apple        9
..     ...      ...
7   Orange        7
```

```
8      Orange      14
9      Banana      40
10     Banana      12
11     Banana     190
```

实际上，你可能有很多需要使用 .melt 方法处理的变量，并且只想指定标识变量。在这种情况下，可以按以下方式调用 .melt 方法，这将产生与步骤（2）相同的结果。在融化单个列时，甚至不需要列表，只要传递其字符串值即可。

```
>>> state_fruit2.melt(id_vars='State')
     State   variable  value
0    Texas      Apple     12
1  Arizona      Apple      9
2  Florida      Apple      0
3    Texas     Orange     10
4  Arizona     Orange      7
5  Florida     Orange     14
6    Texas     Banana     40
7  Arizona     Banana     12
8  Florida     Banana    190
```

10.4 同时堆叠多组变量

一些数据集包含多组变量作为列名称，这些变量需要同时被堆叠到它们自己的列中。我们可以通过一个 movie 数据集的示例来帮助你更清晰地理解这一点。首先，选择包含演员姓名及其对应的 Facebook 平台点赞/喜欢数的所有列。

```
>>> movie = pd.read_csv('data/movie.csv')
>>> actor = movie[[ 'movie_title', 'actor_1_name',
...                 'actor_2_name', 'actor_3_name',
...                 'actor_1_facebook_likes',
...                 'actor_2_facebook_likes',
...                 'actor_3_facebook_likes']]
>>> actor.head()
                                    movie_title ...
0                                        Avatar ...
1    Pirates of the Caribbean: At World's End ...
2                                       Spectre ...
3                         The Dark Knight Rises ...
4     Star Wars: Episode VII - The Force Awakens ...
```

如果将变量分别定义为电影的片名、演员姓名和 Facebook 平台上的点赞/喜欢数，那么我们将需要堆叠两个列的集合，仅调用一次 .stack 或 .melt 方法就完成任务是不可能的。

在本秘笈中，我们将使用 wide_to_long 函数同时将演员姓名及其对应的 Facebook 平台上的点赞/喜欢数堆叠在一起，以规整演员的 DataFrame。

10.4.1 实战操作

（1）如前文所述，我们将使用 wide_to_long 函数将数据重整为规整形式。

当使用 wide_to_long 函数时，需要更改要堆叠的列名称，以使这些列名称以数字结尾。我们首先创建一个用户定义的函数 change_col_name 来更改列名称。

```
>>> def change_col_name(col_name):
...     col_name = col_name.replace('_name', '')
...     if 'facebook' in col_name:
...         fb_idx = col_name.find('facebook')
...         col_name = (col_name[:5] + col_name[fb_idx - 1:]
...                     + col_name[5:fb_idx-1])
...     return col_name
```

（2）将 change_col_name 函数传递给 .rename 方法以转换所有列名称。

```
>>> actor2 = actor.rename(columns=change_col_name)
>>> actor2
         movie_title       actor_1  ...  actor_facebook_likes_2
0             Avatar   CCH Pounder  ...                   936.0
1           Pirates...  Johnny Depp  ...                  5000.0
2            Spectre   Christop...  ...                   393.0
3         The Dark...  Tom Hardy    ...                 23000.0
4         Star War...  Doug Walker  ...                    12.0
...              ...           ...  ...                     ...
4911      Signed S...  Eric Mabius  ...                   470.0
4912      The Foll...  Natalie Zea  ...                   593.0
4913      A Plague...  Eva Boehnke ...                     0.0
4914      Shanghai...  Alan Ruck    ...                   719.0
4915      My Date ...  John August ...                    23.0
```

（3）使用 wide_to_long 函数同时堆叠 actor 和 actor_facebook_likes 列的集合。

```
>>> stubs = ['actor', 'actor_facebook_likes']
>>> actor2_tidy = pd.wide_to_long(actor2,
...     stubnames=stubs,
...     i=['movie_title'],
```

```
...         j='actor_num',
...         sep='_')
>>> actor2_tidy.head()
                            actor   actor_facebook_likes
movie_title    actor_num
Avatar         1          CCH Pounder           1000.0
Pirates o...   1          Johnny Depp          40000.0
Spectre        1          Christop...          11000.0
The Dark ...   1           Tom Hardy           27000.0
Star Wars...   1          Doug Walker            131.0
```

10.4.2 原理解释

wide_to_long 函数以相当特殊的方式工作。该函数的主要参数是 stubnames，这是一个字符串列表。每个字符串代表一个列分组。以该字符串开头的所有列都将被堆叠到一个列中。在此秘笈中，有两组列，即 actor 和 actor_facebook_likes。默认情况下，这些列的每个组都需要以数字结尾。此数字随后将用于标记整形数据。这些列组中的每一个都有一个下画线字符，用于将 stubnames 与结尾数字分开。要解决此问题，必须使用 sep 参数。

原始列名称与 wide_to_long 函数正常工作所需的模式不匹配。列名称可以通过使用列表指定其值来手动更改。这种方式有一个缺陷，那就是需要输入大量的字符，因此可以定义一个函数，该函数自动将列转换为有效的格式。change_col_name 函数可以从 actor 列中删除 *_name* 并重新排列 Facebook 列，以使它们都以数字结尾。

为了完成列的重命名，在步骤（2）中使用了 .rename 方法。接收许多不同类型的参数，其中之一是 change_col_name 函数。将 change_col_name 函数传递给 .rename 方法时，每个列名称都会被隐式地传递给 change_col_name 函数进行处理。

现在，我们已经正确创建了两组列，即分别以 actor 和 actor_facebook_likes 开头的列。wide_to_long 函数需要唯一列参数 i 充当不会堆叠的标识变量。除此之外，wide_to_long 函数还需要参数 j，该参数将重命名从原始列名的末尾去除的标识数字。默认情况下，suffix 参数包含正则表达式 r'\d+'，该正则表达式搜索一位或多位数字。其中，\d 是与数字 0～9 匹配的特殊令牌，+使表达式与这些数字中的一个或多个匹配。

10.4.3 扩展知识

当所有变量分组都具有相同的数字结尾时（本秘笈就是这样做的），wide_to_long 函数可以正常工作。当变量没有相同的结尾或不是以数字结尾时，仍然可以使用 wide_to_long 函数同时堆叠列。例如，来看以下数据集。

```
>>> df = pd.read_csv('data/stackme.csv')
>>> df
  State Country    a1   b2   Test  d  e
0    TX      US  0.45  0.3  Test1  2  6
1    MA      US  0.03  1.2  Test2  9  7
2    ON     CAN  0.70  4.2  Test3  4  2
```

假设我们希望将列 a1 和 b1 以及列 d 和 e 堆叠在一起。另外，我们还想使用 a1 和 b1 作为行的标签。要完成此任务，需要重命名列，以使该列以所需的标签结尾。

```
>>> df.rename(columns = {'a1':'group1_a1', 'b2':'group1_b2',
...                       'd':'group2_a1', 'e':'group2_b2'})
  State Country ...  group2_a1  group2_b2
0    TX      US ...          2          6
1    MA      US ...          9          7
2    ON     CAN ...          4          2
```

然后，我们需要修改 suffix 参数，该参数通常默认为选择数字的正则表达式。在这里，我们通知该参数找到任意数量的字符。

```
>>> pd.wide_to_long(
...     df.rename(columns = { 'a1':'group1_a1',
...                           'b2':'group1_b2',
...                            'd':'group2_a1', 'e':'group2_b2'}),
...     stubnames=['group1', 'group2'],
...     i=['State', 'Country', 'Test'],
...     j='Label',
...     suffix='.+',
...     sep='_')
                              group1  group2
State Country Test  Label
TX    US      Test1 a1         0.45       2
                    b2         0.30       6
MA    US      Test2 a1         0.03       9
                    b2         1.20       7
ON    CAN     Test3 a1         0.70       4
                    b2         4.20       2
```

10.5 反转已堆叠的数据

如前文所述，DataFrame 具有两个类似的方法（即 .stack 和 .melt），用于将水平列名称转换为垂直列值。另外，DataFrame 可以分别使用 .unstack 和 .pivot 方法反转这两个操作。.stack 和 .unstack 方法仅允许控制列和行索引，而 .melt 和 .pivot 方法则提供了更大的灵活性来选择要重塑的列。

在本秘笈中，我们将在数据集上调用 .stack 和 .melt 方法，然后立即使用 .unstack 和 .pivot 方法反转操作。

10.5.1 实战操作

（1）读取 college 数据集，以机构名称（INSTNM 列）作为索引，并且仅包含大学生种族这些列（即以 UGDS_ 开头的列）。

```
>>> def usecol_func(name):
...     return 'UGDS_' in name or name == 'INSTNM'
>>> college = pd.read_csv('data/college.csv',
...         index_col='INSTNM',
...         usecols=usecol_func)
>>> college
                UGDS_WHITE  UGDS_BLACK  ...  UGDS_NRA  UGDS_UNKN
INSTNM                                  ...
Alabama A...    0.0333      0.9353      ...  0.0059    0.0138
Universit...    0.5922      0.2600      ...  0.0179    0.0100
Amridge U...    0.2990      0.4192      ...  0.0000    0.2715
Universit...    0.6988      0.1255      ...  0.0332    0.0350
Alabama S...    0.0158      0.9208      ...  0.0243    0.0137
...             ...         ...         ...  ...       ...
SAE Insti...    NaN         NaN         ...  NaN       NaN
Rasmussen...    NaN         NaN         ...  NaN       NaN
National ...    NaN         NaN         ...  NaN       NaN
Bay Area ...    NaN         NaN         ...  NaN       NaN
Excel Lea...    NaN         NaN         ...  NaN       NaN
```

（2）使用 .stack 方法将每个水平列名称转换为垂直索引级别。

```
>>> college_stacked = college.stack()
>>> college_stacked
INSTNM
Alabama A & M University    UGDS_WHITE  0.0333
                            UGDS_BLACK  0.9353
                            UGDS_HISP   0.0055
                            UGDS_ASIAN  0.0019
                            UGDS_AIAN   0.002
                                        ...
```

```
Coastal Pines Technical College  UGDS_AIAN    0.0034
                                 UGDS_NHPI    0.0017
                                 UGDS_2MOR    0.0191
                                 UGDS_NRA     0.0028
                                 UGDS_UNKN    0.0056
Length: 61866, dtype: float64
```

（3）使用 .unstack 方法将此堆叠数据反转回其原始形式。

```
>>> college_stacked.unstack()
              UGDS_WHITE   UGDS_BLACK   ...   UGDS_NRA   UGDS_UNKN
INSTNM                                  ...
Alabama A...    0.0333      0.9353      ...    0.0059     0.0138
Universit...    0.5922      0.2600      ...    0.0179     0.0100
Amridge U...    0.2990      0.4192      ...    0.0000     0.2715
Universit...    0.6988      0.1255      ...    0.0332     0.0350
Alabama S...    0.0158      0.9208      ...    0.0243     0.0137
...               ...         ...       ...      ...        ...
Hollywood...    0.2182      0.4182      ...    0.0182     0.0909
Hollywood...    0.1200      0.3333      ...    0.0000     0.0667
Coachella...    0.3284      0.1045      ...    0.0000     0.0000
Dewey Uni...    0.0000      0.0000      ...    0.0000     0.0000
Coastal P...    0.6762      0.2508      ...    0.0028     0.0056
```

（4）使用 .melt 和 .pivot 方法可以完成类似的操作。
首先，读取数据但是不将机构名称列放在索引中。

```
>>> college2 = pd.read_csv('data/college.csv',
...             usecols=usecol_func)
>>> college2
         INSTNM       UGDS_WHITE   ...   UGDS_NRA   UGDS_UNKN
0        Alabama ...    0.0333     ...    0.0059     0.0138
1        Universi...    0.5922     ...    0.0179     0.0100
2        Amridge ...    0.2990     ...    0.0000     0.2715
3        Universi...    0.6988     ...    0.0332     0.0350
4        Alabama ...    0.0158     ...    0.0243     0.0137
...        ...            ...      ...      ...        ...
7530     SAE Inst...     NaN       ...     NaN        NaN
7531     Rasmusse...     NaN       ...     NaN        NaN
7532     National...     NaN       ...     NaN        NaN
7533     Bay Area...     NaN       ...     NaN        NaN
7534     Excel Le...     NaN       ...     NaN        NaN
```

（5）使用 .melt 方法将所有种族列转置为单列。

```
>>> college_melted = college2.melt(id_vars='INSTNM',
...       var_name='Race',
...       value_name='Percentage')
>>> college_melted
            INSTNM        Race    Percentage
0          Alabama ...  UGDS_WHITE    0.0333
1          Universi... UGDS_WHITE    0.5922
2          Amridge ... UGDS_WHITE    0.2990
3          Universi... UGDS_WHITE    0.6988
4          Alabama ... UGDS_WHITE    0.0158
...            ...         ...           ...
67810      SAE Inst... UGDS_UNKN       NaN
67811      Rasmusse... UGDS_UNKN       NaN
67812      National... UGDS_UNKN       NaN
67813      Bay Area... UGDS_UNKN       NaN
67814      Excel Le... UGDS_UNKN       NaN
```

（6）使用 .pivot 方法反转之前的结果。

```
>>> melted_inv = college_melted.pivot(index='INSTNM',
...       columns='Race',
...       values='Percentage')
>>> melted_inv
Race         UGDS_2MOR   UGDS_AIAN   ...   UGDS_UNKN   UGDS_WHITE
INSTNM                                ...
A & W Hea...    0.0000     0.0000    ...    0.0000       0.0000
A T Still...      NaN        NaN     ...      NaN          NaN
ABC Beaut...    0.0000     0.0000    ...    0.0000       0.0000
ABC Beaut...    0.0000     0.0000    ...    0.0000       0.2895
AI Miami ...    0.0018     0.0000    ...    0.4644       0.0324
...                ...        ...    ...       ...          ...
Yukon Bea...    0.0000     0.1200    ...    0.0000       0.8000
Z Hair Ac...    0.0211     0.0000    ...    0.0105       0.9368
Zane Stat...    0.0218     0.0029    ...    0.2399       0.6995
duCret Sc...    0.0976     0.0000    ...    0.0244       0.4634
eClips Sc...    0.0000     0.0000    ...    0.0000       0.1446
```

（7）可以看到，机构名称（INSTNM 列）现在已转移到索引中，而不是按其原始顺序排列。列名称不是按其原始顺序。要精确还原到步骤（4）中的起始 DataFrame，请使用 .loc 索引操作符同时选择行和列，然后重置索引。

```
>>> college2_replication = (melted_inv
...       .loc[college2['INSTNM'], college2.columns[1:]]
```

```
...        .reset_index()
... )
>>> college2.equals(college2_replication)
True
```

10.5.2 原理解释

在步骤（1）中，有多种方法可以完成相同的操作。在这里，我们展示了 read_csv 函数的多功能性。usecols 参数接收我们要导入的列的列表，也可以使用一个函数来动态确定它们。我们使用了一个函数来检查列名称是否包含 UGDS_或等于 INSTNM。该函数将每个列名称作为字符串传递，并且必须返回一个布尔值。通过这种方式可以节省大量的内存。

步骤（2）中的 .stack 方法将所有列名称放入最里面的索引级别中，并返回一个 Series。

在步骤（3）中，.unstack 方法通过获取最里面的索引级别中的所有值并将它们转换为列名称来反转此操作。请注意，步骤（1）和步骤（3）中的结果有所不同，因为默认情况下 .stack 方法会删除缺失值。如果传递 dropna = False 参数，那么 .stack 方法将正确返回完全一样的结果。

步骤（4）读取了与步骤（1）相同的数据集，但没有将机构名称放入索引中，因为 .melt 方法无法访问它。

步骤（5）使用 .melt 方法转置了所有 Race（种族）列。.melt 方法通过将 value_vars 参数保留为其默认值 None 来实现此目的。另外，当未指定时，id_vars 参数中不存在的所有列都将得以转置。本示例仅指定了 id_vars='INSTNM'。

步骤（6）使用了 .pivot 方法反转步骤（5）的操作，该方法接收 3 个参数。大多数参数采用单个列作为字符串（values 参数也可以接收列名称的列表）。index 参数引用的列保持垂直，并成为新索引。columns 参数引用的列的值成为列名称。由 values 参数引用的值成为平铺，以便与其之前的索引和列标签的交集相对应。

要使用 .pivot 方法进行复制，需要按照与原始顺序相同的顺序对行和列进行排序。由于机构名称在索引中，因此需要使用 .loc 索引操作符按原始索引对 DataFrame 进行排序。

10.5.3 扩展知识

为了帮助进一步理解 .stack 和 .unstack 方法，可以使用这两个方法来转置 college DataFrame。在这种情况下，我们将使用矩阵转置的精确数学定义，其中的新行就是原始数据矩阵的旧列。

如果查看步骤（2）的输出，则会注意到有两个索引级别。默认情况下，.unstack 方法将使用最里面的索引级别作为新的列值。索引级别从外部起算，从 0 开始编号。Pandas 将.unstack 方法的 level 参数默认为-1，这是指最里面的索引。当然，我们也可以指定 level=0 参数，这样 .unstack 方法就会使用最外面的列。

```
>>> college.stack().unstack(0)
INSTNM       Alaba/rsity   ...   Coast/llege
UGDS_WHITE        0.0333   ...        0.6762
UGDS_BLACK        0.9353   ...        0.2508
UGDS_HISP         0.0055   ...        0.0359
UGDS_ASIAN        0.0019   ...        0.0045
UGDS_AIAN         0.0024   ...        0.0034
UGDS_NHPI         0.0019   ...        0.0017
UGDS_2MOR         0.0000   ...        0.0191
UGDS_NRA          0.0059   ...        0.0028
UGDS_UNKN         0.0138   ...        0.0056
```

有一种不用 .stack 方法或 .unstack 方法就可以转置 DataFrame 的方式，那就是使用 .transpose 方法或 .T 属性，具体如下。

```
>>> college.T
>>> college.transpose()
INSTNM       Alaba/rsity   ...   Coast/llege
UGDS_WHITE        0.0333   ...        0.6762
UGDS_BLACK        0.9353   ...        0.2508
UGDS_HISP         0.0055   ...        0.0359
UGDS_ASIAN        0.0019   ...        0.0045
UGDS_AIAN         0.0024   ...        0.0034
UGDS_NHPI         0.0019   ...        0.0017
UGDS_2MOR         0.0000   ...        0.0191
UGDS_NRA          0.0059   ...        0.0028
UGDS_UNKN         0.0138   ...        0.0056
```

10.6 在 groupby 聚合之后取消堆叠

按单个列对数据进行分组并在单个列上执行聚合将返回易于使用的结果；而按多个列进行分组时，其聚合结果可能反而不易使用。由于默认情况下.groupby 操作会将唯一的分组列放在索引中，因此.unstack 方法可能有利于重新排列数据，这样就可以按更容易解释的方式显示数据。

在此秘笈中，我们将使用 employee（员工）数据集执行聚合，并按多列分组。然后使用 .unstack 方法重塑结果，以获得便于比较不同分组的格式。

10.6.1 实战操作

（1）读取 employee 数据集，并按种族（RACE 列）找到平均薪水。

```
>>> employee = pd.read_csv('data/employee.csv')
>>> (employee
...     .groupby('RACE')
...     ['BASE_SALARY']
...     .mean()
...     .astype(int)
... )
RACE
American Indian or Alaskan Native    60272
Asian/Pacific Islander               61660
Black or African American            50137
Hispanic/Latino                      52345
Others                               51278
White                                64419
Name: BASE_SALARY, dtype: int64
```

（2）这是一个 groupby 操作，可产生易于阅读且无须重塑的 Series。现在让我们按性别（gender）查找所有种族（race）的平均工资。注意结果仍然是一个 Series。

```
>>> (employee
...     .groupby(['RACE', 'GENDER'])
...     ['BASE_SALARY']
...     .mean()
...     .astype(int)
... )
RACE                               GENDER
American Indian or Alaskan Native  Female    60238
                                   Male      60305
Asian/Pacific Islander             Female    63226
                                   Male      61033
Black or African American          Female    48915
                                             ...
Hispanic/Latino                    Male      54782
Others                             Female    63785
                                   Male      38771
```

```
    White                          Female    66793
                                   Male      63940
Name: BASE_SALARY, Length: 12, dtype: int64
```

（3）这种聚合要复杂一些，可以进行调整以使不同的比较更加容易。例如，如果将男性和女性并列而不是像现在这样垂直，则比较男性和女性的工资会更容易。因此，可以在 GENDER（性别）索引级别上调用 .unstack。

```
>>> (employee
...     .groupby(['RACE', 'GENDER'])
...     ['BASE_SALARY']
...     .mean()
...     .astype(int)
...     .unstack('GENDER')
... )
GENDER                             Female    Male
RACE
American Indian or Alaskan Native  60238     60305
Asian/Pacific Islander             63226     61033
Black or African American          48915     51082
Hispanic/Latino                    46503     54782
Others                             63785     38771
White                              66793     63940
```

（4）同理，也可以取消 RACE（种族）索引级别的堆叠。

```
>>> (employee
...     .groupby(['RACE', 'GENDER'])
...     ['BASE_SALARY']
...     .mean()
...     .astype(int)
...     .unstack('RACE')
... )
RACE      American Indian or Alaskan Native   ...   White
GENDER                                        ...
Female                                60238   ...   66793
Male                                  60305   ...   63940
```

10.6.2 原理解释

步骤（1）使用了单个分组列（RACE）、单个聚合列（BASE_SALARY）和单个聚合函数（.mean）进行最简单的聚合。此结果易于使用和解释，因此不需要进行任何处理即可执行评估。

步骤（2）按种族和性别分组。所得的 Series 具有多重索引，在一个维中包含所有值，这使得比较更加困难。为了使信息更易于使用和解读，我们使用了 .unstack 方法将一个（或多个）级别中的值转换为列。

默认情况下，.unstack 方法使用最里面的索引级别作为新列。你也可以使用 level 参数指定要取消堆叠的级别，该参数接收字符串形式的级别名称，也接收使用整数位置指定的级别。最好不要使用整数位置形式的级别名称，以避免产生歧义。

步骤（3）和步骤（4）分别取消了每个级别的堆叠，这使得结果 DataFrame 仅具有单级索引，现在按性别比较每个种族的薪水要容易得多。

10.6.3 扩展知识

如果对来自 DataFrame 的单个列执行分组时有多个聚合函数，则直接结果将是一个 DataFrame 而不是 Series。例如，我们可以计算更多的聚合结果。其方法和步骤（2）一样，但是多了 max 和 min 聚合函数。

```
>>> (employee
...     .groupby(['RACE', 'GENDER'])
...     ['BASE_SALARY']
...     .agg(['mean', 'max', 'min'])
...     .astype(int)
... )
```

RACE	GENDER	mean	max	min
American Indian or Alaskan Native	Female	60238	98536	26125
	Male	60305	81239	26125
Asian/Pacific Islander	Female	63226	130416	26125
	Male	61033	163228	27914
Black or African American	Female	48915	150416	24960
...		...	...	...
Hispanic/Latino	Male	54782	165216	26104
Others	Female	63785	63785	63785
	Male	38771	38771	38771
White	Female	66793	178331	27955
	Male	63940	210588	26125

现在，取消 GENDER（性别）列的堆叠将导致具有多重索引的列。从这里开始，你可以继续使用 .unstack 和 .stack 方法交换行和列级别，直到获得所需的数据结构。

```
>>> (employee
...     .groupby(['RACE', 'GENDER'])
...     ['BASE_SALARY']
...     .agg(['mean', 'max', 'min'])
...     .astype(int)
...     .unstack('GENDER')
... )
                mean           ...    min
GENDER        Female    Male   ... Female    Male
RACE                           ...
American ...   60238   60305   ...  26125   26125
Asian/Pac...   63226   61033   ...  26125   27914
Black or ...   48915   51082   ...  24960   26125
Hispanic/...   46503   54782   ...  26125   26104
Others         63785   38771   ...  63785   38771
White          66793   63940   ...  27955   26125
```

10.7 使用 groupby 聚合复制 .pivot_table 方法的功能

乍一看，.pivot_table 方法似乎提供了一种独特的数据分析方法（Pandas 中的 pivot_table 和 Microsoft Excel 中的数据透视表功能类似），但是，在进行少量调整之后，其实使用 .groupby 方法也可以复制其功能。掌握这种等效操作有助于加深你对 Pandas 功能的理解。

在此秘笈中，我们将使用 flights 航班数据集创建一个数据透视表，然后使用 .groupby 方法重新创建它。

10.7.1 实战操作

（1）读取 flights 数据集，然后使用 .pivot_table 方法查找每家航空公司每个始发机场已取消航班的总数。

```
>>> flights = pd.read_csv('data/flights.csv')
>>> fpt = flights.pivot_table(index='AIRLINE',
...     columns='ORG_AIR',
...     values='CANCELLED',
...     aggfunc='sum',
...     fill_value=0)
>>> fpt
```

```
ORG_AIR   ATL  DEN  DFW  IAH  LAS  LAX  MSP  ORD  PHX  SFO
AIRLINE
AA          3    4   86    3    3   11    3   35    4    2
AS          0    0    0    0    0    0    0    0    0    0
B6          0    0    0    0    0    0    0    0    0    1
DL         28    1    0    0    1    1    4    0    1    2
EV         18    6   27   36    0    0    6   53    0    0
...       ...  ...  ...  ...  ...  ...  ...  ...  ...  ...
OO          3   25    2   10    0   15    4   41    9   33
UA          2    9    1   23    3    6    2   25    3   19
US          0    0    2    2    1    0    0    6    7    3
VX          0    0    0    0    0    3    0    0    0    3
WN          9   13    0    0    7   32    1    0    6   25
```

（2）要使用 .groupby 方法复制此操作，需要对两列进行分组，然后取消堆叠它们。groupby 聚合函数无法复制该表。本示例的诀窍是首先对 index 和 columns 参数中的所有列（即 AIRLINE 和 ORG_AIR 列）进行分组。

```
>>> (flights
...     .groupby(['AIRLINE', 'ORG_AIR'])
...     ['CANCELLED']
...     .sum()
... )
AIRLINE  ORG_AIR
AA       ATL         3
         DEN         4
         DFW        86
         IAH         3
         LAS         3
                    ..
WN       LAS         7
         LAX        32
         MSP         1
         PHX         6
         SFO        25
Name: CANCELLED, Length: 114, dtype: int64
```

（3）使用 .unstack 方法将 ORG_AIR 索引级别旋转到列名称中。

```
>>> fpg = (flights
...     .groupby(['AIRLINE', 'ORG_AIR'])
...     ['CANCELLED']
...     .sum()
```

```
...         .unstack('ORG_AIR', fill_value=0)
... )

>>> fpt.equals(fpg)
True
```

10.7.2 原理解释

.pivot_table 方法非常通用且很灵活,但是该方法执行的操作其实与 groupby 聚合很相似,步骤(1)就清楚显示了这一点。其中:index 参数采用一列(或列的列表),该列将不会被透视(旋转),并且其唯一值将被放置在索引中;columns 参数同样采用了一列(或列的列表),但是该列将被透视(旋转),并且其唯一值将作为列名称;values 参数采用的则是将要聚合的一列(或列的列表)。

还存在一个 aggfunc 参数,该参数采用一个聚合函数(或函数列表),该函数确定如何对 values 参数中的列进行聚合。aggfunc 参数默认为字符串 mean(平均值),在此示例中,我们将其更改为计算 sum(总和)。

此外,AIRLINE 和 ORG_AIR 列的某些唯一组合是不存在的,这些缺失的组合在结果 DataFrame 中将默认为缺失值。在这里,我们使用了 fill_value 参数将其更改为 0 值。

步骤(2)使用了 index 和 columns 参数中的所有列作为分组列开始复制过程。这是使此秘笈生效的关键。数据透视表其实就是分组列的所有唯一组合的交集。

步骤(3)通过使用 .unstack 方法将最里面的索引级别转换为列名称来完成复制操作。就像 .pivot_table 方法一样,AIRLINE 和 ORG_AIR 列的所有组合并非都是存在的,因此我们同样使用了 fill_value 参数将这些缺失的交集强制为 0 值。

10.7.3 扩展知识

还可以使用 .groupby 方法复制更复杂的数据透视表。
首先使用 .pivot_table 方法获取以下结果。

```
>>> flights.pivot_table(index=['AIRLINE', 'MONTH'],
...     columns=['ORG_AIR', 'CANCELLED'],
...     values=['DEP_DELAY', 'DIST'],
...     aggfunc=['sum', 'mean'],
...     fill_value=0)
                    sum       ...    mean
                  DEP_DELAY   ...    DIST
ORG_AIR             ATL       ...    SFO
```

```
                CANCELLED          0      1    ...         0         1
AIRLINE         MONTH                          ...
AA              1                -13      0    ...  1860.166667     0.0
                2                -39      0    ...  1337.916667  2586.0
                3                 -2      0    ...  1502.758621     0.0
                4                  1      0    ...  1646.903226     0.0
                5                 52      0    ...  1436.892857     0.0
...             ...              ...    ...    ...         ...       ...
WN              7               2604      0    ...   636.210526     0.0
                8               1718      0    ...   644.857143   392.0
                9               1033      0    ...   731.578947   354.5
                11               700      0    ...   580.875000   392.0
                12              1679      0    ...   782.256410     0.0
```

要使用 .groupby 方法复制此代码,可遵循秘笈中的相同模式,将 index 和 columns 参数中的所有列放入 .groupby 方法中,然后调用 .unstack 方法将索引级别拉出至各列中。

```
>>> (flights
...     .groupby(['AIRLINE', 'MONTH', 'ORG_AIR', 'CANCELLED'])
...     [['DEP_DELAY', 'DIST']]
...     .agg(['mean', 'sum'])
...     .unstack(['ORG_AIR', 'CANCELLED'], fill_value=0)
...     .swaplevel(0, 1, axis='columns')
... )
                             mean            ...      sum
                          DEP_DELAY          ...     DIST
ORG_AIR                      ATL             ...      SFO
CANCELLED                   0       1        ...     0         1
AIRLINE         MONTH                        ...
AA              1        -3.250000   NaN     ...   33483.0     NaN
                2        -3.000000   NaN     ...   32110.0  2586.0
                3        -0.166667   NaN     ...   43580.0     NaN
                4         0.071429   NaN     ...   51054.0     NaN
                5         5.777778   NaN     ...   40233.0     NaN
...             ...           ...    ...     ...       ...     ...
WN              7        21.700000   NaN     ...   24176.0     NaN
                8        16.207547   NaN     ...   18056.0   784.0
                9         8.680672   NaN     ...   27800.0   709.0
                11        5.932203   NaN     ...   23235.0   784.0
                12       15.691589   NaN     ...   30508.0     NaN
```

可以看到,列级别的顺序有所不同,.pivot_table 方法将聚合函数置于 values 参数的列之前的级别中。可以使用 .swaplevel 方法来解决此问题。.swaplevel 方法将最外面的列

（级别 0）交换为里面的列（级别 1）。

另外还值得一提的是，这两种方法获得的列的顺序也是不同的。

10.8　重命名轴的级别以方便数据的重塑

当每个轴（索引和列）级别都有名称时，使用 .stack 和 .unstack 方法进行数据的重塑要容易得多。如前文所述，Pandas 允许用户按整数位置或名称引用每个轴级别。由于整数位置是隐式的而不是显式的，因此应尽可能考虑使用级别名称。6.2 节"选择 Series 数据"中已经介绍过，The Zen of Python（Python 之禅）里面的第二句就是 Explicit is better than implicit（显式优于隐式）。如果你想知道全部的 Python 之禅（也就是 Python 代码编写指导原则），可在 Python 中新建文件，然后输入 import this，按 F5 键运行。

当对多列进行分组或聚合时，所得的 Pandas 对象将在一个或两个轴上具有多个级别。在此秘笈中，我们将命名每个轴的每个级别，然后分别使用 .stack 和 .unstack 方法将数据重塑为所需的形式。

10.8.1　实战操作

（1）读取 college 数据集，并按机构和宗教信仰找到一些关于本科生总体人数和 SAT 数学成绩的基本摘要统计数据。

```
>>> college = pd.read_csv('data/college.csv')
>>> (college
...     .groupby(['STABBR', 'RELAFFIL'])
...     [['UGDS', 'SATMTMID']]
...     .agg(['size', 'min', 'max'])
... )
                    UGDS                      SATMTMID
                    size    min      max      size    min      max
STABBR  RELAFFIL
AK      0           7       109.0    12865.0  7       NaN      NaN
        1           3       27.0     275.0    3       503.0    503.0
AL      0           72      12.0     29851.0  72      420.0    590.0
        1           24      13.0     3033.0   24      400.0    560.0
AR      0           68      18.0     21405.0  68      427.0    565.0
...     ...         ...     ...      ...      ...     ...      ...
WI      0           87      20.0     29302.0  87      480.0    680.0
        1           25      4.0      8212.0   25      452.0    605.0
```

```
WV              0           65   20.0   44924.0         65   430.0   530.0
                1            8   63.0    1375.0          8   455.0   510.0
WY              0           11   52.0    9910.0         11   540.0   540.0
```

（2）可以看到：一方面，步骤（1）中的两个索引级别都有名称，并且都是旧的列名称；另一方面，列级别则没有名称，可以使用 .rename_axis 方法为其指定级别名称。

```
>>> (college
...     .groupby(['STABBR', 'RELAFFIL'])
...     [['UGDS', 'SATMTMID']]
...     .agg(['size', 'min', 'max'])
...     .rename_axis(['AGG_COLS', 'AGG_FUNCS'], axis='columns')
... )
AGG_COLS              UGDS                    SATMTMID
AGG_FUNCS             size    min     max     size    min     max
STABBR    RELAFFIL
AK        0              7  109.0  12865.0       7    NaN    NaN
          1              3   27.0    275.0       3  503.0  503.0
AL        0             72   12.0  29851.0      72  420.0  590.0
          1             24   13.0   3033.0      24  400.0  560.0
AR        0             68   18.0  21405.0      68  427.0  565.0
...                    ...    ...      ...     ...    ...    ...
WI        0             87   20.0  29302.0      87  480.0  680.0
          1             25    4.0   8212.0      25  452.0  605.0
WV        0             65   20.0  44924.0      65  430.0  530.0
          1              8   63.0   1375.0       8  455.0  510.0
WY        0             11   52.0   9910.0      11  540.0  540.0
```

（3）现在每个轴级别都有一个名称，重塑数据将变得轻而易举。使用 .stack 方法将 AGG_FUNCS 列移至索引级别中。

```
>>> (college
...     .groupby(['STABBR', 'RELAFFIL'])
...     [['UGDS', 'SATMTMID']]
...     .agg(['size', 'min', 'max'])
...     .rename_axis(['AGG_COLS', 'AGG_FUNCS'], axis='columns')
...     .stack('AGG_FUNCS')
... )
AGG_COLS                              UGDS   SATMTMID
STABBR   RELAFFIL   AGG_FUNCS
AK       0          size                7.0       7.0
                    min               109.0       NaN
                    max             12865.0       NaN
```

	1	size	3.0	3.0
		min	27.0	503.0
...			...	...
WV	1	min	63.0	455.0
		max	1375.0	510.0
WY	0	size	11.0	11.0
		min	52.0	540.0
		max	9910.0	540.0

（4）默认情况下，堆叠会将新列级别放置在最里面的索引位置处。使用 .swaplevel 方法可以将 AGG_FUNCS 列从最内层移动到最外层。

```
>>> (college
...     .groupby(['STABBR', 'RELAFFIL'])
...     [['UGDS', 'SATMTMID']]
...     .agg(['size', 'min', 'max'])
...     .rename_axis(['AGG_COLS', 'AGG_FUNCS'], axis='columns')
...     .stack('AGG_FUNCS')
...     .swaplevel('AGG_FUNCS', 'STABBR',
...         axis='index')
... )
```

AGG_COLS			UGDS	SATMTMID
AGG_FUNCS	RELAFFIL	STABBR		
size	0	AK	7.0	7.0
min	0	AK	109.0	NaN
max	0	AK	12865.0	NaN
size	1	AK	3.0	3.0
min	1	AK	27.0	503.0
...			...	...
		WV	63.0	455.0
max	1	WV	1375.0	510.0
size	0	WY	11.0	11.0
min	0	WY	52.0	540.0
max	0	WY	9910.0	540.0

（5）通过使用 .sort_index 方法对级别进行排序，可以继续使用轴级别名称。

```
>>> (college
...     .groupby(['STABBR', 'RELAFFIL'])
...     [['UGDS', 'SATMTMID']]
...     .agg(['size', 'min', 'max'])
...     .rename_axis(['AGG_COLS', 'AGG_FUNCS'], axis='columns')
...     .stack('AGG_FUNCS')
...     .swaplevel('AGG_FUNCS', 'STABBR', axis='index')
```

```
...         .sort_index(level='RELAFFIL', axis='index')
...         .sort_index(level='AGG_COLS', axis='columns')
... )
AGG_COLS                           SATMTMID       UGDS
AGG_FUNCS    RELAFFIL    STABBR
max          0           AK             NaN    12865.0
                         AL           590.0    29851.0
                         AR           565.0    21405.0
                         AS             NaN     1276.0
                         AZ           580.0   151558.0
...                                    ...        ...
size         1           VI             1.0        1.0
                         VT             5.0        5.0
                         WA            17.0       17.0
                         WI            25.0       25.0
                         WV             8.0        8.0
```

（6）为了完全重塑数据，你可能还需要堆叠一些列，而取消堆叠另一些列。对此，可将这两个方法链接在一起。

```
>>> (college
...     .groupby(['STABBR', 'RELAFFIL'])
...     [['UGDS', 'SATMTMID']]
...     .agg(['size', 'min', 'max'])
...     .rename_axis(['AGG_COLS', 'AGG_FUNCS'], axis='columns')
...     .stack('AGG_FUNCS')
...     .unstack(['RELAFFIL', 'STABBR'])
... )
AGG_COLS        UGDS              ...   SATMTMID
RELAFFIL          0       1       ...       1       0
STABBR           AK      AK       ...      WV      WY
AGG_FUNCS                         ...
size            7.0     3.0       ...     8.0    11.0
min           109.0    27.0       ...   455.0   540.0
max         12865.0   275.0       ...   510.0   540.0
```

（7）一次堆叠所有列以返回一个 Series。

```
>>> (college
...     .groupby(['STABBR', 'RELAFFIL'])
...     [['UGDS', 'SATMTMID']]
...     .agg(['size', 'min', 'max'])
...     .rename_axis(['AGG_COLS', 'AGG_FUNCS'], axis='columns')
...     .stack(['AGG_FUNCS', 'AGG_COLS'])
```

```
...   )
       STABBR  RELAFFIL  AGG_FUNCS  AGG_COLS
       AK      0         size       UGDS          7.0
                                    SATMTMID      7.0
                         min        UGDS        109.0
                         max        UGDS      12865.0
               1         size       UGDS          3.0
                                                  ...
       WY      0         size       SATMTMID     11.0
                         min        UGDS         52.0
                                    SATMTMID    540.0
                         max        UGDS       9910.0
                                    SATMTMID    540.0
Length: 640, dtype: float64
```

（8）我们还可以将索引中的所有内容都取消堆叠。在这种情况下，它会展开产生一个非常宽的结果，Pandas 将该结果显示为一个 Series。

```
>>> (college
...   .groupby(['STABBR', 'RELAFFIL'])
...   [['UGDS', 'SATMTMID']]
...   .agg(['size', 'min', 'max'])
...   .rename_axis(['AGG_COLS', 'AGG_FUNCS'], axis='columns')
...   .unstack(['STABBR', 'RELAFFIL'])
... )
AGG_COLS   AGG_FUNCS  STABBR  RELAFFIL
UGDS       size       AK      0            7.0
                              1            3.0
                      AL      0           72.0
                              1           24.0
                      AR      0           68.0
                                           ...
SATMTMID   max        WI      1          605.0
                      WV      0          530.0
                              1          510.0
                      WY      0          540.0
                              1            NaN
Length: 708, dtype: float64
```

10.8.2 原理解释

调用 .groupby 方法的结果通常会产生具有多个轴级别的 DataFrame 或 Series。

步骤（1）中的 groupby 操作生成的 DataFrame 每个轴具有多个级别。这些列级别均未命名，因此只能按它们的整数位置对其进行引用。为了更轻松引用列级别，可以使用 .rename_axis 方法重命名它们。

.rename_axis 方法有点奇怪，因为该方法可以根据传递给其自身的第一个参数的类型来修改级别名称和级别值。将一个列表传递给 .rename_axis 方法即可更改多个级别的名称（如果只有一个级别，则只需要传递一个标量）。

在步骤（2）中，我们向 .rename_axis 方法传递了一个列表，并返回一个已命名所有轴级别的 DataFrame。

一旦所有轴级别都有名称，就可以控制数据的结构。步骤（3）将 AGG_FUNCS 列堆叠到最里面的索引级别中。

步骤（4）中的 .swaplevel 方法接收要交换的级别的名称或位置作为前两个参数。

在步骤（5）中，.sort_index 方法被调用两次，并对每个级别的值进行了排序。请注意，列级别的值是列名称 SATMTMID 和 UGDS。

在步骤（6）中使用了 .stack 方法堆叠一些列，同时又使用了 .unstack 方法取消堆叠另一些列。通过这两个方法可以得到截然不同的输出。也可以对每个单列或索引级别进行堆叠或取消堆叠，并且这两个方法都可以展开成一个 Series。

10.8.3 扩展知识

如果你希望完全丢弃级别值，则可以将其设置为 None。当你想要减少视觉上的混乱，或者列的级别已经很明显并且不需要做进一步处理时，可以执行以下操作。

```
>>> (college
...     .groupby(['STABBR', 'RELAFFIL'])
...     [['UGDS', 'SATMTMID']]
...     .agg(['size', 'min', 'max'])
...     .rename_axis([None, None], axis='index')
...     .rename_axis([None, None], axis='columns')
... )
        UGDS                    SATMTMID
        size   min      max     size   min      max
AK  0      7  109.0  12865.0       7    NaN      NaN
    1      3   27.0    275.0       3  503.0    503.0
AL  0     72   12.0  29851.0      72  420.0    590.0
    1     24   13.0   3033.0      24  400.0    560.0
AR  0     68   18.0  21405.0      68  427.0    565.0
...      ...   ...      ...     ...    ...      ...
```

```
WI   0    87   20.0   29302.0      87   480.0   680.0
     1    25    4.0    8212.0      25   452.0   605.0
WV   0    65   20.0   44924.0      65   430.0   530.0
     1     8   63.0    1375.0       8   455.0   510.0
WY   0    11   52.0    9910.0      11   540.0   540.0
```

10.9 对多个变量存储为列名称的情况进行规整

每当列名称本身包含多个不同的变量时，就会出现一种特殊的混乱数据。例如，当年龄和性别被连接在一起时，就会出现这种情况。要规整这样的数据集，必须使用 Pandas .str 属性操作列。该属性包含用于字符串处理的更多方法。

在本秘笈中，我们将首先确定所有变量（其中一些变量被串联在一起作为列名），然后对数据进行重塑并解析文本以提取正确的变量值。

10.9.1 实战操作

（1）读取男子举重数据集 weightlifting，并识别变量。

```
>>> weightlifting = pd.read_csv('data/weightlifting_men.csv')
>>> weightlifting
   Weight Category   M35 35-39   ...   M75 75-79   M80 80+
0              56          137   ...          62        55
1              62          152   ...          67        57
2              69          167   ...          75        60
3              77          182   ...          82        65
4              85          192   ...          87        70
5              94          202   ...          90        75
6             105          210   ...          95        80
7            105+          217   ...         100        85
```

（2）本示例中包含了变量 Weight Category（它其实已经组合了性别和年龄）和 Qual Total（合格总数）。年龄和性别变量已被合并到一个单元格中。在将它们分开之前，需要使用 .melt 方法将年龄和性别列名称转置为单个垂直列。

```
>>> (weightlifting
...     .melt(id_vars='Weight Category',
...           var_name='sex_age',
...           value_name='Qual Total')
... )
```

```
     Weight Category     sex_age    Qual Total
0                56       M35 35-39        137
1                62       M35 35-39        152
2                69       M35 35-39        167
3                77       M35 35-39        182
4                85       M35 35-39        192
..              ...             ...        ...
75               77       M80 80+           65
76               85       M80 80+           70
77               94       M80 80+           75
78              105       M80 80+           80
79             105+       M80 80+           85
```

（3）选择 sex_age 列，然后使用.str 属性中可用的.split 方法将其拆分为两个不同的列。

```
>>> (weightlifting
...     .melt(id_vars='Weight Category',
...         var_name='sex_age',
...         value_name='Qual Total')
...     ['sex_age']
...     .str.split(expand=True)
... )
        0       1
0     M35    35-39
1     M35    35-39
2     M35    35-39
3     M35    35-39
4     M35    35-39
..    ...     ...
75    M80    80+
76    M80    80+
77    M80    80+
78    M80    80+
79    M80    80+
```

（4）上述操作返回了一个包含无意义列名称的 DataFrame，下面重命名这些列。

```
>>> (weightlifting
...     .melt(id_vars='Weight Category',
...         var_name='sex_age',
...         value_name='Qual Total')
...     ['sex_age']
...     .str.split(expand=True)
...     .rename(columns={0:'Sex', 1:'Age Group'})
... )
```

```
     Sex  Age Group
0    M35  35-39
1    M35  35-39
2    M35  35-39
3    M35  35-39
4    M35  35-39
..   ...  ...
75   M80  80+
76   M80  80+
77   M80  80+
78   M80  80+
79   M80  80+
```

（5）在 .str 属性后使用索引操作创建一个 Sex（性别）列，以从重命名的 Sex 列中选择第一个字符。

```
>>> (weightlifting
...     .melt(id_vars='Weight Category',
...           var_name='sex_age',
...           value_name='Qual Total')
...     ['sex_age']
...     .str.split(expand=True)
...     .rename(columns={0:'Sex', 1:'Age Group'})
...     .assign(Sex=lambda df_: df_.Sex.str[0])
... )
     Sex  Age Group
0    M    35-39
1    M    35-39
2    M    35-39
3    M    35-39
4    M    35-39
..   ..   ...
75   M    80+
76   M    80+
77   M    80+
78   M    80+
79   M    80+
```

（6）使用 pd.concat 函数将此 DataFrame 与 Weight Category（重量分类）和 Qual Total（合格总数）列连接起来。

```
>>> melted = (weightlifting
...     .melt(id_vars='Weight Category',
...           var_name='sex_age',
```

第 10 章 将数据重组为规整形式 · 353 ·

```
...             value_name='Qual Total')
... )
>>> tidy = pd.concat([melted
...         ['sex_age']
...         .str.split(expand=True)
...         .rename(columns={0:'Sex', 1:'Age Group'})
...         .assign(Sex=lambda df_: df_.Sex.str[0]),
...         melted[['Weight Category', 'Qual Total']]],
...         axis='columns'
... )
>>> tidy
    Sex Age Group  Weight Category  Qual Total
0   M   35-39                  56         137
1   M   35-39                  62         152
2   M   35-39                  69         167
3   M   35-39                  77         182
4   M   35-39                  85         192
..  ..  ...                   ...         ...
75  M   80+                    77          65
76  M   80+                    85          70
77  M   80+                    94          75
78  M   80+                   105          80
79  M   80+                  105+          85
```

（7）可以使用以下方法创建相同的结果。

```
>>> melted = (weightlifting
...     .melt(id_vars='Weight Category',
...           var_name='sex_age',
...           value_name='Qual Total')
... )
>>> (melted
...     ['sex_age']
...     .str.split(expand=True)
...     .rename(columns={0:'Sex', 1:'Age Group'})
...     .assign(Sex=lambda df_: df_.Sex.str[0],
...             Category=melted['Weight Category'],
...             Total=melted['Qual Total'])
... )
    Sex Age Group  Category  Total
0   M   35-39           56    137
1   M   35-39           62    152
2   M   35-39           69    167
3   M   35-39           77    182
```

4	M	35-39	85	192
...	...	...	...	...
75	M	80+	77	65
76	M	80+	85	70
77	M	80+	94	75
78	M	80+	105	80
79	M	80+	105+	85

10.9.2 原理解释

像许多数据集一样，本示例中的举重数据集具有一些原始形式的易于理解的信息，但是从技术上讲，此数据集仍然是混乱的，因为除了一个 Weight Category（重量分类）列之外，其他所有列名称都包含性别和年龄信息。

一旦确定了变量，就可以开始规整数据集。只要列名称包含变量，就需要使用.melt（或.stack）方法。Weight Category 变量已经在正确的位置，因此可以通过将其传递给 id_vars 参数来将其保留为已标识的变量。请注意，我们不需要显式地命名要使用 value_vars 融化的所有列。默认情况下，id_vars 中不存在的所有列都会被.melt 方法融化。

sex_age 列需要被解析，并将其拆分为两个变量。为此，我们转向.str 属性提供的额外功能，该功能仅适用于 Series（单个 DataFrame 列）或索引（不能是分层结构）。.split 方法是这种情况下较常见的方法之一，因为该方法可以将字符串的不同部分拆分成各自的列。默认情况下，.split 方法在一个空白空格处分割，但是你也可以使用 pat 参数指定一个字符串或正则表达式。当 expand 参数被设置为 True 时，将为每个独立的分割字符段形成一个新列；如果为 False，则返回单个列，其中包含所有段的列表。

在步骤（4）中重命名列之后，还需要再次使用.str 属性。此属性使我们可以像对字符串一样对列名称进行索引或分割。在这里，我们选择了第一个字符，这是性别变量。我们可以进一步将年龄分为最小和最大年龄两个单独的列，但是举重专业通常以这种方式指代整个年龄组，因此保持不变即可。

步骤（6）显示了将所有数据连接在一起的两种不同方法之一。其中，concat 函数接收 DataFrame 的集合，并将其垂直（axis='index'）或水平（axis='columns'）连接起来。由于两个 DataFrame 的索引相同，因此可以将一个 DataFrame 的值分配给另一个 DataFrame 中的新列，步骤（7）就是这样做的。

10.9.3 扩展知识

还有一种方法可以完成此秘笈。从步骤（2）开始，可以在不使用.split 方法的情况下从 sex_age 列中分配新列。.assign 方法可用于动态地添加这些新列。

```
>>> tidy2 = (weightlifting
...     .melt(id_vars='Weight Category',
...         var_name='sex_age',
...         value_name='Qual Total')
...     .assign(Sex=lambda df_:df_.sex_age.str[0],
...         **{'Age Group':(lambda df_: (df_
...             .sex_age
...             .str.extract(r'(\d{2}[-+](?:\d{2})?)',
...                 expand=False)))})
...     .drop(columns='sex_age')
... )

>>> tidy2
    Weight Category  Qual Total  Sex  Age Group
0                56         137   M      35-39
1                62         152   M      35-39
2                69         167   M      35-39
3                77         182   M      35-39
4                85         192   M      35-39
..              ...         ...  ..        ...
75               77          65   M        80+
76               85          70   M        80+
77               94          75   M        80+
78              105          80   M        80+
79             105+          85   M        80+

>>> tidy.sort_index(axis=1).equals(tidy2.sort_index(axis=1))
True
```

Sex（性别）列是按照与步骤（5）中相同的方式找到的。由于这里没有使用 .split 方法，因此 Age Group（年龄组）列必须以不同的方式被提取。.extract 方法可使用复杂的正则表达式来提取字符串的特定部分。

要正确使用.extract 方法，你的模式必须包含捕获组。可以通过将圆括号括在模式的一部分周围来形成捕获组。在此示例中，整个表达式是一个大捕获组。它以\d{2}开头，这将精确地搜索两位数，然后是文字的正负号，或者是可选的后两位。尽管表达式的最后部分(?:\d{2})?通过括号被括起来，但?:表示它不是捕获组。从技术上讲，它是一个非捕获组，用于表示两个数字（可选）。由于不再需要 sex_age 列，因此将其删除。

最后，将两个规整 DataFrame 相互比较，发现它们是等效的。

10.10 对多个变量存储为单个列的情况进行规整

如前文所述,规整数据集中的每个变量必须有一个单独的列。有时也会出现多个变量名被放在一列中,而其对应的值被放在另一列中的情况。

在此秘笈中,我们将确定包含错误结构的变量的列,并将其旋转以创建规整数据。

10.10.1 实战操作

(1)读取 restaurant inspections(餐厅检查)数据集,并将 Date(日期)列的数据类型转换为 datetime64。

```
>>> inspections = pd.read_csv('data/restaurant_inspections.csv',
...     parse_dates=['Date'])
>>> inspections
                          Name    ...
0             E & E Grill House   ...
1             E & E Grill House   ...
2             E & E Grill House   ...
3             E & E Grill House   ...
4             E & E Grill House   ...
..                          ...   ...
495    PIER SIXTY ONE-THE LIGHTHOUSE   ...
496    PIER SIXTY ONE-THE LIGHTHOUSE   ...
497    PIER SIXTY ONE-THE LIGHTHOUSE   ...
498    PIER SIXTY ONE-THE LIGHTHOUSE   ...
499    PIER SIXTY ONE-THE LIGHTHOUSE   ...
```

(2)restaurant inspections 数据集有两列,即 Name(名称)和 Date(日期),但是这两列被纳入一个列中。Info(信息)列具有 5 个不同的变量,即 Borough(城区)、Cuisine(菜品)、Description(说明)、Grade(等级)和 Score(分数)。我们可以尝试使用 .pivot 方法使 Name 和 Date 列保持垂直,从 Info 列的所有值中创建新列,并将 Value 列用作它们的交集。

```
>>> inspections.pivot(index=['Name', 'Date'],
...     columns='Info', values='Value')
Traceback (most recent call last):
...
NotImplementedError: > 1 ndim Categorical are not supported at
```

第 10 章　将数据重组为规整形式

this time

（3）遗憾的是，Pandas 开发人员尚未实现此功能。幸运的是，在大多数情况下，Pandas 有多种完成同一任务的方法。在本示例中，可以将 Name、Date 和 Info 放入索引中。

```
>>> inspections.set_index(['Name','Date', 'Info'])
                                           Value
Name          Date        Info
E & E Gri...  2017-08-08  Borough          MANHATTAN
                         Cuisine           American
                         Description      Non-food...
                         Grade                    A
                         Score                  9.0
...                                              ...
PIER SIXT...  2017-09-01  Borough          MANHATTAN
                         Cuisine           American
                         Description       Filth fl...
                         Grade                    Z
                         Score                 33.0
```

（4）使用 .unstack 方法旋转 Info 列中的所有值。

```
>>> (inspections
...     .set_index(['Name','Date', 'Info'])
...     .unstack('Info')
... )
                         Value                      ...
Info                     Borough       Cuisine      ...  Grade    Score
Name          Date                                  ...
3 STAR JU...  2017-05-10  BROOKLYN     Juice, S...  ...      A     12.0
A & L PIZ...  2017-08-22  BROOKLYN         Pizza   ...      A      9.0
AKSARAY T...  2017-07-25  BROOKLYN        Turkish  ...      A     13.0
ANTOJITOS...  2017-06-01  BROOKLYN     Latin (C... ...      A     10.0
BANGIA        2017-06-16  MANHATTAN        Korean  ...      A      9.0
...                          ...             ...   ...    ...      ...
VALL'S PI...  2017-03-15  STATEN I...  Pizza/It... ...      A      9.0
VIP GRILL     2017-06-12  BROOKLYN     Jewish/K... ...      A     10.0
WAHIZZA       2017-04-13  MANHATTAN        Pizza   ...      A     10.0
WANG MAND...  2017-08-29  QUEENS           Korean  ...      A     12.0
XIAOYAN Y...  2017-08-29  QUEENS           Korean  ...      Z     49.0
```

（5）使用 .reset_index 方法将索引级别分为若干列。

```
>>> (inspections
```

```
...         .set_index(['Name','Date', 'Info'])
...         .unstack('Info')
...         .reset_index(col_level=-1)
... )
.                        ...    Value
Info        Name    Date  ...    Grade    Score
0    3 STAR J...  2017-05-10 ...    A     12.0
1    A & L PI...  2017-08-22 ...    A      9.0
2    AKSARAY ...  2017-07-25 ...    A     13.0
3    ANTOJITO...  2017-06-01 ...    A     10.0
4        BANGIA   2017-06-16 ...    A      9.0
..           ...         ... ...   ...     ...
95   VALL'S P...  2017-03-15 ...    A      9.0
96    VIP GRILL   2017-06-12 ...    A     10.0
97      WAHIZZA   2017-04-13 ...    A     10.0
98   WANG MAN...  2017-08-29 ...    A     12.0
99   XIAOYAN ...  2017-08-29 ...    Z     49.0
```

（6）restaurant inspections 数据集现在已经很规整了，但是还有一些烦人的剩余 Pandas 残骸需要清除。这里不妨使用 .droplevel 方法删除顶级列级别，然后将索引级别重命名为 None。

```
>>> (inspections
...         .set_index(['Name','Date', 'Info'])
...         .unstack('Info')
...         .reset_index(col_level=-1)
...         .droplevel(0, axis=1)
...         .rename_axis(None, axis=1)
... )
            Name    Date  ...    Grade    Score
0    3 STAR J...  2017-05-10 ...    A     12.0
1    A & L PI...  2017-08-22 ...    A      9.0
2    AKSARAY ...  2017-07-25 ...    A     13.0
3    ANTOJITO...  2017-06-01 ...    A     10.0
4        BANGIA   2017-06-16 ...    A      9.0
..           ...         ... ...   ...     ...
95   VALL'S P...  2017-03-15 ...    A      9.0
96    VIP GRILL   2017-06-12 ...    A     10.0
97      WAHIZZA   2017-04-13 ...    A     10.0
98   WANG MAN...  2017-08-29 ...    A     12.0
99   XIAOYAN ...  2017-08-29 ...    Z     49.0
```

（7）通过使用 .squeeze 方法将步骤（3）中的一列 DataFrame 转换为 Series，可以避免在步骤（4）中创建的多重索引列。以下代码可产生与步骤（6）相同的结果。

```
>>> (inspections
...     .set_index(['Name','Date', 'Info'])
...     .squeeze()
...     .unstack('Info')
...     .reset_index()
...     .rename_axis(None, axis='columns')
... )
              Name          Date    ...  Grade   Score
0     3 STAR J...    2017-05-10    ...      A    12.0
1     A & L PI...    2017-08-22    ...      A     9.0
2     AKSARAY  ...   2017-07-25    ...      A    13.0
3     ANTOJITO...    2017-06-01    ...      A    10.0
4          BANGIA    2017-06-16    ...      A     9.0
..            ...           ...    ...    ...     ...
95    VALL'S P...    2017-03-15    ...      A     9.0
96      VIP GRILL    2017-06-12    ...      A    10.0
97        WAHIZZA    2017-04-13    ...      A    10.0
98    WANG MAN...    2017-08-29    ...      A    12.0
99    XIAOYAN  ...   2017-08-29    ...      Z    49.0
```

10.10.2 原理解释

在步骤（1）中，我们注意到在 Info 列中垂直放置了 5 个变量，在 Value 列中有它们的对应值。因为我们需要将这 5 个变量中的每一个作为水平列名称进行透视（旋转），所以看起来似乎可以使用 .pivot 方法。遗憾的是，当有多个非透视列时，该方法不能正常工作，Pandas 开发人员尚未实现处理这种特殊情况的功能。我们必须考虑使用另一种方法。

.unstack 方法也可以透视垂直数据，但仅适用于索引中的数据。步骤（3）使用了 .set_index 方法，将不会旋转的两列移入索引中。将这些列放入索引之后，即可使用 .unstack 方法，如步骤（4）所示。

请注意，当我们取消堆叠一个 DataFrame 时，Pandas 会保留原始列名（在这里，它只是单个列，即 Value 列），并创建一个以旧列名为上层的多重索引。

数据集现在基本上是规整的，但是我们还需要继续使用 .reset_index 方法将非透视的列设置为普通列。因为有多重索引列，所以可使用 col_level 参数选择新的列名称所属的级别。默认情况下，名称会被插入最高级别（级别 0）中。我们使用-1 表示最底层。

在此之后，还有一些多余的 DataFrame 名称和索引需要删除。在步骤（6）中，使用了.droplevel 和.rename_axis 执行此操作。这些列仍然具有无用的.name 属性 Info，它被重命名为 None。

通过将步骤（3）中的结果 DataFrame 强制为 Series，可以避免清理多重索引列。.squeeze 方法适用于单列 DataFrame，并可以将其转换为 Series。

10.10.3 扩展知识

.pivot_table 方法是可以使用的，该方法对允许的非透视列数没有限制。.pivot_table 方法与.pivot 方法的不同之处在于，它可以对与 index 和 columns 参数中的列之间的交点相对应的所有值执行聚合。

由于此交叉点可能有多个值，因此 .pivot_table 要求用户向其传递一个聚合函数以输出单个值。我们使用 first 聚合函数（该函数采用组中的第一个值）。在此特定示例中，每个交叉点都只有一个值，因此没有要聚合的内容。默认的聚合函数是 mean，在这里会产生错误，因为某些值是字符串。

```
>>> (inspections
...     .pivot_table(index=['Name', 'Date'],
...                  columns='Info',
...                  values='Value',
...                  aggfunc='first')
...     .reset_index()
...     .rename_axis(None, axis='columns')
... )
          Name        Date  ...  Grade  Score
0   3 STAR J...  2017-05-10  ...      A   12.0
1   A & L PI...  2017-08-22  ...      A    9.0
2   AKSARAY ...  2017-07-25  ...      A   13.0
3   ANTOJITO...  2017-06-01  ...      A   10.0
4       BANGIA  2017-06-16  ...      A    9.0
..          ...         ...  ...    ...    ...
95  VALL'S P...  2017-03-15  ...      A    9.0
96    VIP GRILL  2017-06-12  ...      A   10.0
97      WAHIZZA  2017-04-13  ...      A   10.0
98  WANG MAN...  2017-08-29  ...      A   12.0
99  XIAOYAN ...  2017-08-29  ...      Z   49.0
```

10.11 对多个值存储在同一单元格中的情况进行规整

表格数据本质上是二维的，因此，可以在单个单元格中显示的信息量有限。在这种情况下，你偶尔会看到有些数据集在同一单元格中存储的不止一个值。但是，规整数据仅允许每个单元格使用一个值。因此，要纠正这些情况，通常需要使用.str 属性中的方法将字符串数据解析为多列。

在本秘笈中，我们将检查一个数据集，该数据集有一列，其每个单元格中都包含多个不同的变量。可以使用.str 属性将这些字符串解析为单独的列以规整数据。

10.11.1 实战操作

（1）读取德克萨斯州 cities（城市）数据集。

```
>>> cities = pd.read_csv('data/texas_cities.csv')
>>> cities
      City              Geolocation
0  Houston    29.7604° N, 95.3698° W
1   Dallas    32.7767° N, 96.7970° W
2   Austin    30.2672° N, 97.7431° W
```

（2）City（城市）列看起来良好，并且仅包含一个值。另外，Geolocation（地理位置）列却包含 4 个变量，即 latitude（纬度）、latitude direction（纬度方向）、longitude（经度）、longitude direction（经度方向）。因此，可以考虑将 Geolocation 列分为 4 个单独的列。我们将使用正则表达式匹配任何字符，后面跟着一个空格。

```
>>> geolocations = cities.Geolocation.str.split(pat='. ',
...         expand=True)
>>> geolocations.columns = ['latitude', 'latitude direction',
...         'longitude', 'longitude direction']
```

（3）因为 Geolocation 列的原始数据类型是 object，所以所有新列也都是 object。可以将 latitude 和 longitude 列更改为 float 类型。

```
>>> geolocations = geolocations.astype({'latitude':'float',
...         'longitude':'float'})
>>> geolocations.dtypes
latitude                float64
latitude direction       object
longitude               float64
longitude direction      object
dtype: object
```

（4）将这些新列与原始列中的 City 列合并。

```
>>> (geolocations
...     .assign(city=cities['City'])
... )
    latitude  latitude direction  ...  longitude  direction     city
0    29.7604          N           ...          W              Houston
1    32.7767          N           ...          W               Dallas
2    30.2672          N           ...          W               Austin
```

10.11.2 原理解释

读取数据之后，我们将决定数据集中有多少个变量。在这里，我们选择将 Geolocation 列拆分为 4 个变量，但是我们也可以只选择 latitude 和 longitude 两个变量，并使用负号来区分东方和西方、南方和北方。

有若干种方式可以使用 .str 属性的方法解析 Geolocation 列。最简单的方式是使用 .split 方法。我们给 .split 方法传递由任何字符（句点）和空格定义的正则表达式。当空格跟随任何字符时，将进行拆分，并形成一个新列。在纬度的末尾，第一次出现了该模式。空格跟随度数字符（°），从而形成一次拆分。拆分字符将被丢弃，而不保留在结果列中。下一次拆分与逗号和空格匹配，紧跟在纬度方向之后。

这里总共进行了 3 次拆分，得到了 4 列。步骤（2）的第 2 行为其提供了有意义的名称。即使所得的 latitude 和 longitude 列似乎是浮点类型，但事实并非如此。它们最初是从 object 列进行解析的，因此仍然是 object 数据类型。

步骤（3）使用字典将列名称映射到其新类型。

如果字典有很多列名称，则需要大量输入代码，因此，也可以使用 to_numeric 函数尝试将每一列转换为整数或浮点数，而不是使用字典。

要在每个列上迭代应用此函数，可使用 .apply 方法。

```
>>> geolocations.apply(pd.to_numeric, errors='ignore')
    latitude  latitude direction  longitude  longitude direction
0    29.7604          N            95.3698                   W
1    32.7767          N            96.7970                   W
2    30.2672          N            97.7431                   W
```

步骤（4）将 City 列连接到 DataFrame，以完成规整数据的过程。

10.11.3 扩展知识

在此示例中，.split 方法使用了正则表达式，并且工作得很好。对于其他示例，某些

列可能会要求你根据若干种不同的模式创建拆分。要搜索多个正则表达式，可使用管道字符（|）。例如，如果只想分割度数符号和逗号，并在其后跟一个空格，则可以执行以下操作。

```
>>> cities.Geolocation.str.split(pat=r'° |, ', expand=True)
       0  1       2  3
0  29.7604  N  95.3698  W
1  32.7767  N  96.7970  W
2  30.2672  N  97.7431  W
```

这将从步骤（2）返回相同的 DataFrame。可以使用管道字符将任意数量的其他拆分模式附加到前面的字符串模式。

.extract 方法是另一种允许你提取每个单元格中特定组的方法。这些捕获组必须用括号括起来。结果中不包含任何括号外匹配的内容。以下代码将产生与步骤（2）相同的输出。

''' {.sourceCode .pycon}
```
>>> cities.Geolocation.str.extract(r'([0-9.]+). (N|S), ([0-9.]+). (E|W)',
...     expand=True)
       0  1       2  3
0  29.7604  N  95.3698  W
1  32.7767  N  96.7970  W
2  30.2672  N  97.7431  W
```

'''

此正则表达式具有 4 个捕获组。第一组和第三组至少搜索一个或多个带小数的连续数字。第二组和第四组搜索单个字符（方向）。第一个和第三个捕获组由任何字符分隔，后跟一个空格。第二个捕获组用逗号分隔，然后是一个空格。

10.12 对变量存储在列名称和值中的情况进行规整

当变量以水平方式存储在列名称中或以垂直方式存储在值中时，就会出现一种特别难以诊断的混乱数据形式。

一般来说，在数据库中不会找到这种类型的数据集，但是在其他人已经生成的汇总报告中很可能找到这种数据集。

10.12.1 实战操作

在此秘笈中，使用 .melt 和 .pivot_table 方法将数据重塑为规整数据。

（1）读取 sensors（传感器）数据集。

```
>>> sensors = pd.read_csv('data/sensors.csv')
>>> sensors
   Group     Property  2012  2013  2014  2015  2016
0      A     Pressure   928   873   814   973   870
1      A  Temperature  1026  1038  1009  1036  1042
2      A         Flow   819   806   861   882   856
3      B     Pressure   817   877   914   806   942
4      B  Temperature  1008  1041  1009  1002  1013
5      B         Flow   887   899   837   824   873
```

（2）可以看到，正确放置在垂直列中的唯一变量是 Group。在 Property（属性）列中，似乎具有 3 个唯一变量，即 Pressure（压力）、Temperature（温度）和 Flow（流量）。其余各列（2012—2016 年）本身只是一个变量，可以考虑将其命名为 Year。用单个 DataFrame 方法不可能重组这种混乱的数据。下面从 .melt 方法开始，将各个年份旋转到它们自己的列中（使用 Year 作为列名称）。

```
>>> sensors.melt(id_vars=['Group', 'Property'], var_name='Year')
   Group     Property  Year  value
0      A     Pressure  2012    928
1      A  Temperature  2012   1026
2      A         Flow  2012    819
3      B     Pressure  2012    817
4      B  Temperature  2012   1008
..   ...          ...   ...    ...
25     A  Temperature  2016   1042
26     A         Flow  2016    856
27     B     Pressure  2016    942
28     B  Temperature  2016   1013
29     B         Flow  2016    873
```

（3）上述操作解决了部分问题。接下来，可以使用 .pivot_table 方法将 Property 列转换为新的列名称。

```
>>> (sensors
```

```
...         .melt(id_vars=['Group', 'Property'], var_name='Year')
...         .pivot_table(index=['Group', 'Year'],
...                      columns='Property', values='value')
...         .reset_index()
...         .rename_axis(None, axis='columns')
... )
    Group  Year  Flow  Pressure  Temperature
0     A    2012   819    928        1026
1     A    2013   806    873        1038
2     A    2014   861    814        1009
3     A    2015   882    973        1036
4     A    2016   856    870        1042
5     B    2012   887    817        1008
6     B    2013   899    877        1041
7     B    2014   837    914        1009
8     B    2015   824    806        1002
9     B    2016   873    942        1013
```

10.12.2 原理解释

一旦在步骤（1）中确定了变量，就可以开始重组。Pandas 没有同时旋转数据列的方法，因此我们必须一步一步来。我们可以将 Group 和 Property 列传递给 .melt 方法中的 id_vars 参数，使它们保持垂直，从而仅旋转年份。

在执行上述操作后，获得的数据就是在 10.12 节"对变量存储在列名称和值中的情况进行规整"秘笈中讨论的混乱数据模式，即多个值被存储在同一单元格中的情况。当在 index 参数中使用多个列时，必须使用 .pivot_table 方法来旋转 DataFrame。

在经过数据透视处理之后，Group 和 Year 变量被卡在索引中，可以使用.reset_index 方法将它们拉回来（作为正常列）。.pivot_table 方法将 columns 参数中使用的列名称保留为列索引的名称。重置索引之后，此名称变得毫无意义，可以使用 .rename_axis 方法将其删除。

10.12.3 扩展知识

每当涉及.melt、.pivot_table 或.pivot 方法的解决方案时，你都可以确认存在使用.stack 和.unstack 的替代方法。技巧是首先要移动当前未旋转到索引中的列。

```
>>> (sensors
...     .set_index(['Group', 'Property'])
...     .rename_axis('Year', axis='columns')
...     .stack()
...     .unstack('Property')
...     .rename_axis(None, axis='columns')
...     .reset_index()
... )
   Group  Year  Flow  Pressure  Temperature
0      A  2012   819       928         1026
1      A  2013   806       873         1038
2      A  2014   861       814         1009
3      A  2015   882       973         1036
4      A  2016   856       870         1042
5      B  2012   887       817         1008
6      B  2013   899       877         1041
7      B  2014   837       914         1009
8      B  2015   824       806         1002
9      B  2016   873       942         1013
```

第 11 章　组合 Pandas 对象

本章包含以下秘笈。
- 将新行追加到 DataFrame。
- 将多个 DataFrame 连接在一起。
- 了解 concat 函数、join 和 merge 方法之间的区别。
- 连接到 SQL 数据库。

11.1　介　　绍

在 Pandas 中，要将两个或多个 DataFrame 或 Series 组合在一起，可以有多种选择。
- append 方法：灵活性最差，仅允许将新行追加到 DataFrame。
- concat 方法：用途广泛，可以在任一轴上组合任意数量的 DataFrame 或 Series。
- join 方法：可以通过将一个 DataFrame 的列与其他 DataFrame 的索引对齐来提供快速查找功能。
- merge 方法：提供了类似 SQL 的功能，可以将两个 DataFrame 连接在一起。

11.2　将新行追加到 DataFrame

执行数据分析时，创建新列比创建新行更为常见。这是因为新的数据行通常表示新的观察结果，而作为数据分析人员，连续捕获新数据通常不是你的工作。数据捕获通常留给其他平台，如关系数据库管理系统。无论如何，这是一个必不可少的功能，因为新的数据总是会不时出现。

在本秘笈中，我们将使用 .loc 属性将新的数据行追加到一个小型数据集，然后过渡到使用 .append 方法。

11.2.1　实战操作

（1）读取 names 数据集，并查看其输出。

```
>>> import pandas as pd
>>> import numpy as np
>>> names = pd.read_csv('data/names.csv')
>>> names
       Name  Age
0   Cornelia   70
1      Abbas   69
2   Penelope    4
3       Niko    2
```

（2）下面创建一个包含一些新数据的列表，然后使用 .loc 属性设置一个行标签，让它等于这个新数据。

```
>>> new_data_list = ['Aria', 1]
>>> names.loc[4] = new_data_list
>>> names
       Name  Age
0   Cornelia   70
1      Abbas   69
2   Penelope    4
3       Niko    2
4       Aria    1
```

（3）.loc 属性使用标签来引用行。在这种情况下，行标签与整数位置完全匹配。也可以使用非整数标签追加更多行。

```
>>> names.loc['five'] = ['Zach', 3]
>>> names
          Name  Age
0     Cornelia   70
1        Abbas   69
2     Penelope    4
3         Niko    2
4         Aria    1
five      Zach    3
```

（4）为了更明确地将变量与值相关联，可以使用字典。在此步骤中，可以动态选择新的索引标签作为 DataFrame 的长度。

```
>>> names.loc[len(names)] = {'Name':'Zayd', 'Age':2}
>>> names
       Name  Age
0   Cornelia   70
1      Abbas   69
```

第 11 章　组合 Pandas 对象

```
2       Penelope    4
3       Niko        2
4       Aria        1
five    Zach        3
6       Zayd        2
```

（5）也可以使用 Series 保存新数据，其使用方式与字典完全相同。

```
>>> names.loc[len(names)] = pd.Series({'Age':32, 'Name':'Dean'})
>>> names
        Name        Age
0       Cornelia    70
1       Abbas       69
2       Penelope    4
3       Niko        2
4       Aria        1
five    Zach        3
6       Zayd        2
7       Dean        32
```

（6）上面的操作全部使用.loc 属性来就地更改名称 DataFrame，这种操作方式不会返回 DataFrame 的单独副本。在接下来的步骤中，我们将讨论.append 方法的应用，该方法不会修改调用它的 DataFrame，而是返回带有已追加新行的 DataFrame 的新副本。

我们可从最初的 names 这个 DataFrame 开始，并尝试追加一行。.append 方法的第一个参数必须是另一个 DataFrame、Series、字典或它们的列表，但不能是像步骤（2）中那样的列表。下面看看当尝试将字典与.append 方法一起使用时会发生什么。

```
>>> names = pd.read_csv('data/names.csv')
>>> names.append({'Name':'Aria', 'Age':1})
Traceback (most recent call last):
...
TypeError: Can only append a Series if ignore_index=True or if the
Series has a name
```

（7）步骤（6）中的错误消息似乎有点不正确。我们传递的是字典而不是一个 Series，但是尽管如此，它还是为我们提供了如何纠正的说明，即我们需要传递 ignore_index=True 参数。

```
>>> names.append({'Name':'Aria', 'Age':1}, ignore_index=True)
        Name        Age
0       Cornelia    70
1       Abbas       69
```

```
2     Penelope     4
3        Niko      2
4        Aria      1
```

（8）这种方法可行，但是 ignore_index 是一个喜欢偷偷"干活"的参数。当将其设置为 True 时，旧索引将被完全删除，并替换为 0～n-1 的 RangeIndex。例如，我们可以尝试为 names DataFrame 指定一个索引。

```
>>> names.index = ['Canada', 'Canada', 'USA', 'USA']
>>> names
             Name    Age
Canada    Cornelia   70
Canada     Abbas     69
USA      Penelope    4
USA        Niko      2
```

（9）重新运行步骤（7）中的代码，你将获得相同的结果。原始索引将被完全忽略。

（10）现在继续使用这个在索引中包含国家名称字符串的 names DataFrame。下面用 .append 方法追加一个具有 name 属性的 Series。

```
>>> s = pd.Series({'Name': 'Zach', 'Age': 3}, name=len(names))
>>> s
Name    Zach
Age       3
Name: 4, dtype: object

>>> names.append(s)
             Name    Age
Canada    Cornelia   70
Canada     Abbas     69
USA      Penelope    4
USA        Niko      2
4          Zach      3
```

（11）.append 方法比 .loc 属性更灵活，该方法支持同时添加多行。实现此操作的一种方式是传递一个 Series 列表。

```
>>> s1 = pd.Series({'Name': 'Zach', 'Age': 3}, name=len(names))
>>> s2 = pd.Series({'Name': 'Zayd', 'Age': 2}, name='USA')
>>> names.append([s1, s2])
             Name    Age
Canada    Cornelia   70
Canada     Abbas     69
```

```
USA        Penelope    4
USA        Niko        2
4          Zach        3
USA        Zayd        2
```

（12）仅具有两列的小型 DataFrame 足够简单，我们可以手动写出所有列名称和值。但是，当它们变得很大时，此过程将非常困难。例如，下面查看 2016 年的棒球数据集。

```
>>> bball_16 = pd.read_csv('data/baseball16.csv')
>>> bball_16
    playerID   yearID  stint  teamID  ...   HBP   SH    SF    GIDP
0   altuv...   2016    1      HOU     ...   7.0   3.0   7.0   15.0
1   bregm...   2016    1      HOU     ...   0.0   0.0   1.0   1.0
2   castr...   2016    1      HOU     ...   1.0   1.0   0.0   9.0
3   corre...   2016    1      HOU     ...   5.0   0.0   3.0   12.0
4   gatti...   2016    1      HOU     ...   4.0   0.0   5.0   12.0
..  ...        ...     ...    ...     ...   ...   ...   ...   ...
11  reedaj01   2016    1      HOU     ...   0.0   0.0   1.0   1.0
12  sprin...   2016    1      HOU     ...   11.0  0.0   1.0   12.0
13  tucke...   2016    1      HOU     ...   2.0   0.0   0.0   2.0
14  valbu...   2016    1      HOU     ...   1.0   3.0   2.0   5.0
15  white...   2016    1      HOU     ...   2.0   0.0   2.0   6.0
```

（13）当前数据集包含 22 列，如果你手动输入新的数据行，则很容易输错列名称或完全忘记其中的一个。为了防止发生这些错误，可以选择一行作为 Series，并将.to_dict 方法链接到该行，以获取一个示例行作为字典。

```
>>> data_dict = bball_16.iloc[0].to_dict()
>>> data_dict
{'playerID': 'altuvjo01', 'yearID': 2016, 'stint': 1, 'teamID':
'HOU', 'lgID': 'AL', 'G': 161, 'AB': 640, 'R': 108, 'H': 216,
'2B': 42, '3B': 5, 'HR': 24, 'RBI': 96.0, 'SB': 30.0, 'CS': 10.0,
'BB': 60, 'SO': 70.0, 'IBB': 11.0, 'HBP': 7.0, 'SH': 3.0, 'SF':
7.0, 'GIDP': 15.0}
```

（14）使用字典推导式（dictionary comprehension）清除旧值，将任何先前的字符串值分配为空字符串，而所有其他字符串值则分配为缺失值。现在，该字典可以用作你要输入的任何新数据的模板。

```
>>> new_data_dict = {k: '' if isinstance(v, str) else
...     np.nan for k, v in data_dict.items()}
>>> new_data_dict
```

```
{'playerID': '', 'yearID': nan, 'stint': nan, 'teamID': '', 'lgID': '',
 'G': nan, 'AB': nan, 'R': nan, 'H': nan, '2B': nan, '3B': nan, 'HR': nan,
 'RBI': nan, 'SB': nan, 'CS': nan, 'BB': nan, 'SO': nan, 'IBB': nan,
 'HBP': nan, 'SH': nan, 'SF': nan, 'GIDP': nan}
```

11.2.2 原理解释

.loc 属性用于根据行和列标签选择和分配数据。传递给.loc 属性的第一个值表示行标签。

在步骤（2）中，names.loc[4]引用的是标签整数位置为 4 的行。此标签当前在 DataFrame 中不存在。赋值语句使用列表提供的数据创建新行。如果以前存在标签等于整数 4 的行，则该命令将覆盖该行。

如前文所述，此操作将修改 names DataFrame 本身而不是返回一个 DataFrame 的副本。这种就地修改方式会使得该索引操作符的使用风险比.append 方法更高，后者永远不会修改调用它的原始 DataFrame。

任何有效的标签都可以与.loc 属性一起使用，如步骤（3）所示。无论新标签的值是多少，新行始终追加在末尾。在本示例中，即使列表可以正常工作，为清晰起见，最好还是使用字典，以便我们准确地知道哪些列与每个值相关联，步骤（4）就是这样做的。

步骤（4）和步骤（5）显示了一种技巧，可以动态地将新标签设置为 DataFrame 中的当前行数。只要索引标签与列名称匹配，存储在 Series 中的数据也将得到正确分配。

接下来的步骤都使用了.append 方法，该方法仅将新行追加到 DataFrame 中。大多数 DataFrame 方法都允许通过 axis 参数指定行和列操作。.append 方法是一个例外，该方法只能将行追加到 DataFrame 中。

如步骤（6）中的错误消息所示，使用映射到值的列名称的字典不足以使.append 方法正常工作。要正确附加不带行名的字典，必须将.ignore_index 参数设置为 True。

步骤（10）展示了如何通过将字典转换为 Series 来保留旧索引。确保使用 name 参数，然后将其用作新的索引标签。通过这种方式，我们可以使用.append 方法添加任意数量的行（需要传递 Series 列表作为第一个参数）。

当需要处理更大的 DataFrame 并且以这种方式附加行时，有必要避免大量输入和随之产生的错误。此时可以使用 .to_dict 方法将单行转换为字典，然后使用字典推导式清除所有旧值，使用一些默认值取代它们，这可以用作新行的模板。

11.2.3 扩展知识

将单个行添加到 DataFrame 中是成本相当昂贵的操作。如果你发现自己编写了一个循环，试图将单个数据行添加到 DataFrame 中，那么这样做是有问题的。为什么这么说？
下面首先创建 1000 行的新数据作为 Series 列表。

```
>>> random_data = []
>>> for i in range(1000):
...     d = dict()
...     for k, v in data_dict.items():
...         if isinstance(v, str):
...             d[k] = np.random.choice(list('abcde'))
...         else:
...             d[k] = np.random.randint(10)
...     random_data.append(pd.Series(d, name=i + len(bball_16)))
>>> random_data[0]
2B    3
3B    9
AB    3
BB    9
CS    4
Name: 16, dtype: object
```

现在来计算循环遍历每一项进行一次追加需要多长时间。

```
>>> %%timeit
>>> bball_16_copy = bball_16.copy()
>>> for row in random_data:
...     bball_16_copy = bball_16_copy.append(row)
4.88 s ± 190 ms per loop (mean ± std. dev. of 7 runs, 1 loop each)
```

可以看到，仅 1000 行的数据就花了将近 5s。如果改为传递整个 Series 列表，则速度会大大提高。

```
>>> %%timeit
>>> bball_16_copy = bball_16.copy()
>>> bball_16_copy = bball_16_copy.append(random_data)
78.4 ms ± 6.2 ms per loop (mean ± std. dev. of 7 runs, 10 loops each)
```

可以看到，如果传递 Series 对象列表，则时间已减少到 100ms（0.1s）以下。在内部，Pandas 会将 Series 列表转换为单个 DataFrame，然后追加数据。

11.3　将多个 DataFrame 连接在一起

concat 函数可将两个或多个 DataFrame（或 Series）以垂直或水平方式连接在一起。

一般来说，当同时处理多个 Pandas 对象时，连接（concatenation）并不是无序发生的，而是通过它们的索引对齐每个对象。

在此秘笈中，我们将按水平和垂直方式使用 concat 函数组合 DataFrame，然后更改参数值以产生不同的结果。

11.3.1 实战操作

（1）读取 2016 年和 2017 年的股票数据集，并将其股票代码作为索引。

```
>>> stocks_2016 = pd.read_csv('data/stocks_2016.csv',
...        index_col='Symbol')
>>> stocks_2017 = pd.read_csv('data/stocks_2017.csv',
...        index_col='Symbol')

>>> stocks_2016
        Shares   Low   High
Symbol
AAPL        80    95    110
TSLA        50    80    130
WMT         40    55     70


>>> stocks_2017
        Shares   Low   High
Symbol
AAPL        50   120    140
GE         100    30     40
IBM         87    75     95
SLB         20    55     85
TXN        500    15     23
TSLA       100   100    300
```

（2）将所有股票数据集放置在一个列表中，然后调用 concat 函数，以将这些数据集沿着默认轴（0）连接在一起。

```
>>> s_list = [stocks_2016, stocks_2017]
>>> pd.concat(s_list)
        Shares   Low   High
Symbol
AAPL        80    95    110
TSLA        50    80    130
WMT         40    55     70
AAPL        50   120    140
```

```
GE          100    30    40
IBM          87    75    95
SLB          20    55    85
TXN         500    15    23
TSLA        100   100   300
```

（3）可以看到，默认情况下，concat 函数以垂直方式连接 DataFrame，其中一个 DataFrame 在另一个 DataFrame 的上面。上面获得的这个 DataFrame 的一个问题是无法识别每一行的年份。concat 函数允许使用 key 参数标记结果 DataFrame 的每个片段。该标签将显示在已连接的 DataFrame 的最外层索引级别中，并强制创建 MultiIndex（多重索引）。另外，为了清晰起见，还可以使用 names 参数重命名每个索引级别。

```
>>> pd.concat(s_list, keys=['2016', '2017'],
...     names=['Year', 'Symbol'])
              Shares   Low   High
Year Symbol
2016 AAPL        80    95    110
     TSLA        50    80    130
     WMT         40    55     70
2017 AAPL        50   120    140
     GE         100    30     40
     IBM         87    75     95
     SLB         20    55     85
     TXN        500    15     23
     TSLA       100   100    300
```

（4）也可以通过将 axis 参数更改为 columns 或 1 来进行水平方向的连接。

```
>>> pd.concat(s_list, keys=['2016', '2017'],
...     axis='columns', names=['Year', None])
Year      2016                    2017
        Shares   Low   High    Shares   Low   High
AAPL      80.0  95.0  110.0     50.0  120.0  140.0
GE         NaN   NaN    NaN    100.0   30.0   40.0
IBM        NaN   NaN    NaN     87.0   75.0   95.0
SLB        NaN   NaN    NaN     20.0   55.0   85.0
TSLA      50.0  80.0  130.0    100.0  100.0  300.0
TXN        NaN   NaN    NaN    500.0   15.0   23.0
WMT       40.0  55.0   70.0      NaN    NaN    NaN
```

（5）可以看到，当 2017 年有的股票代号出现在 2016 年时，由于该年没有该股票，所以出现了缺失值（GE、IBM、SLB 和 TXN 都是这种情况）。同样，2017 年没有 WMT

股票，所以该年也出现了缺失值。

默认情况下，concat 函数使用外连接（outer join）将列表中每个 DataFrame 的所有行保留在列表中。但是，concat 函数也为我们提供了一个选项，可以在两个 DataFrame 中仅保留具有相同索引值的行，这称为内连接（inner join）。

现在可以将 join 参数设置为 inner 来改变 concat 函数的行为。

```
>>> pd.concat(s_list, join='inner', keys=['2016', '2017'],
...        axis='columns', names=['Year', None])
Year      2016              2017
       Shares  Low  High  Shares  Low  High
Symbol
AAPL      80    95   110     50   120   140
TSLA      50    80   130    100   100   300
```

11.3.2 原理解释

concat 函数接收列表作为第一个参数。该列表必须是一系列的 Pandas 对象，通常是 DataFrame 或 Series 的列表。默认情况下，所有这些对象都将被垂直堆叠，一个被堆叠在另一个之上。在此秘笈中，仅连接了两个 DataFrame，但是任何数量的 Pandas 对象都可以被堆叠在一起。当垂直连接时，DataFrame 通过其列名称对齐。

在此数据集中，所有列的名称都相同，因此 2017 年数据中的每个列都在 2016 年数据中的同一列名称下精确对齐。但是，当将它们以水平方式连接时，如在步骤（4）就是这样，两个年份中只有两个索引标签是匹配的（即 AAPL 和 TSLA）。因此，这些股票代号在任何一年都没有缺失值，而其他股票代号则会产生缺失值。

使用 concat 函数可能有两种对齐方式，即外连接（join='outer'，这也是默认设置）和内连接（需指定 join='inner'）。

11.3.3 扩展知识

.append 方法是 concat 函数的简化版本，只能将新行追加到 DataFrame 中。在内部，.append 方法调用的就是 concat 函数。例如，此秘笈中的步骤（2）可使用以下方法重复。

```
>>> stocks_2016.append(stocks_2017)
       Shares  Low  High
Symbol
AAPL      80    95   110
TSLA      50    80   130
```

```
WMT      40    55    70
AAPL     50   120   140
GE      100    30    40
IBM      87    75    95
SLB      20    55    85
TXN     500    15    23
TSLA    100   100   300
```

11.4 了解 concat 函数、.join 和 .merge 方法之间的区别

concat 函数以及 .merge 和 .join 这两个 DataFrame（而不是 Series）方法都提供了非常相似的功能，可以将多个 Pandas 对象组合在一起。由于 concat 函数、.join 和 .merge 方法是如此相似，并且它们在某些情况下可以相互复制操作，因此分析人员往往会对需要在何时使用它们以及如何正确使用它们感到非常困惑。

为了帮助理解 concat 函数、.join 和 .merge 方法之间的差异，请看以下列表介绍。

- concat 函数。
    - Pandas 函数。
    - 以垂直或水平方式组合两个或更多 Pandas 对象。
    - 仅在索引上对齐。
    - 每当索引中出现重复项时将出错。
    - 默认为外连接，提供内连接选项。
- .join 方法。
    - DataFrame 方法。
    - 以水平方式组合两个或多个 Pandas 对象。
    - 将调用它的 DataFrame 的列或索引与另一个对象的索引（而不是列）对齐。
    - 通过执行笛卡儿积处理连接列或索引上的重复值。
    - 默认为左连接，提供内连接、外连接和右连接选项。
- .merge 方法。
    - DataFrame 方法。
    - 以水平方式准确地合并两个 DataFrame。
    - 将调用它的 DataFrame 的列或索引与其他 DataFrame 的列或索引对齐。
    - 通过执行笛卡儿积处理连接列或索引上的重复值。
    - 默认为内连接，提供左连接、外连接和右连接选项。

在本秘笈中，我们将组合 DataFrame。第一种情况使用 concat 函数会更简单，而第二

种情况则使用.merge 方法会更简单。

11.4.1 实战操作

（1）使用循环将 2016 年、2017 年和 2018 年的股票数据读入一个 DataFrame 列表中。请注意，这里没有分 3 次调用 read_csv 函数。

```
>>> years = 2016, 2017, 2018
>>> stock_tables = [pd.read_csv(
...     f'data/stocks_{year}.csv', index_col='Symbol')
...     for year in years]
>>> stocks_2016, stocks_2017, stocks_2018 = stock_tables
>>> stocks_2016
        Shares  Low  High
Symbol
AAPL        80   95   110
TSLA        50   80   130
WMT         40   55    70

>>> stocks_2017
        Shares  Low  High
Symbol
AAPL        50  120   140
GE         100   30    40
IBM         87   75    95
SLB         20   55    85
TXN        500   15    23
TSLA       100  100   300

>>> stocks_2018
        Shares  Low  High
Symbol
AAPL        40  135   170
AMZN         8  900  1125
TSLA        50  220   400
```

（2）concat 函数是唯一能够以垂直方式组合 DataFrame 的 Pandas 方法。将 stock_tables 列表传递给 concat 函数，即可做到这一点。

```
>>> pd.concat(stock_tables, keys=[2016, 2017, 2018])
             Shares  Low  High
     Symbol
2016 AAPL        80   95   110
```

2017	TSLA	50	80	130
	WMT	40	55	70
	AAPL	50	120	140
	GE	100	30	40
...	...	...	...	...
2018	TXN	500	15	23
	TSLA	100	100	300
	AAPL	40	135	170
	AMZN	8	900	1125
	TSLA	50	220	400

（3）要以水平方式组合 DataFrame，可以将 axis 参数更改为 columns。

```
>>> pd.concat(dict(zip(years, stock_tables)), axis='columns')
```

	2016			...	2018		
	Shares	Low	High	...	Shares	Low	High
AAPL	80.0	95.0	110.0	...	40.0	135.0	170.0
AMZN	NaN	NaN	NaN	...	8.0	900.0	1125.0
GE	NaN	NaN	NaN	...	NaN	NaN	NaN
IBM	NaN	NaN	NaN	...	NaN	NaN	NaN
SLB	NaN	NaN	NaN	...	NaN	NaN	NaN
TSLA	50.0	80.0	130.0	...	50.0	220.0	400.0
TXN	NaN	NaN	NaN	...	NaN	NaN	NaN
WMT	40.0	55.0	70.0	...	NaN	NaN	NaN

（4）现在开始以水平方式组合 DataFrame，我们可以使用.join 和.merge 方法复制 concat 函数的功能。接下来，我们将使用.join 方法组合 stock_2016 和 stock_2017 两个 DataFrame。默认情况下，DataFrame 按其索引对齐。如果任何列具有相同的名称，那么必须给 lsuffix 或 rsuffix 参数提供一个值，以在结果中区分它们。

```
>>> stocks_2016.join(stocks_2017, lsuffix='_2016',
...     rsuffix='_2017', how='outer')
```

	Shares_2016	Low_2016	...	Low_2017	High_2017
Symbol			...		
AAPL	80.0	95.0	...	120.0	140.0
GE	NaN	NaN	...	30.0	40.0
IBM	NaN	NaN	...	75.0	95.0
SLB	NaN	NaN	...	55.0	85.0
TSLA	50.0	80.0	...	100.0	300.0
TXN	NaN	NaN	...	15.0	23.0
WMT	40.0	55.0	...	NaN	NaN

（5）要复制步骤（3）中 concat 函数的输出，可以将一个 DataFrame 列表传递给 .join

方法，具体如下。

```
>>> other = [stocks_2017.add_suffix('_2017'),
...     stocks_2018.add_suffix('_2018')]
>>> stocks_2016.add_suffix('_2016').join(other, how='outer')
      Shares_2016  Low_2016  ...  Low_2018  High_2018
AAPL         80.0      95.0  ...     135.0      170.0
TSLA         50.0      80.0  ...     220.0      400.0
WMT          40.0      55.0  ...       NaN        NaN
GE            NaN       NaN  ...       NaN        NaN
IBM           NaN       NaN  ...       NaN        NaN
SLB           NaN       NaN  ...       NaN        NaN
TXN           NaN       NaN  ...       NaN        NaN
AMZN          NaN       NaN  ...     900.0     1125.0
```

（6）现在来检查它们是否相等。

```
>>> stock_join = stocks_2016.add_suffix('_2016').join(other,
...     how='outer')

>>> stock_concat = (
...     pd.concat(
...         dict(zip(years, stock_tables)), axis="columns")
...     .swaplevel(axis=1)
...     .pipe(lambda df_:
...         df_.set_axis(df_.columns.to_flat_index(), axis=1))
...     .rename(lambda label:
...         "_".join([str(x) for x in label]), axis=1)
... )
>>> stock_join.equals(stock_concat)
True
```

（7）现在来讨论.merge 方法的示例。与 concat 函数和.join 方法不同，.merge 方法只能将两个 DataFrame 组合在一起。默认情况下，.merge 方法会尝试对齐每个 DataFrame 中具有相同名称的列中的值。但是，你也可以通过将布尔参数 left_index 和 right_index 设置为 True 来选择使其与索引对齐。下面将 2016 年和 2017 年的股票数据合并在一起。

```
>>> stocks_2016.merge(stocks_2017, left_index=True,
...     right_index=True)
        Shares_x  Low_x  High_x  Shares_y  Low_y  High_y
Symbol
AAPL          80     95     110        50    120     140
TSLA          50     80     130       100    100     300
```

（8）默认情况下，.merge 方法使用内连接，并自动为名称相同的列提供后缀。可以更改为外连接，然后对 2018 年的股票数据执行另一个外连接以复制 concat 函数的行为。

请注意，在 Pandas 1.0 中，merge 索引将被排序，而 concat 版本则不会。

```
>>> stock_merge = (stocks_2016
...     .merge(stocks_2017, left_index=True,
...            right_index=True, how='outer',
...            suffixes=('_2016', '_2017'))
...     .merge(stocks_2018.add_suffix('_2018'),
...            left_index=True, right_index=True,
...            how='outer')
... )
>>> stock_concat.sort_index().equals(stock_merge)
True
```

（9）现在可以换一个比较方式。来查看希望将列值对齐而不是索引或列标签本身对齐的数据集。.merge 方法就是针对这种情况而构建的。首先查看两个新的小型数据集——food_prices（食品价格）和 food_transactions（食品交易）。

```
>>> names = ['prices', 'transactions']
>>> food_tables = [pd.read_csv('data/food_{}.csv'.format(name))
...     for name in names]
>>> food_prices, food_transactions = food_tables
>>> food_prices
     item  store  price  Date
0    pear      A   0.99  2017
1    pear      B   1.99  2017
2   peach      A   2.99  2017
3   peach      B   3.49  2017
4  banana      A   0.39  2017
5  banana      B   0.49  2017
6   steak      A   5.99  2017
7   steak      B   6.99  2017
8   steak      B   4.99  2015


>>> food_transactions
   custid    item  store  quantity
0       1    pear      A         5
1       1  banana      A        10
2       2   steak      B         3
3       2    pear      B         1
```

```
4        2        peach         B           2
5        2        steak         B           1
6        2        coconut       B           4
```

（10）如果想查找每笔交易的总金额，则需要将这些表按 item 和 store 列连接。

```
>>> food_transactions.merge(food_prices, on=['item', 'store'])
   custid    item   store   quantity   price   Date
0       1    pear       A          5    0.99   2017
1       1  banana       A         10    0.39   2017
2       2   steak       B          3    6.99   2017
3       2   steak       B          3    4.99   2015
4       2   steak       B          1    6.99   2017
5       2   steak       B          1    4.99   2015
6       2    pear       B          1    1.99   2017
7       2   peach       B          2    3.49   2017
```

（11）现在，price（价格）已正确与其对应的 item（商品）和 store（商店）对齐，但是这里也存在一个问题。客户 2 总共有 4 个 steak（牛排）购买记录。由于牛排在商店 B 的每个表中出现两次，因此在它们之间产生了笛卡儿积，导致出现 4 行。另外还可以看到，缺少 coconut（椰子）商品，因为它没有相应的价格。

现在来修改这两个问题。

```
>>> food_transactions.merge(food_prices.query('Date == 2017'),
...     how='left')
   custid    item   store   quantity   price    Date
0       1    pear       A          5    0.99  2017.0
1       1  banana       A         10    0.39  2017.0
2       2   steak       B          3    6.99  2017.0
3       2    pear       B          1    1.99  2017.0
4       2   peach       B          2    3.49  2017.0
5       2   steak       B          1    6.99  2017.0
6       2 coconut       B          4     NaN     NaN
```

（12）可以使用 .join 方法复制上述操作，但是必须首先将 food_prices DataFrame 的连接列放入索引中。

```
>>> food_prices_join = food_prices.query('Date == 2017') \
...     .set_index(['item', 'store'])
>>> food_prices_join
              price   Date
item   store
pear   A       0.99   2017
```

```
          B           1.99    2017
peach     A           2.99    2017
          B           3.49    2017
banana    A           0.39    2017
          B           0.49    2017
steak     A           5.99    2017
          B           6.99    2017
```

(13).join 方法仅与传递给它的 DataFrame 的索引对齐,但可以使用调用它的 DataFrame 的索引或列。要使用调用.join 方法的 DataFrame 的列进行对齐,需要将此列传递给 on 参数。

```
>>> food_transactions.join(food_prices_join, on=['item', 'store'])
   custid      item  store  quantity  price    Date
0       1      pear      A         5   0.99  2017.0
1       1    banana      A        10   0.39  2017.0
2       2     steak      B         3   6.99  2017.0
3       2      pear      B         1   1.99  2017.0
4       2     peach      B         2   3.49  2017.0
5       2     steak      B         1   6.99  2017.0
6       2   coconut      B         4    NaN     NaN
```

可以看到,此输出结果与步骤(11)的结果是匹配的。

要使用 concat 函数复制上述操作,需要将 item 和 store 列放置到两个 DataFrames 的索引中。但是,在这种特定情况下,由于在至少一个 DataFrame 中出现重复的索引值(包含牛排商品和商店 B 的 DataFrame),因此该操作将产生错误。

```
>>> pd.concat([ food_transactions.set_index(['item', 'store']),
...             food_prices.set_index(['item', 'store'])],
...             axis='columns')
Traceback (most recent call last):
...
ValueError: cannot handle a non-unique multi-index!
```

11.4.2 原理解释

同时导入多个 DataFrame 时,重复编写 read_csv 函数可能很麻烦。自动执行此过程的一种方法是将所有文件名放在一个列表中,并使用 for 循环遍历它们。在步骤(1)中就使用了这种列表推导式(list comprehension)来完成此操作。

在步骤(1)的末尾,我们将 DataFrame 列表解压缩到它们自己的相应命名变量中,以便可以轻松、清晰地引用每个表。

使用 DataFrames 列表的好处是，它可以满足 concat 函数的确切要求，如步骤（2）所示。请注意步骤（2）如何使用 keys 参数命名每个数据块。

如步骤（3）所示，也可以将字典传递给 concat 函数。

在步骤（4）中，必须将 .join 的类型更改为 outer，以包括在传递给它的 DataFrame 中的所有行（这些行在调用它的 DataFrame 中不存在索引）。

在步骤（5）中，传递的 DataFrame 列表不能有任何共同的列。尽管有一个 rsuffix 参数，但是它仅在传递单个 DataFrame 时才起作用，如果传递的是 DataFrame 列表，则不起作用。要解决该问题，可以预先使用.add_suffix 方法更改列的名称，然后调用.join 方法。

在步骤（7）中使用了.merge 方法，该方法默认可以对齐两个 DataFrame 中相同的所有列名称。要更改此默认行为，并对齐一个 DataFrame 或两个 DataFrame 的索引，可以将 left_index 或 right_index 参数设置为 True。

步骤（8）通过两次调用.merge 方法复制了前面的操作。可以看到，当要对齐多个 DataFrame 的索引时，一般来说 concat 函数比.merge 方法要好得多。

在步骤（9）中讨论了.merge 方法具有优势的情况。.merge 方法是唯一能够按列值对齐调用它的 DataFrame 和传递给它的 DataFrame 的方法。

步骤（10）演示了两个 DataFrame 的合并操作。使用.merge 方法可以轻松完成该操作。on 参数并不是必需的，本示例为清晰起见而使用了它。

遗憾的是，在组合 DataFrame 时很容易复制或删除数据，在步骤（10）中可以看到这一点。因此，在组合数据后花一些时间进行完整性检查至关重要。在本示例中，food_prices 数据集在商店 B 中的牛排具有重复的价格，因此在步骤（11）中仅查询当前年份，从而消除了重复的行。此外，我们还更改为左连接方式，以确保无论价格是否存在，每一笔交易都将被保留。

在这些实例中可以使用.join 方法，但是必须首先将传递给它的 DataFrame 中的所有列移入索引中。最后，每当你打算按数据列中的值对齐数据时，concat 函数都不是一个好的选择。

总而言之，除非知道索引是对齐的，否则都可以使用.merge 方法。

11.4.3 扩展知识

可以在不知道文件名的情况下将所有文件从特定目录读取到 DataFrames 中。Python 提供了若干种遍历目录的方法，其中 glob 模块是一种流行的选择。

在本书配套资源包的 gas_prices 目录中包含 5 个不同的 CSV 文件，每个文件都包含从 2007 年开始的特定等级汽油的每周价格。每个文件只有两列，即日期和价格。比较理想的情况是，遍历所有文件，将它们读入 DataFrames 中，然后使用 concat 函数将它们组合在一起。

glob 模块包含了 glob 函数，该函数具有一个参数，该参数是一个字符串，指示要迭代的目录的位置。要获取目录中的所有文件，可使用星号（*）。在以下示例中，'*.csv' 将返回所有以 .csv 结尾的文件。glob 函数返回的结果是一个字符串文件名列表，可以将其传递给 read_csv 函数。

```
>>> import glob
>>> df_list = []
>>> for filename in glob.glob('data/gas prices/*.csv'):
...     df_list.append(pd.read_csv(filename, index_col='Week',
...         parse_dates=['Week']))
>>> gas = pd.concat(df_list, axis='columns')
>>> gas
            Midgrade  Premium  Diesel  All Grades  Regular
Week
2017-09-25     2.859    3.105   2.788       2.701    2.583
2017-09-18     2.906    3.151   2.791       2.750    2.634
2017-09-11     2.953    3.197   2.802       2.800    2.685
2017-09-04     2.946    3.191   2.758       2.794    2.679
2017-08-28     2.668    2.901   2.605       2.513    2.399
...              ...      ...     ...         ...      ...
2007-01-29     2.277    2.381   2.413       2.213    2.165
2007-01-22     2.285    2.391   2.430       2.216    2.165
2007-01-15     2.347    2.453   2.463       2.280    2.229
2007-01-08     2.418    2.523   2.537       2.354    2.306
2007-01-01     2.442    2.547   2.580       2.382    2.334
```

11.5 连接到 SQL 数据库

对于数据分析人员来说，学习和掌握 SQL 是一项必备的技能。世界上许多数据都存储在接收 SQL 语句的数据库中。关系数据库管理系统有许多种，其中 SQLite 是最受欢迎和易于使用的系统之一。

我们将探索由 SQLite 提供的 chinook 示例数据库，该数据库包含一个假想的音乐商店的 11 个数据表。当进入某个关系数据库中时，首先要做的事情之一就是研究数据库图（有时也称为实体关系图），以了解表之间的关系。图 11-1 对于理解此秘笈中的操作非常有帮助。

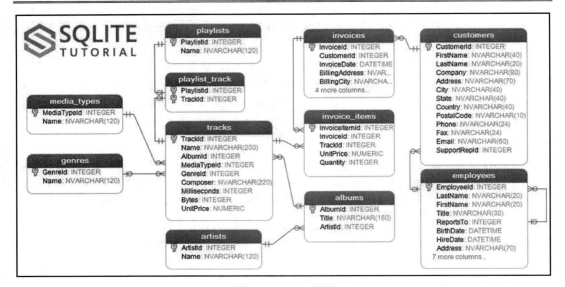

图 11-1　SQL 关系

要跟随此秘笈操作，需要安装 sqlalchemy Python 软件包。如果安装的是 Anaconda 发行版，则应该已经可以使用它。与数据库建立连接时，SQLAlchemy 是首选的 Pandas 工具。此秘笈将介绍如何连接到 SQLite 数据库，然后提出两个不同的查询，并通过使用.merge 方法将表连接在一起来进行响应。

11.5.1　实战操作

（1）在开始从 chinook 数据库中读取表之前，需要设置 SQLAlchemy 引擎。

```
>>> from sqlalchemy import create_engine
>>> engine = create_engine('sqlite:///data/chinook.db')
```

（2）现在回到 Pandas 中。使用 read_sql_table 函数读取 tracks（曲目）表。该表的名称是第一个参数，SQLAlchemy 引擎是第二个参数。

```
>>> tracks = pd.read_sql_table('tracks', engine)
>>> tracks
     TrackId  ...  UnitPrice
0          1  ...       0.99
1          2  ...       0.99
2          3  ...       0.99
3          4  ...       0.99
4          5  ...       0.99
```

第 11 章 组合 Pandas 对象

...	...	...	...
3498	3499	...	0.99
3499	3500	...	0.99
3500	3501	...	0.99
3501	3502	...	0.99
3502	3503	...	0.99

（3）在本秘笈的其余部分中，我们将在数据库图的帮助下响应几个不同的特定查询。首先，查找每种 genre（音乐类型）的平均歌曲长度。

```
>>> (pd.read_sql_table('genres', engine)
...     .merge(tracks[['GenreId', 'Milliseconds']],
...            on='GenreId', how='left')
...     .drop('GenreId', axis='columns')
... )
           Name  Milliseconds
0          Rock        343719
1          Rock        342562
2          Rock        230619
3          Rock        252051
4          Rock        375418
...         ...           ...
3498  Classical        286741
3499  Classical        139200
3500  Classical         66639
3501  Classical        221331
3502      Opera        174813
```

（4）现在可以轻松找到每种音乐类型的每首歌曲的平均长度。为了帮助简化解释，还可以将 Milliseconds（毫秒）列转换为 timedelta 数据类型。

```
>>> (pd.read_sql_table('genres', engine)
...     .merge(tracks[['GenreId', 'Milliseconds']],
...            on='GenreId', how='left')
...     .drop('GenreId', axis='columns')
...     .groupby('Name')
...     ['Milliseconds']
...     .mean()
...     .pipe(lambda s_: pd.to_timedelta(s_, unit='ms')
...                        .rename('Length'))
...     .dt.floor('s')
...     .sort_values()
... )
Name
Rock And Roll         00:02:14
```

```
Opera                    00:02:54
Hip Hop/Rap              00:02:58
Easy Listening           00:03:09
Bossa Nova               00:03:39
                           ...
Comedy                   00:26:25
TV Shows                 00:35:45
Drama                    00:42:55
Science Fiction          00:43:45
Sci Fi & Fantasy         00:48:31
Name: Length, Length: 25, dtype: timedelta64[ns]
```

(5) 现在，查找每个客户花费的总金额。这需要将 customers、invoices 和 invoice_items 表相互连接起来。

```
>>> cust = pd.read_sql_table('customers', engine,
...     columns=['CustomerId','FirstName',
...     'LastName'])
>>> invoice = pd.read_sql_table('invoices', engine,
...     columns=['InvoiceId','CustomerId'])
>>> invoice_items = pd.read_sql_table('invoice_items', engine,
...     columns=['InvoiceId', 'UnitPrice', 'Quantity'])
>>> (cust
...     .merge(invoice, on='CustomerId')
...     .merge(invoice_items, on='InvoiceId')
... )
      CustomerId  FirstName  ...  UnitPrice  Quantity
0              1       Luís  ...       1.99         1
1              1       Luís  ...       1.99         1
2              1       Luís  ...       0.99         1
3              1       Luís  ...       0.99         1
4              1       Luís  ...       0.99         1
...          ...        ...  ...        ...       ...
2235          59       Puja  ...       0.99         1
2236          59       Puja  ...       0.99         1
2237          59       Puja  ...       0.99         1
2238          59       Puja  ...       0.99         1
2239          59       Puja  ...       0.99         1
```

(6) 现在可以将数量乘以单价，然后查找每个客户花费的总金额。

```
>>> (cust
...     .merge(invoice, on='CustomerId')
...     .merge(invoice_items, on='InvoiceId')
...     .assign(Total=lambda df_:df_.Quantity * df_.UnitPrice)
```

```
...         .groupby(['CustomerId', 'FirstName', 'LastName'])
...         ['Total']
...         .sum()
...         .sort_values(ascending=False)
... )
CustomerId  FirstName   LastName
6           Helena      Holý          49.62
26          Richard     Cunningham    47.62
57          Luis        Rojas         46.62
46          Hugh        O'Reilly      45.62
45          Ladislav    Kovács        45.62
                                       ...
32          Aaron       Mitchell      37.62
31          Martha      Silk          37.62
29          Robert      Brown         37.62
27          Patrick     Gray          37.62
59          Puja        Srivastava    36.64
Name: Total, Length: 59, dtype: float64
```

11.5.2 原理解释

create_engine 函数需要一个连接字符串才能正常工作。SQLite 的连接字符串是数据库的位置，它位于 data 目录中。其他关系数据库管理系统具有更复杂的连接字符串。你将需要提供用户名、密码、主机名、端口以及（可选）数据库，另外可能还需要提供 SQL 方言（Dialect）和驱动程序（Driver）。连接字符串的一般格式如下。

```
dialect+driver://username:password@host:port/database
```

此外，你可能还需要单独安装特定的关系数据库的驱动程序。

在创建引擎之后，即可使用步骤（2）中的 read_sql_table 函数，轻松选择整个表并读入 DataFrames 中。数据库中的每个表都有一个主键（primary key）来标识每一行。在图 11-1 中使用了钥匙符号来标识它。

在步骤（3）中，通过 GenreId 将 genres 链接到了 tracks（曲目）。由于我们只关心曲目长度，因此在执行合并操作之前，将 tracks DataFrame 修剪为仅包含需要的列。在合并表之后，即可使用基本的.groupby 操作来响应查询。

在找到每种音乐类型的每首歌曲的平均长度之后，我们还更进一步，将以毫秒为单位的整数值转换为更容易阅读和理解的 Timedelta 对象。这里的关键是以字符串形式传递正确的度量单位。在有了一个 Timedelta Series 之后，可以使用.dt 属性访问.floor 方法，该方法会将时间向下取整到最接近的秒数。

响应步骤（5）的查询涉及 3 个表。这里同样可以将表修剪为仅包含需要的列，方式是将这些列名称传递给 columns 参数。使用.merge 方法时，具有相同名称的连接列将不被保留。在步骤（6）中，还可以专门为价格乘以数量的结果值分配一列，代码如下。

```
cust_inv['Total'] = cust_inv['Quantity'] * cust_inv['UnitPrice']
```

正如本书所强调的那样，我们在可能的情况下更喜欢链接操作，因此在这里你又看到了经常使用的.assign 方法。

11.5.3 扩展知识

如果你精通 SQL，则可以将 SQL 查询编写为字符串，然后将其传递给 read_sql_query 函数。例如，以下将重现步骤（4）的输出。

```
>>> sql_string1 = '''
... SELECT
...     Name,
...     time(avg(Milliseconds) / 1000, 'unixepoch') as avg_time
... FROM (
...     SELECT
...         g.Name,
...         t.Milliseconds
...     FROM
...         genres as g
...     JOIN
...         tracks as t on
...         g.genreid == t.genreid
...     )
... GROUP BY Name
... ORDER BY avg_time'''
>>> pd.read_sql_query(sql_string1, engine)
              Name    avg_time
0     Rock And Roll  00:02:14
1             Opera  00:02:54
2       Hip Hop/Rap  00:02:58
3    Easy Listening  00:03:09
4        Bossa Nova  00:03:39
..              ...       ...
20           Comedy  00:26:25
21         TV Shows  00:35:45
22            Drama  00:42:55
```

```
23       Science Fiction    00:43:45
24       Sci Fi & Fantasy   00:48:31
```

要重现步骤（6）的响应结果，可使用以下 SQL 查询。

```
>>> sql_string2 = '''
...     SELECT
...         c.customerid,
...         c.FirstName,
...         c.LastName,
...         sum(ii.quantity * ii.unitprice) as Total
...     FROM
...         customers as c
...     JOIN
...         invoices as i
...         on c.customerid = i.customerid
...     JOIN
...         invoice_items as ii
...         on i.invoiceid = ii.invoiceid
...     GROUP BY
...         c.customerid, c.FirstName, c.LastName
...     ORDER BY
...         Total desc'''
>>> pd.read_sql_query(sql_string2, engine)
    CustomerId  FirstName   LastName    Total
0            6     Helena       Holý    49.62
1           26    Richard Cunningham    47.62
2           57       Luis      Rojas    46.62
3           45   Ladislav     Kovács    45.62
4           46       Hugh    O'Reilly    45.62
..         ...        ...        ...      ...
54          53       Phil     Hughes    37.62
55          54      Steve     Murray    37.62
56          55       Mark     Taylor    37.62
57          56      Diego   Gutiérrez    37.62
58          59       Puja  Srivastava    36.64
```

第 12 章　时间序列分析

本章包含以下秘笈。
- 了解 Python 和 Pandas 日期工具之间的区别。
- 智能分割时间序列。
- 用时间数据过滤列。
- 使用仅适用于 DatetimeIndex 的方法。
- 计算每周犯罪数。
- 分别汇总每周犯罪和交通事故。
- 按星期和年份衡量犯罪情况。
- 使用匿名函数进行分组。
- 按 Timestamp 和其他列分组。

12.1　介　　绍

Pandas 最初是作为金融数据分析工具而开发出来的，因此，Pandas 为时间序列分析提供了很好的支持。时间序列是随着时间的推移而收集的数据点。一般来说，时间在每个数据点之间是平均间隔的。当然，观察结果也可能存在差距。Pandas 的功能包括操纵日期、在不同时间段内聚合、在不同时间段内采样等。

12.2　了解 Python 和 Pandas 日期工具之间的区别

在深入介绍 Pandas 时间序列分析功能之前，了解 Python 核心的日期和时间功能可能会有所帮助。datetime 模块提供了 3 种数据类型，即 date、time 和 datetime。
- date：从形式上来看，它是一个由年、月和日组成的时刻。例如，2020 年 12 月 31 日就是一个 date。
- time：由小时、分钟、秒和微秒（百万分之一秒）组成，并且与任何日期无关。例如，12 点 30 分就是一个 time。
- datetime：同时包含 date 和 time 这两个元素。

另外，Pandas 有一个封装日期和时间的对象，称为 Timestamp（时间戳）。Timestamp 具有纳秒级（十亿分之一秒）的精度，并且是从 NumPy 的 datetime64 数据类型派生而来的。

Python 和 Pandas 都有一个 timedelta 对象，在进行日期加减时非常有用。

在本秘笈中，我们将首先讨论 Python 的 datetime 模块，然后转向讨论 Pandas 中相应的日期工具。

12.2.1　实战操作

（1）将 datetime 模块导入我们的命名空间中，然后分别创建一个 date、time 和 datetime 对象。

```
>>> import pandas as pd
>>> import numpy as np
>>> import datetime
>>> date = datetime.date(year=2013, month=6, day=7)
>>> time = datetime.time(hour=12, minute=30,
...     second=19, microsecond=463198)
>>> dt = datetime.datetime(year=2013, month=6, day=7,
...     hour=12, minute=30, second=19,
...     microsecond=463198)
>>> print(f"date is {date}")
date is 2013-06-07

>>> print(f"time is {time}")
time is 12:30:19.463198

>>> print(f"datetime is {dt}")
datetime is 2013-06-07 12:30:19.463198
```

（2）构造并输出一个 timedelta 对象，这是 datetime 模块中的另一种主要数据类型。

```
>>> td = datetime.timedelta(weeks=2, days=5, hours=10,
...     minutes=20, seconds=6.73,
...     milliseconds=99, microseconds=8)
>>> td
datetime.timedelta(days=19, seconds=37206, microseconds=829008)
```

（3）将此 td 添加到步骤（1）的 date 和 dt 对象中。

```
>>> print(f'new date is {date+td}')
new date is 2013-06-26
```

```
>>> print(f'new datetime is {dt+td}')
new datetime is 2013-06-26 22:50:26.292206
```

(4)尝试将 timedelta 添加到 time 对象中,结果出现了错误。

```
>>> time + td
Traceback (most recent call last):
...
TypeError: unsupported operand type(s) for +: 'datetime.time' and
'datetime.timedelta'
```

(5)现在来看看 Pandas 及其 Timestamp 对象,该对象是一个具有纳秒级精度的时刻。Timestamp 构造函数非常灵活,可以处理各种输入。

```
>>> pd.Timestamp(year=2012, month=12, day=21, hour=5,
...     minute=10, second=8, microsecond=99)
Timestamp('2012-12-21 05:10:08.000099')

>>> pd.Timestamp('2016/1/10')
Timestamp('2016-01-10 00:00:00')

>>> pd.Timestamp('2014-5/10')
Timestamp('2014-05-10 00:00:00')

>>> pd.Timestamp('Jan 3, 2019 20:45.56')
Timestamp('2019-01-03 20:45:33')

>>> pd.Timestamp('2016-01-05T05:34:43.123456789')
Timestamp('2016-01-05 05:34:43.123456789')
```

(6)还可以将单个整数或浮点数传递给 Timestamp 构造函数,该构造函数返回的日期等于 UNIX 纪元(1970 年 1 月 1 日)之后的纳秒数。

```
>>> pd.Timestamp(500)
Timestamp('1970-01-01 00:00:00.000000500')

>>> pd.Timestamp(5000, unit='D')
Timestamp('1983-09-10 00:00:00')
```

(7)Pandas 提供了 to_datetime 函数,该函数的功能与 Timestamp 构造函数相似,但是该函数还包含一些不同的参数以应对特殊情况。这对于将 DataFrames 中的字符串列转换为日期数据很有用。

to_datetime 函数也可以用于标量日期。请参考以下示例。

```
>>> pd.to_datetime('2015-5-13')
Timestamp('2015-05-13 00:00:00')

>>> pd.to_datetime('2015-13-5', dayfirst=True)
Timestamp('2015-05-13 00:00:00')

>>> pd.to_datetime('Start Date: Sep 30, 2017 Start Time: 1:30 pm',
...     format='Start Date: %b %d, %Y Start Time: %I:%M %p')
Timestamp('2017-09-30 13:30:00')

>>> pd.to_datetime(100, unit='D', origin='2013-1-1')
Timestamp('2013-04-11 00:00:00')
```

（8）to_datetime 函数具有更多功能，它能够将整个列表或字符串 Series 或整数转换为 Timestamp 对象。由于我们更有可能与 Series 或 DataFrames 交互，而不是与单个标量值交互，因此使用 to_datetime 函数比使用 Timestamp 的可能性更大。

```
>>> s = pd.Series([10, 100, 1000, 10000])
>>> pd.to_datetime(s, unit='D')
0   1970-01-11
1   1970-04-11
2   1972-09-27
3   1997-05-19
dtype: datetime64[ns]

>>> s = pd.Series(['12-5-2015', '14-1-2013',
...     '20/12/2017', '40/23/2017'])

>>> pd.to_datetime(s, dayfirst=True, errors='coerce')
0   2015-05-12
1   2013-01-14
2   2017-12-20
3          NaT
dtype: datetime64[ns]

>>> pd.to_datetime(['Aug 3 1999 3:45:56', '10/31/2017'])
DatetimeIndex(['1999-08-03 03:45:56', '2017-10-31 00:00:00'],
    dtype='datetime64[ns]', freq=None)
```

（9）Pandas 具有 Timedelta 构造函数和 to_timedelta 函数来表示时间量（与 Timestamp 构造函数和 to_datetime 函数类似）。

Timedelta 构造函数和 to_timedelta 函数都可以创建单个 Timedelta 对象。

与 to_datetime 函数一样，to_timedelta 函数也具有更多功能，并且可以将整个列表或 Series 转换为 Timedelta 对象。

```
>>> pd.Timedelta('12 days 5 hours 3 minutes 123456789 nanoseconds')
Timedelta('12 days 05:03:00.123456')

>>> pd.Timedelta(days=5, minutes=7.34)
Timedelta('5 days 00:07:20.400000')

>>> pd.Timedelta(100, unit='W')
Timedelta('700 days 00:00:00')

>>> pd.to_timedelta('67:15:45.454')
Timedelta('2 days 19:15:45.454000')

>>> s = pd.Series([10, 100])
>>> pd.to_timedelta(s, unit='s')
0   00:00:10
1   00:01:40
dtype: timedelta64[ns]

>>> time_strings = ['2 days 24 minutes 89.67 seconds',
...                 '00:45:23.6']
>>> pd.to_timedelta(time_strings)
TimedeltaIndex(['2 days 00:25:29.670000', '0 days
00:45:23.600000'], dtype='timedelta64[ns]', freq=None)
```

（10）可以将 Timedelta 构造函数添加到另一个 Timestamp 构造函数中，或从中减去一个 Timedelta 构造函数。Timedelta 构造函数甚至可以彼此相除以返回一个浮点数。

```
>>> pd.Timedelta('12 days 5 hours 3 minutes') * 2
Timedelta('24 days 10:06:00')

>>> (pd.Timestamp('1/1/2017') +
...      pd.Timedelta('12 days 5 hours 3 minutes') * 2)
Timestamp('2017-01-25 10:06:00')

>>> td1 = pd.to_timedelta([10, 100], unit='s')
>>> td2 = pd.to_timedelta(['3 hours', '4 hours'])
>>> td1 + td2
TimedeltaIndex(['03:00:10', '04:01:40'], dtype='timedelta64[ns]',
```

```
freq=None)

>>> pd.Timedelta('12 days') / pd.Timedelta('3 days')
4.0
```

（11）Timestamp 构造函数和 Timedelta 构造函数都具有大量可用作属性和方法的功能。以下是一些示例。

```
>>> ts = pd.Timestamp('2016-10-1 4:23:23.9')
>>> ts.ceil('h')
Timestamp('2016-10-01 05:00:00')

>>> ts.year, ts.month, ts.day, ts.hour, ts.minute, ts.second
(2016, 10, 1, 4, 23, 23)

>>> ts.dayofweek, ts.dayofyear, ts.daysinmonth
(5, 275, 31)

>>> ts.to_pydatetime()
datetime.datetime(2016, 10, 1, 4, 23, 23, 900000)

>>> td = pd.Timedelta(125.8723, unit='h')
>>> td
Timedelta('5 days 05:52:20.280000')

>>> td.round('min')
Timedelta('5 days 05:52:00')

>>> td.components
Components(days=5, hours=5, minutes=52, seconds=20,
milliseconds=280, microseconds=0, nanoseconds=0)

>>> td.total_seconds()
453140.28
```

12.2.2 原理解释

datetime 模块是 Python 标准库的一部分。数据分析人员最好能够熟悉该模块，因为你可能会在很多情况下应用到该模块。此外，datetime 模块只有 6 种类型的对象，即 date、time、datetime、timedelta、timezone 和 tzinfo。

Pandas Timestamp 和 Timedelta 对象具有 Python datetime 模块所具有的全部功能，甚

至还包括更多其他功能。因此，在处理时间序列时，完全可以停留在 Pandas 中。

步骤（1）和步骤（2）显示了如何使用 datetime 模块创建 datetime、date、time 和 timedelta。只有整数可以用作日期或时间的参数。

可以将此操作与步骤（5）进行比较。在步骤（5）中，Pandas Timestamp 构造函数可以接收相同的参数以及各种日期字符串。除了整数部分和字符串，步骤（6）还显示了如何将单个数字标量用作日期。此标量的单位默认为纳秒（ns），但在第二条语句中更改为天（D），其他可选项还包括小时（h）、分钟（min）、秒（s）、毫秒（ms）和微秒（μs）。

步骤（2）详细说明了 datetime 模块的 timedelta 对象及其所有参数的构造。同样，也可以将其与步骤（9）中显示的 Pandas Timedelta 构造函数进行比较，该构造函数可接收相同的参数以及字符串和标量数字。

除了只能创建单个对象的 Timestamp 和 Timedelta 构造函数，to_datetime 和 to_timedelta 函数还可以将整数或字符串的整个序列转换为所需的类型。这些函数还提供了构造函数不可用的其他几个参数。其中之一是 errors 参数，默认情况下是字符串值 raise，但也可以将其设置为 ignore 或 coerce。

每当无法转换字符串日期时，errors 参数将确定要执行的操作。当将该参数设置为 raise 时，将抛出异常，程序执行将停止；当将该参数设置为 ignore 时，将返回原始序列，就像进入函数之前一样；当将该参数设置为 coerce 时，将使用 NaT（not a time，不是时间）对象来表示新值。

步骤（8）中对 to_datetime 函数的第二次调用可以将所有值正确转换为 Timestamp 对象，但最后一个除外，它被迫变为 NaT 对象。

在只有 to_datetime 函数可用的参数中，另一个参数是 format，这在字符串包含 Pandas 无法自动识别的特定日期模式时特别有用。在步骤（7）的第三条语句中，我们在其他一些字符中嵌入了 datetime，然后使用其各自的格式指令替换了字符串的日期和时间，这就是通过 format 参数实现的。

日期格式指令以单个百分号（%）开头，后跟单个字符。每个指令都将指定日期或时间的某些部分。有关所有指令的表格，可访问 Python 官方说明文档，其网址如下。

http://bit.ly/2kePoRe

12.3 智能分割时间序列

在本书第 6 章"选择数据子集"中，已经详细介绍过 DataFrame 的选择和切片。当 DataFrame 具有 DatetimeIndex 时，将出现更多选择和切片的机会。

在本秘笈中,我们将使用部分日期匹配(partial date matching)来选择和分割具有 DatetimeIndex 的 DataFrame。

12.3.1 实战操作

(1)从 hdf5 文件 Crimes.h5 中读入丹佛犯罪数据集,并输出列数据类型和前几行。hdf5 文件格式允许有效存储大量数据,并且与 CSV 文本文件不同。

```
>>> crime = pd.read_hdf('data/crime.h5', 'crime')
>>> crime.dtypes
OFFENSE_TYPE_ID              category
OFFENSE_CATEGORY_ID          category
REPORTED_DATE          datetime64[ns]
GEO_LON                       float64
GEO_LAT                       float64
NEIGHBORHOOD_ID              category
IS_CRIME                        int64
IS_TRAFFIC                      int64
dtype: object
```

(2)请注意,这里有 3 个 category(分类)列和 1 个 Timestamp(由 NumPy 的 datetime64 对象表示)。这些数据类型是在创建数据文件时存储的,这与仅存储原始文本的 CSV 文件不同。将 REPORTED_DATE(报告日期)列设置为索引,以使智能 Timestamp 切片成为可能。

```
>>> crime = crime.set_index('REPORTED_DATE')
>>> crime
                              OFFENSE_TYPE_ID  ...
REPORTED_DATE                                  ...
2014-06-29 02:01:00   traffic-accident-dui-duid  ...
2014-06-29 01:54:00   vehicular-eluding-no-chase ...
2014-06-29 02:00:00        disturbing-the-peace  ...
2014-06-29 02:18:00                      curfew  ...
2014-06-29 04:17:00          aggravated-assault  ...
...                                         ...  ...
2017-09-13 05:48:00   burglary-business-by-force ...
2017-09-12 20:37:00  weapon-unlawful-discharge-of ...
2017-09-12 16:32:00        traf-habitual-offender ...
2017-09-12 13:04:00      criminal-mischief-other  ...
2017-09-12 09:30:00                 theft-other  ...
```

(3)像以前一样,可以通过将值传递给.loc 属性来选择等于单个索引的所有行。

```
>>> crime.loc['2016-05-12 16:45:00']
                    OFFENSE_TYPE_ID  OFFENSE_CATEGORY_ID    GEO_LON
                                     OFFENSE_TYPE_ID    ...  IS_TRAFFIC
REPORTED_DATE                                           ...
2016-05-12 16:45:00        traffic-accident        ...     1
2016-05-12 16:45:00        traffic-accident        ...     1
2016-05-12 16:45:00     fraud-identity-theft       ...     0
```

(4)使用索引中的 Timestamp,可以选择部分匹配索引值的所有行。例如,如果想要查找 2016 年 5 月 5 日以后的所有犯罪记录,可以按以下方式选择。

```
>>> crime.loc['2016-05-12']
                          OFFENSE_TYPE_ID    ...  IS_TRAFFIC
REPORTED_DATE                                ...
2016-05-12 23:51:00    criminal-mischief-other   ...     0
2016-05-12 18:40:00       liquor-possession      ...     0
2016-05-12 22:26:00       traffic-accident       ...     1
2016-05-12 20:35:00        theft-bicycle         ...     0
2016-05-12 09:39:00    theft-of-motor-vehicle    ...     0
...                          ...               ...    ...
2016-05-12 17:55:00      public-peace-other      ...     0
2016-05-12 19:24:00      threats-to-injure       ...     0
2016-05-12 22:28:00         sex-aslt-rape        ...     0
2016-05-12 15:59:00    menacing-felony-w-weap    ...     0
2016-05-12 16:39:00          assault-dv          ...     0
```

(5)我们不仅可以模糊选择一个日期,而且可以选择整月、整年,甚至也可以细化到一天中的某个小时。

```
>>> crime.loc['2016-05'].shape
(8012, 7)
>>> crime.loc['2016'].shape
(91076, 7)
>>> crime.loc['2016-05-12 03'].shape
(4, 7)
```

(6)选择字符串还可以包含月份名称。

```
>>> crime.loc['Dec 2015'].sort_index()
                            OFFENSE_TYPE_ID    ...
REPORTED_DATE                                  ...
2015-12-01 00:48:00      drug-cocaine-possess  ...
```

```
2015-12-01  00:48:00              theft-of-motor-vehicle   ...
2015-12-01  01:00:00             criminal-mischief-other   ...
2015-12-01  01:10:00                          traf-other   ...
2015-12-01  01:10:00               traf-habitual-offender   ...
       ...                                            ...  ...
2015-12-31  23:35:00                drug-cocaine-possess   ...
2015-12-31  23:40:00                     traffic-accident   ...
2015-12-31  23:44:00                drug-cocaine-possess   ...
2015-12-31  23:45:00         violation-of-restraining-order ...
2015-12-31  23:50:00         weapon-poss-illegal-dangerous  ...
```

（7）包含月份名称的许多其他字符串模式也可以使用。

```
>>> crime.loc['2016 Sep, 15'].shape
(252, 7)
>>> crime.loc['21st October 2014 05'].shape
(4, 7)
```

（8）除了选择，你还可以使用切片符号选择精确的数据范围。以下示例将包含从 2015 年 3 月 4 日—2016 年 1 月 1 日结束的所有值。

```
>>> crime.loc['2015-3-4':'2016-1-1'].sort_index()
                                OFFENSE_TYPE_ID   ...
REPORTED_DATE                                     ...
2015-03-04  00:11:00                  assault-dv  ...
2015-03-04  00:19:00                  assault-dv  ...
2015-03-04  00:27:00            theft-of-services ...
2015-03-04  00:49:00      traffic-accident-hit-and-run ...
2015-03-04  01:07:00        burglary-business-no-force ...
       ...                                        ...  ...
2016-01-01  23:15:00      traffic-accident-hit-and-run ...
2016-01-01  23:16:00               traffic-accident   ...
2016-01-01  23:40:00              robbery-business    ...
2016-01-01  23:45:00           drug-cocaine-possess    ...
2016-01-01  23:48:00         drug-poss-paraphernalia   ...
```

（9）可以看到，在返回的结果中包含了从开始日期到结束日期（不考虑具体的时间）所提交的所有犯罪记录。使用基于标签的 .loc 属性，还可以为切片的开始或结束部分提供更多的精度，示例如下。

```
>>> crime.loc['2015-3-4 22':'2016-1-1 11:22:00'].sort_index()
                                OFFENSE_TYPE_ID   ...
REPORTED_DATE                                     ...
2015-03-04  22:25:00   traffic-accident-hit-and-run ...
```

```
2015-03-04  22:30:00              traffic-accident  ...
2015-03-04  22:32:00  traffic-accident-hit-and-run  ...
2015-03-04  22:33:00  traffic-accident-hit-and-run  ...
2015-03-04  22:36:00      theft-unauth-use-of-ftd  ...
    ...                                        ...  ...
2016-01-01  11:10:00       theft-of-motor-vehicle  ...
2016-01-01  11:11:00              traffic-accident  ...
2016-01-01  11:11:00  traffic-accident-hit-and-run  ...
2016-01-01  11:16:00                    traf-other  ...
2016-01-01  11:22:00              traffic-accident  ...
```

12.3.2 原理解释

hdf5 文件的功能之一是能够保留每一列的数据类型，从而减少了所需的内存。在本示例中，有 3 列被存储为 Pandas 数据类型中的 category 而不是作为 object。将这 3 列存储为 object 将导致内存使用量增加 4 倍。

```
>>> mem_cat = crime.memory_usage().sum()
>>> mem_obj = (crime
...     .astype({'OFFENSE_TYPE_ID':'object',
...              'OFFENSE_CATEGORY_ID':'object',
...              'NEIGHBORHOOD_ID':'object'})
...     .memory_usage(deep=True)
...     .sum()
... )
>>> mb = 2 ** 20
>>> round(mem_cat / mb, 1), round(mem_obj / mb, 1)
(29.4, 122.7)
```

要使用索引操作符按日期选择和切片行，则索引必须包含日期值。所以，在步骤（2）中，将 REPORTED_DATE 列移到索引中并创建一个 DatetimeIndex 作为新索引。

```
>>> crime.index[:2]
DatetimeIndex(['2014-06-29 02:01:00', '2014-06-29 01:54:00'],
dtype='datetime64[ns]', name='REPORTED_DATE', freq=None)
```

在有了 DatetimeIndex 之后，可以使用 .loc 属性，通过各种各样的字符串来选择行。实际上，所有可以发送到 Pandas Timestamp 构造函数的字符串都可以正常工作。

令人惊讶的是，对于此秘笈中的任何选择或切片，并不是必须使用 .loc 属性，索引操作符本身也可以按相同的方式工作。例如，步骤（7）的第二个语句也可以编写为如下形式。

```
crime['21st October 2014 05']
```

就个人而言，在选择行时笔者更喜欢使用.loc属性，而不是使用DataFrame的索引操作符。究其原因，.loc索引器是显式的（Python之禅说过，显式优于隐式），并且毫无疑问，传递给它的第一个值始终用于选择行。

步骤（8）和步骤（9）演示了如何对Timestamp使用切片。结果中将包括与切片的开始或结束值部分匹配的任何日期。

12.3.3　扩展知识

最初的犯罪数据DataFrame虽然未经过排序，但是切片仍然可以按预期进行。当然，如果能够先对索引进行排序，则将大幅提高性能。来看看以两种不同的方式重复步骤（8）中的切片操作，结果会有什么不同。

```
>>> %timeit crime.loc['2015-3-4':'2016-1-1']
12.2 ms ± 1.93 ms per loop (mean ± std. dev. of 7 runs, 100 loops each)

>>> crime_sort = crime.sort_index()
>>> %timeit crime_sort.loc['2015-3-4':'2016-1-1']
14.4 ms ± 41.9 µs per loop (mean ± std. dev. of 7 runs, 1000 loops each)
```

由此可见，排序后的DataFrame的性能是原始数据的8倍。

12.4　用时间数据过滤列

本节讨论如何过滤具有DatetimeIndex的数据。一般来说，你会有一些在其中包含日期的列，有些列作为索引是没有意义的。在本节中，我们将用列复制12.3节"智能分割时间序列"的切片。遗憾的是，切片构造无法在列上使用，因此我们将不得不采取其他措施。

12.4.1　实战操作

（1）从hdf5文件Crimes.h5中读入丹佛犯罪数据集，并检查列类型。

```
>>> crime = pd.read_hdf('data/crime.h5', 'crime')
>>> crime.dtypes
OFFENSE_TYPE_ID            category
OFFENSE_CATEGORY_ID        category
REPORTED_DATE              datetime64[ns]
```

```
GEO_LON                         float64
GEO_LAT                         float64
NEIGHBORHOOD_ID                 category
IS_CRIME                          int64
IS_TRAFFIC                        int64
dtype: object
```

（2）选择 REPORTED_DATE 列具有特定值的所有行。我们将使用布尔数组进行过滤。请注意，这里可以将 datetime 列与字符串进行比较。

```
>>> (crime
...     [crime.REPORTED_DATE == '2016-05-12 16:45:00']
... )
            OFFEN/PE_ID      ...    IS_TRAFFIC
300905     traffic-accident  ...         1
302354     traffic-accident  ...         1
302373     fraud-identity-theft ...      0
```

（3）选择部分日期匹配的所有行。如果使用相等运算符（==）尝试此操作，则此操作将失败。虽然没有收到错误，但是没有返回任何行。

```
>>> (crime
...     [crime.REPORTED_DATE == '2016-05-12']
... )
Empty DataFrame
Columns: [OFFENSE_TYPE_ID, OFFENSE_CATEGORY_ID, REPORTED_DATE,
GEO_LON, GEO_LAT, NEIGHBORHOOD_ID, IS_CRIME, IS_TRAFFIC]
Index: []
```

如果我们尝试与 .dt.date 属性进行比较，则此操作也会失败。这是因为这是一系列 Python datetime.date 对象，并且这些对象不支持这种比较。

```
>>> (crime
...     [crime.REPORTED_DATE.dt.date == '2016-05-12']
... )
Empty DataFrame
Columns: [OFFENSE_TYPE_ID, OFFENSE_CATEGORY_ID, REPORTED_DATE,
GEO_LON, GEO_LAT, NEIGHBORHOOD_ID, IS_CRIME, IS_TRAFFIC]
Index: []
```

（4）如果要部分日期匹配，可以使用 .between 方法，该方法支持部分日期字符串。请注意，默认情况下包含开始日期和结束日期（参数名称分别为 left 和 right）。如果某行的日期为 2016 年 5 月 13 日午夜，则可以按以下方式将其包含在内。

```
>>> (crime
...     [crime.REPORTED_DATE.between(
...         '2016-05-12', '2016-05-13')]
... )
                    OFFEN/PE_ID   ...   IS_TRAFFIC
295715    criminal-mischief-other   ...        0
296474         liquor-possession    ...        0
297204           traffic-accident   ...        1
299383             theft-bicycle   ...        0
299389       theft-of-motor-vehicle ...        0
...                           ...   ...       ...
358208         public-peace-other   ...        0
358448          threats-to-injure   ...        0
363134              sex-aslt-rape   ...        0
365959       menacing-felony-w-weap ...        0
378711                 assault-dv   ...        0
```

（5）因为.between 方法支持部分日期字符串，所以可使用该方法复制 12.3 节"智能分割时间序列"中的大多数切片功能。例如，可以使用 .between 方法匹配一个月、一年或一天中的某个小时。

```
>>> (crime
...     [crime.REPORTED_DATE.between(
...         '2016-05', '2016-06')]
...     .shape
... )
(8012, 8)

>>> (crime
...     [crime.REPORTED_DATE.between(
...         '2016', '2017')]
...     .shape
... )
(91076, 8)

>>> (crime
...     [crime.REPORTED_DATE.between(
...         '2016-05-12 03', '2016-05-12 04')]
...     .shape
... )
(4, 8)
```

（6）还可以使用其他字符串模式。

```
>>> (crime
...     [crime.REPORTED_DATE.between(
...         '2016 Sep, 15', '2016 Sep, 16')]
...     .shape
... )
(252, 8)

>>> (crime
...     [crime.REPORTED_DATE.between(
...         '21st October 2014 05', '21st October 2014 06')]
...     .shape
... )
(4, 8)
```

（7）因为.loc 是封闭的，并且包括开始和结束值，所以 .between 方法其实是模仿了它的功能。但是，在部分日期字符串中会稍有不同。例如，在以 2016-1-1 结束的切片中，将包含 2016 年 1 月 1 日的所有值。而 .between 方法在使用该值作为结束值时，将仅包括直到当天开始的值。因此，要准确复制切片['2015-3-4' : '2016-1-1']，需要添加结束日期的最后时间。

```
>>> (crime
...     [crime.REPORTED_DATE.between(
...         '2015-3-4','2016-1-1 23:59:59')]
...     .shape
... )
(75403, 8)
```

（8）可以根据需要调整上述日期。以下语句复制了 12.3 节 "智能分割时间序列" 秘笈步骤（9）的操作。

```
>>> (crime
...     [crime.REPORTED_DATE.between(
...         '2015-3-4 22','2016-1-1 11:22:00')]
...     .shape
... )
(75071, 8)
```

12.4.2 原理解释

Pandas 库可以切片索引值，但不能切片列。要在列上复制 DatetimeIndex 切片，需要使用 .between 方法。该方法的主体只有以下 7 行代码。

```
def between(self, left, right, inclusive=True):
if inclusive:
lmask = self >= left
rmask = self <= right
else:
lmask = self > left
rmask = self < right

return lmask & rmask
```

这使我们了解到，可以构建 mask 并根据需要对其进行组合。例如，可以使用两个 mask 来复制上面步骤（8）中的操作。

```
>>> lmask = crime.REPORTED_DATE >= '2015-3-4 22'
>>> rmask = crime.REPORTED_DATE <= '2016-1-1 11:22:00'
>>> crime[lmask & rmask].shape
(75071, 8)
```

12.4.3 扩展知识

现在来比较在索引上执行的 .loc 和在列上执行的 .between 方法的计时。

```
>>> ctseries = crime.set_index('REPORTED_DATE')
>>> %timeit ctseries.loc['2015-3-4':'2016-1-1']
11 ms ± 3.1 ms per loop (mean ± std. dev. of 7 runs, 100 loops each)

>>> %timeit crime[crime.REPORTED_DATE.between('2015-3-4','2016-1-1')]
20.1 ms ± 525 µs per loop (mean ± std. dev. of 7 runs, 10 loops each)
```

可以看到，将日期信息包含在索引中可以略微提高一些速度。如果需要在单个列上执行日期切片，则将索引设置为日期列可能很有意义。

另外需要注意的是，将索引设置为列也是有开销的，并且如果仅切片一次，则索引开销会使这两个操作的时间大致相同。

12.5　使用仅适用于 DatetimeIndex 的方法

在 DataFrame 和 Series 的方法中，有许多仅适用于 DatetimeIndex 的方法。如果索引是其他类型，则这些方法将失败。

在本秘笈中，我们将首先使用方法按照时间成分选择数据行。然后，将学习功能强大的 DateOffset 对象及其别名。

12.5.1 实战操作

（1）读取 hdf5 犯罪数据集，将索引设置为 REPORTED_DATE 列，并确保获得的是 DatetimeIndex。

```
>>> crime = (pd.read_hdf('data/crime.h5', 'crime')
...          .set_index('REPORTED_DATE')
...         )

>>> type(crime.index)
<class 'pandas.core.indexes.datetimes.DatetimeIndex'>
```

（2）使用 .between_time 方法选择凌晨 2 点～凌晨 5 点发生的所有犯罪，无论日期是在哪一天。

```
>>> crime.between_time('2:00', '5:00', include_end=False)
                              OFFENSE_TYPE_ID   ...
REPORTED_DATE                                   ...
2014-06-29 02:01:00   traffic-accident-dui-duid ...
2014-06-29 02:00:00        disturbing-the-peace ...
2014-06-29 02:18:00                      curfew ...
2014-06-29 04:17:00          aggravated-assault ...
2014-06-29 04:22:00   violation-of-restraining-order ...
...                                         ... ...
2017-08-25 04:41:00       theft-items-from-vehicle ...
2017-09-13 04:17:00        theft-of-motor-vehicle ...
2017-09-13 02:21:00                assault-simple ...
2017-09-13 03:21:00     traffic-accident-dui-duid ...
2017-09-13 02:15:00  traffic-accident-hit-and-run ...
```

（3）使用 .at_time 方法选择所有日期的特定时间。

```
>>> crime.at_time('5:47')
                              OFFENSE_TYPE_ID   ...
REPORTED_DATE                                   ...
2013-11-26 05:47:00       criminal-mischief-other ...
2017-04-09 05:47:00     criminal-mischief-mtr-veh ...
2017-02-19 05:47:00       criminal-mischief-other ...
2017-02-16 05:47:00            aggravated-assault ...
2017-02-12 05:47:00           police-interference ...
...                                           ... ...
2013-09-10 05:47:00              traffic-accident ...
```

```
2013-03-14  05:47:00                    theft-other  ...
2012-10-08  05:47:00       theft-items-from-vehicle  ...
2013-08-21  05:47:00       theft-items-from-vehicle  ...
2017-08-23  05:47:00   traffic-accident-hit-and-run  ...
```

（4）.first 方法提供了一种选择前 n 个时间段的好方法，其中 n 是整数。这些时间段由 pd.offsets 模块中的 DateOffset 对象表示。DataFrame 必须按其索引进行排序，以确保此方法可以工作。例如，可以选择前 6 个月的犯罪数据。

```
>>> crime_sort = crime.sort_index()
>>> crime_sort.first(pd.offsets.MonthBegin(6))
                                  OFFENSE_TYPE_ID   ...
REPORTED_DATE                                       ...
2012-01-02  00:06:00             aggravated-assault ...
2012-01-02  00:06:00   violation-of-restraining-order ...
2012-01-02  00:16:00       traffic-accident-dui-duid ...
2012-01-02  00:47:00                traffic-accident ...
2012-01-02  01:35:00             aggravated-assault  ...
         ...                                    ...  ...
2012-06-30  23:40:00       traffic-accident-dui-duid ...
2012-06-30  23:44:00                traffic-accident ...
2012-06-30  23:50:00         criminal-mischief-mtr-veh ...
2012-06-30  23:54:00    traffic-accident-hit-and-run ...
2012-07-01  00:01:00                  robbery-street ...
```

（5）这捕获了从 1～6 月的数据，但令人惊讶的是，它还在 7 月选择了一行。原因是 Pandas 使用了索引中 first 元素的时间分量，在此示例中为 6min。下面使用 MonthEnd，这是一个稍微不同的偏移量（offset）。

```
>>> crime_sort.first(pd.offsets.MonthEnd(6))
                                  OFFENSE_TYPE_ID   ...
REPORTED_DATE                                       ...
2012-01-02  00:06:00             aggravated-assault ...
2012-01-02  00:06:00   violation-of-restraining-order ...
2012-01-02  00:16:00       traffic-accident-dui-duid ...
2012-01-02  00:47:00                traffic-accident ...
2012-01-02  01:35:00             aggravated-assault  ...
         ...                                    ...  ...
2012-06-29  23:01:00             aggravated-assault  ...
2012-06-29  23:11:00                traffic-accident ...
2012-06-29  23:41:00                  robbery-street ...
2012-06-29  23:57:00                  assault-simple ...
2012-06-30  00:04:00                traffic-accident ...
```

（6）这捕获了几乎相同数量的数据，但是如果仔细观察就能发现，6 月 30 日之后仅捕获了一行。这同样是因为 first 索引的时间分量的问题。确切的搜索结果应该还能找到 2012-06-30 00:06:00 之类的记录。

那么，如何才能准确地获得前 6 个月的数据呢？有两种方法。所有 DateOffset 对象都有一个 normalize 参数，当将其设置为 True 时，会将所有时间分量设置为 0。以下操作应该使我们非常接近想要的结果。

```
>>> crime_sort.first(pd.offsets.MonthBegin(6, normalize=True))
                              OFFENSE_TYPE_ID      ...
REPORTED_DATE                                      ...
2012-01-02   00:06:00          aggravated-assault  ...
2012-01-02   00:06:00  violation-of-restraining-order ...
2012-01-02   00:16:00      traffic-accident-dui-duid ...
2012-01-02   00:47:00              traffic-accident ...
2012-01-02   01:35:00          aggravated-assault  ...
        ...                               ...      ...
2012-06-30   23:40:00   traffic-accident-hit-and-run ...
2012-06-30   23:40:00      traffic-accident-dui-duid ...
2012-06-30   23:44:00              traffic-accident ...
2012-06-30   23:50:00      criminal-mischief-mtr-veh ...
2012-06-30   23:54:00   traffic-accident-hit-and-run ...
```

（7）此方法已成功捕获了 2012 年前 6 个月的所有数据。将 normalize 参数设置为 True 后，搜索将截止到 2012-07-01 00:00:00，这将包括在该日期和时间报告的所有犯罪记录。也就是说，无法使用 .first 方法来确保仅捕获 1~6 月的数据。在上面的输出结果中没有 2012-07-01 00:00:00 的记录，只是因为该 DataFrame 中恰好没有该时间节点的记录而已。

要产生确切的结果，可使用以下切片。

```
>>> crime_sort.loc[:'2012-06']
                              OFFENSE_TYPE_ID      ...
REPORTED_DATE                                      ...
2012-01-02   00:06:00          aggravated-assault  ...
2012-01-02   00:06:00  violation-of-restraining-order ...
2012-01-02   00:16:00      traffic-accident-dui-duid ...
2012-01-02   00:47:00              traffic-accident ...
2012-01-02   01:35:00          aggravated-assault  ...
        ...                               ...      ...
2012-06-30   23:40:00   traffic-accident-hit-and-run ...
2012-06-30   23:40:00      traffic-accident-dui-duid ...
```

```
2012-06-30  23:44:00          traffic-accident  ...
2012-06-30  23:50:00   criminal-mischief-mtr-veh  ...
2012-06-30  23:54:00  traffic-accident-hit-and-run  ...
```

（8）有 12 个 DateOffset 对象，可用于向前或向后移动到下一个最近的偏移量。你可以使用被称为偏移别名（offset alias）的字符串代替在 pd.offsets 中查找 DateOffset 对象。例如，MonthEnd 的字符串是 M，MonthBegin 的字符串是 MS。要表示这些偏移别名的数量，可在其前面放置一个整数。有关所有别名的列表，可访问以下地址。

https://pandas.pydata.org/pandas-docs/stable/user_guide/timeseries.html#timeseries-offset-aliases

现在来看偏移别名的一些示例，在注释中包含对所选偏移别名的描述。

```
>>> crime_sort.first('5D')  # 5 天（Day）
                              OFFENSE_TYPE_ID  ...
REPORTED_DATE                                   ...
2012-01-02  00:06:00            aggravated-assault  ...
2012-01-02  00:06:00  violation-of-restraining-order  ...
2012-01-02  00:16:00       traffic-accident-dui-duid  ...
2012-01-02  00:47:00              traffic-accident  ...
2012-01-02  01:35:00            aggravated-assault  ...
       ...                              ...         ...
2012-01-06  23:11:00        theft-items-from-vehicle  ...
2012-01-06  23:23:00  violation-of-restraining-order  ...
2012-01-06  23:30:00                    assault-dv  ...
2012-01-06  23:44:00          theft-of-motor-vehicle  ...
2012-01-06  23:55:00             threats-to-injure  ...

>>> crime_sort.first('5B')  # 5 工作日（Business day）
                              OFFENSE_TYPE_ID  ...
REPORTED_DATE                                   ...
2012-01-02  00:06:00            aggravated-assault  ...
2012-01-02  00:06:00  violation-of-restraining-order  ...
2012-01-02  00:16:00       traffic-accident-dui-duid  ...
2012-01-02  00:47:00              traffic-accident  ...
2012-01-02  01:35:00            aggravated-assault  ...
       ...                              ...         ...
2012-01-08  23:46:00        theft-items-from-vehicle  ...
2012-01-08  23:51:00    burglary-residence-no-force  ...
2012-01-08  23:52:00                   theft-other  ...
2012-01-09  00:04:00  traffic-accident-hit-and-run  ...
```

```
2012-01-09  00:05:00       fraud-criminal-impersonation  ...

>>> crime_sort.first('7W')  # 7周（week），周的最后一天是星期日
                                       OFFENSE_TYPE_ID  ...
REPORTED_DATE                                           ...
2012-01-02  00:06:00               aggravated-assault  ...
2012-01-02  00:06:00      violation-of-restraining-order  ...
2012-01-02  00:16:00          traffic-accident-dui-duid  ...
2012-01-02  00:47:00                   traffic-accident  ...
2012-01-02  01:35:00               aggravated-assault  ...
     ...                                          ...    ...
2012-02-18  21:57:00                   traffic-accident  ...
2012-02-18  22:19:00          criminal-mischief-graffiti  ...
2012-02-18  22:20:00          traffic-accident-dui-duid  ...
2012-02-18  22:44:00           criminal-mischief-mtr-veh  ...
2012-02-18  23:27:00            theft-items-from-vehicle  ...

>>> crime_sort.first('3QS')  # 前3个季度（Quarter Start）
                                       OFFENSE_TYPE_ID  ...
REPORTED_DATE                                           ...
2012-01-02  00:06:00               aggravated-assault  ...
2012-01-02  00:06:00      violation-of-restraining-order  ...
2012-01-02  00:16:00          traffic-accident-dui-duid  ...
2012-01-02  00:47:00                   traffic-accident  ...
2012-01-02  01:35:00               aggravated-assault  ...
     ...                                          ...    ...
2012-09-30  23:17:00           drug-hallucinogen-possess  ...
2012-09-30  23:29:00                      robbery-street  ...
2012-09-30  23:29:00               theft-of-motor-vehicle  ...
2012-09-30  23:41:00        traffic-accident-hit-and-run  ...
2012-09-30  23:43:00                    robbery-business  ...

>>> crime_sort.first('A')  # 一整年
                                       OFFENSE_TYPE_ID  ...
REPORTED_DATE                                           ...
2012-01-02  00:06:00               aggravated-assault  ...
2012-01-02  00:06:00      violation-of-restraining-order  ...
2012-01-02  00:16:00          traffic-accident-dui-duid  ...
2012-01-02  00:47:00                   traffic-accident  ...
2012-01-02  01:35:00               aggravated-assault  ...
     ...                                          ...    ...
2012-12-30  23:13:00                   traffic-accident  ...
```

```
2012-12-30  23:14:00      burglary-residence-no-force    ...
2012-12-30  23:39:00          theft-of-motor-vehicle     ...
2012-12-30  23:41:00                  traffic-accident   ...
2012-12-31  00:05:00                    assault-simple   ...
```

12.5.2 原理解释

一旦确认索引为 DatetimeIndex，就可以利用本秘笈中的所有方法。

- 要想对 Timestamp 的时间分量进行选择或切片，只能使用 .loc 属性。
- 要按时间范围选择所有日期的记录，必须使用 .between_time 方法。
- 要选择确切的时间，可使用 .at_time 方法。

请确保为开始时间和结束时间传递的字符串至少包含小时和分钟。也可以使用 datetime 模块中的 time 对象。例如，以下命令将产生与步骤（2）相同的结果。

```
>>> import datetime
>>> crime.between_time(datetime.time(2,0), datetime.time(5,0),
...                    include_end=False)
                                   OFFENSE_TYPE_ID    ...
REPORTED_DATE                                         ...
2014-06-29  02:01:00       traffic-accident-dui-duid  ...
2014-06-29  02:00:00          disturbing-the-peace    ...
2014-06-29  02:18:00                        curfew    ...
2014-06-29  04:17:00             aggravated-assault   ...
2014-06-29  04:22:00   violation-of-restraining-order ...
        ...                                           ... ...
2017-08-25  04:41:00          theft-items-from-vehicle ...
2017-09-13  04:17:00          theft-of-motor-vehicle   ...
2017-09-13  02:21:00                  assault-simple   ...
2017-09-13  03:21:00       traffic-accident-dui-duid  ...
2017-09-13  02:15:00    traffic-accident-hit-and-run  ...
```

在步骤（4）中，我们开始使用 .first 方法，但是使用了复杂的参数 offset。offset 参数必须是一个 DateOffset 对象或一个字符串形式的偏移别名。为了帮助理解 DateOffset 对象，最好了解此类对象对单个 Timestamp 的作用。例如，可以采用索引的第一个元素，并以两种不同的方式为其添加 6 个月的时间。

```
>>> first_date = crime_sort.index[0]
>>> first_date
Timestamp('2012-01-02 00:06:00')

>>> first_date + pd.offsets.MonthBegin(6)
```

```
Timestamp('2012-07-01 00:06:00')

>>> first_date + pd.offsets.MonthEnd(6)
Timestamp('2012-06-30 00:06:00')
```

MonthBegin 偏移量和 MonthEnd 偏移量都不会增加或减少确切的时间量，而是有效地舍入到该月的下一个开始或结束，而不管它是在哪一天。在内部，.first 方法使用 DataFrame 的第一个索引元素，并添加传递给它的 DateOffset。然后切片直到这个新日期。例如，步骤（4）等效于以下操作。

```
>>> step4 = crime_sort.first(pd.offsets.MonthEnd(6))
>>> end_dt = crime_sort.index[0] + pd.offsets.MonthEnd(6)
>>> step4_internal = crime_sort[:end_dt]
>>> step4.equals(step4_internal)
True
```

在步骤（8）中，偏移别名使引用 DateOffsets 的方法更加紧凑。

12.5.3 扩展知识

当可用的 DateOffset 不能满足你的需求时，可以构建一个自定义的 DateOffset。

```
>>> dt = pd.Timestamp('2012-1-16 13:40')
>>> dt + pd.DateOffset(months=1)
Timestamp('2012-02-16 13:40:00')
```

请注意，此自定义 DateOffset 将 Timestamp 刚好增加了一个月。下面看一个使用更多日期和时间分量的示例。

```
>>> do = pd.DateOffset(years=2, months=5, days=3,
...         hours=8, seconds=10)
>>> pd.Timestamp('2012-1-22 03:22') + do
Timestamp('2014-06-25 11:22:10')
```

12.6 计算每周犯罪数

丹佛犯罪数据集非常庞大，有超过 460000 行，每一行都标有报告日期。统计每周犯罪的数量，这是可以通过时间分组来响应的许多查询之一。.resample 方法提供了一个简单的接口，可以按任何可能的时间跨度进行分组。

在此秘笈中，将同时使用 .resample 和 .groupby 方法来计算每周犯罪的数量。

12.6.1 实战操作

(1) 读取丹佛犯罪 hdf5 数据集,将索引设置为 REPORTED_DATE,然后对其进行排序以提高此秘笈其余操作的性能。

```
>>> crime_sort = (pd.read_hdf('data/crime.h5', 'crime')
...     .set_index('REPORTED_DATE')
...     .sort_index()
... )
```

(2) 要计算每周的犯罪数量,需要按周分组。.resample 方法可采用 DateOffset 对象或别名作为参数,并返回准备对所有组执行操作的对象。从 .resample 方法返回的对象与调用 .groupby 方法后产生的对象非常相似。

```
>>> crime_sort.resample('W')
<pandas.core.resample.DatetimeIndexResampler object at 0x10f07acf8>
```

(3) 上面的偏移别名 W 用于告诉 Pandas,我们要按周分组。在步骤(2)中并没有发生更多操作。Pandas 已经验证了我们的偏移量,并返回了一个对象,该对象已准备好将每周作为一组执行操作。

在调用 .resample 方法返回一些数据后,可以链接若干种方法。现在让我们链接 .size 方法以计算每周犯罪的数量。

```
>>> (crime_sort
...     .resample('W')
...     .size()
... )
REPORTED_DATE
2012-01-08      877
2012-01-15     1071
2012-01-22      991
2012-01-29      988
2012-02-05      888
                ...
2017-09-03     1956
2017-09-10     1733
2017-09-17     1976
2017-09-24     1839
2017-10-01     1059
Freq: W-SUN, Length: 300, dtype: int64
```

(4) 现在可以将每周犯罪统计数据作为一个 Series,新的索引一次增加一周。

在不同国家/地区，一星期的开始时间并不完全一致。本示例选择星期日作为一周的最后一天，并且该日期也是用来标记结果 Series 中每个元素的日期。

例如，第一个索引值 2012 年 1 月 8 日是星期日。在截至 8 日的那一周内，共发生了 877 起犯罪。1 月 9 日星期一～1 月 15 日星期日这周记录了 1071 起犯罪。下面进行一些完整性检查，以确保重采样正确执行了此操作。

```
>>> len(crime_sort.loc[:'2012-1-8'])
877
>>> len(crime_sort.loc['2012-1-9':'2012-1-15'])
1071
```

（5）也可以选择除星期日外的另一天，以锚定的偏移量作为一周的最后一天。

```
>>> (crime_sort
...     .resample('W-THU')
...     .size()
... )
REPORTED_DATE
2012-01-05     462
2012-01-12    1116
2012-01-19     924
2012-01-26    1061
2012-02-02     926
              ...
2017-09-07    1803
2017-09-14    1866
2017-09-21    1926
2017-09-28    1720
2017-10-05      28
Freq: W-THU, Length: 301, dtype: int64
```

（6）.resample 方法的几乎所有功能都可以通过 .groupby 方法重现。唯一的区别是必须将偏移量传递到 pd.Grouper 对象中。

```
>>> weekly_crimes = (crime_sort
...     .groupby(pd.Grouper(freq='W'))
...     .size()
... )
>>> weekly_crimes
REPORTED_DATE
2012-01-08     877
2012-01-15    1071
2012-01-22     991
```

```
2012-01-29      988
2012-02-05      888
                ...
2017-09-03     1956
2017-09-10     1733
2017-09-17     1976
2017-09-24     1839
2017-10-01     1059
Freq: W-SUN, Length: 300, dtype: int64
```

12.6.2 原理解释

默认情况下，.resample 方法隐式使用 DatetimeIndex 工作，这就是为什么要在步骤（1）中将索引设置为 REPORTED_DATE 的原因。

在步骤（2）中，我们创建了一个中间对象，该对象可以帮助我们了解如何在数据内形成组。.resample 方法的第一个参数是确定将索引中的 Timestamp 分组的规则。在本示例中，使用了偏移别名 W 形成长度为一周的组，该组在星期日结束。默认的结束日期是星期天，但是也可以使用锚定偏移量（anchored offset）进行修改，方法是追加一个短横，然后输入星期的前 3 个字母。例如，W-THU 表示以星期四作为每周的最后一天。

使用 .resample 方法形成组之后，必须链接一个方法以对每个组进行操作。在步骤（3）中，使用了 .size 方法计算每周的犯罪数量。你可能想知道调用 .resample 方法之后可使用的所有可能的属性和方法。下面即列出了 .resample 对象。

```
>>> r = crime_sort.resample('W')
>>> [attr for attr in dir(r) if attr[0].islower()]
['agg', 'aggregate', 'apply', 'asfreq', 'ax', 'backfill', 'bfill',
'count', 'ffill', 'fillna', 'first', 'get_group', 'groups', 'indices',
'interpolate', 'last', 'max', 'mean', 'median', 'min', 'ndim', 'ngroups',
'nunique', 'obj', 'ohlc', 'pad', 'plot', 'prod', 'sem', 'size', 'std',
'sum', 'transform', 'var']
```

步骤（4）通过按周对数据进行切片并计算行数来验证步骤（3）中计数的准确性。

按 Timestamp 分组并不是必须要使用 .resample 方法，因为该功能也可从 .groupby 方法本身获得。但是，你必须使用 freq 参数作为偏移量，将 pd.Grouper 的实例传递给 groupby 方法，步骤（6）就是这样做的。

12.6.3 扩展知识

即使索引不包含 Timestamp，也可以使用 .resample 方法。例如，可以使用 on 参数来选择包含 Timestamp 的列，它们可用于形成组。

```
>>> crime = pd.read_hdf('data/crime.h5', 'crime')
>>> weekly_crimes2 = crime.resample('W', on='REPORTED_DATE').size()
>>> weekly_crimes2.equals(weekly_crimes)
True
```

也可以将 groupby 与 pd.Grouper 一起使用，方法是使用 key 参数选择 Timestamp 列。

```
>>> weekly_crimes_gby2 = (crime
...     .groupby(pd.Grouper(key='REPORTED_DATE', freq='W'))
...     .size()
... )
>>> weekly_crimes2.equals(weekly_crimes)
True
```

还可以通过获得的每周犯罪 Series 调用 .plot 方法，以绘制丹佛所有犯罪（包括交通事故）的折线图。

```
>>> import matplotlib.pyplot as plt
>>> fig, ax = plt.subplots(figsize=(16, 4))
>>> weekly_crimes.plot(title='All Denver Crimes', ax=ax)
>>> fig.savefig('c12-crimes.png', dpi=300)
```

其输出结果如图 12-1 所示。

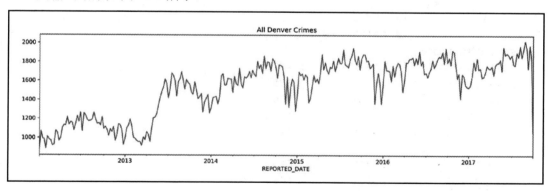

图 12-1　每周犯罪情节

12.7　分别汇总每周犯罪和交通事故

丹佛犯罪数据集将所有犯罪和交通事故汇总在一个表格中，并通过二进制列

IS_CRIME（犯罪记录）和 IS_TRAFFIC（交通事故）将它们分开。.resample 方法允许按时间段进行分组并分别汇总特定的列。

在此秘笈中，我们将使用.resample方法对一年中的每个季度进行分组，然后分别汇总犯罪和交通事故的数量。

12.7.1 实战操作

（1）读取犯罪 hdf5 数据集，将索引设置为 REPORTED_DATE，然后对其进行排序以提高此秘笈其余操作的性能。

```
>>> crime = (pd.read_hdf('data/crime.h5', 'crime')
...     .set_index('REPORTED_DATE')
...     .sort_index()
... )
```

（2）使用 .resample 方法对一年中的每个季度进行分组，然后对每个组的 IS_CRIME 和 IS_TRAFFIC 列求和。

```
>>> (crime
...     .resample('Q')
...     [['IS_CRIME', 'IS_TRAFFIC']]
...     .sum()
... )
                IS_CRIME   IS_TRAFFIC
REPORTED_DATE
2012-03-31          7882         4726
2012-06-30          9641         5255
2012-09-30         10566         5003
2012-12-31          9197         4802
2013-03-31          8730         4442
...                  ...          ...
2016-09-30         17427         6199
2016-12-31         15984         6094
2017-03-31         16426         5587
2017-06-30         17486         6148
2017-09-30         17990         6101
```

（3）请注意，日期全部显示为该季度的最后一天。这是因为偏移别名 Q 代表季度末。也可以使用偏移别名 QS 表示季度的开始。

```
>>> (crime
...     .resample('QS')
...     [['IS_CRIME', 'IS_TRAFFIC']]
...     .sum()
... )
                IS_CRIME    IS_TRAFFIC
REPORTED_DATE
2012-01-01          7882          4726
2012-04-01          9641          5255
2012-07-01         10566          5003
2012-10-01          9197          4802
2013-01-01          8730          4442
...                  ...           ...
2016-07-01         17427          6199
2016-10-01         15984          6094
2017-01-01         16426          5587
2017-04-01         17486          6148
2017-07-01         17990          6101
```

（4）可以通过检查第二季度的数据是否正确来验证这些结果。

```
>>> (crime
...     .loc['2012-4-1':'2012-6-30', ['IS_CRIME', 'IS_TRAFFIC']]
...     .sum()
... )
IS_CRIME      9641
IS_TRAFFIC    5255
dtype: int64
```

（5）可以使用 .groupby 方法重复上述操作。

```
>>> (crime
...     .groupby(pd.Grouper(freq='Q'))
...     [['IS_CRIME', 'IS_TRAFFIC']]
...     .sum()
... )
                IS_CRIME    IS_TRAFFIC
REPORTED_DATE
2012-03-31          7882          4726
2012-06-30          9641          5255
2012-09-30         10566          5003
2012-12-31          9197          4802
2013-03-31          8730          4442
```

```
        ...              ...              ...
2016-09-30             17427             6199
2016-12-31             15984             6094
2017-03-31             16426             5587
2017-06-30             17486             6148
2017-09-30             17990             6101
```

（6）现在可以绘图以直观显示随着时间的推移犯罪和交通事故的趋势。

```
>>> fig, ax = plt.subplots(figsize=(16, 4))
>>> (crime
...     .groupby(pd.Grouper(freq='Q'))
...     [['IS_CRIME', 'IS_TRAFFIC']]
...     .sum()
...     .plot(color=['black', 'lightgrey'], ax=ax,
...           title='Denver Crimes and Traffic Accidents')
... )
>>> fig.savefig('c12-crimes2.png', dpi=300)
```

其输出结果如图 12-2 所示。

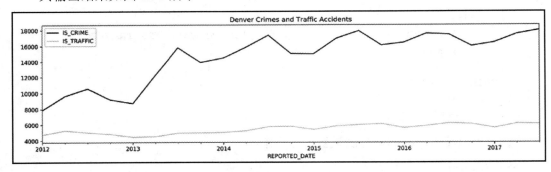

图 12-2　按季度绘制的犯罪和交通事故趋势图

12.7.2　原理解释

在步骤（1）中读取并准备好数据后，在步骤（2）中开始分组和聚合。调用.resample 方法后，可以通过链接方法或选择一组要聚合的列来继续进行操作。

本示例选择了 IS_CRIME 和 IS_TRAFFIC 列进行汇总。如果不只是选择这两列，那么所有数字列的总和将输出以下结果。

```
>>> (crime
...     .resample('Q')
```

```
...         .sum()
... )
                    GEO_LON    ...    IS_TRAFFIC
REPORTED_DATE                  ...
2012-03-31      -1.313006e+06  ...          4726
2012-06-30      -1.547274e+06  ...          5255
2012-09-30      -1.615835e+06  ...          5003
2012-12-31      -1.458177e+06  ...          4802
2013-03-31      -1.368931e+06  ...          4442
...                       ...  ...           ...
2016-09-30      -2.459343e+06  ...          6199
2016-12-31      -2.293628e+06  ...          6094
2017-03-31      -2.288383e+06  ...          5587
2017-06-30      -2.453857e+06  ...          6148
2017-09-30      -2.508001e+06  ...          6101
```

默认情况下，偏移别名 Q 在技术上使用 12 月 31 日作为一年的最后一天。代表一个季度的日期范围全部使用此结束日期进行计算。汇总结果使用该季度的最后一天作为标签。

步骤（3）使用了偏移别名 QS。默认情况下，偏移别名 QS 使用 1 月 1 日作为一年的第一天来计算季度。

大多数公共企业都会报告季度收入，但这些季度收入并不都是从一月开始的。很多国家和企业都具有不同的财年和财季。例如，如果我们希望季度开始于 3 月 1 日，则可以使用 QS-MAR 来锚定偏移别名。

```
>>> (crime_sort
...     .resample('QS-MAR')
...     [['IS_CRIME', 'IS_TRAFFIC']]
...     .sum()
... )
                IS_CRIME    IS_TRAFFIC
REPORTED_DATE
2011-12-01          5013          3198
2012-03-01          9260          4954
2012-06-01         10524          5190
2012-09-01          9450          4777
2012-12-01          9003          4652
...                  ...           ...
2016-09-01         16932          6202
2016-12-01         15615          5731
2017-03-01         17287          5940
```

2017-06-01	18545	6246
2017-09-01	5417	1931

与 12.6 节"计算每周犯罪数"秘笈一样，步骤（4）也通过手动切片验证了结果。

在步骤（5）中，通过 .groupby 方法（使用 pd.Grouper 设置组长度）复制了步骤（3）的操作结果。

在步骤（6）中，调用了 DataFrame .plot 方法绘图。默认情况下，为每列数据绘制一条折线。该图清楚地表明，在每年的前 3 个季度，报告的犯罪数量急剧增加。犯罪和交通事故似乎都与季节有关，在气温较冷的月份数字较低，在较为暖和的月份则数字较高。

12.7.3 扩展知识

为了获得不同的观察角度，我们还可以绘制犯罪和交通事故的增加百分比，而不是原始计数。这可以将所有数据除以第一行并再次绘图。

```
>>> crime_begin = (crime
...     .resample('Q')
...     [['IS_CRIME', 'IS_TRAFFIC']]
...     .sum()
...     .iloc[0]
... )

>>> fig, ax = plt.subplots(figsize=(16, 4))
>>> (crime
...     .resample('Q')
...     [['IS_CRIME', 'IS_TRAFFIC']]
...     .sum()
...     .div(crime_begin)
...     .sub(1)
...     .round(2)
...     .mul(100)
...     .plot.bar(color=['black', 'lightgrey'], ax=ax,
...         title='Denver Crimes and Traffic Accidents % Increase')
... )

>>> fig.autofmt_xdate()
>>> fig.savefig('c12-crimes3.png', dpi=300, bbox_inches='tight')
```

其输出结果如图 12-3 所示。

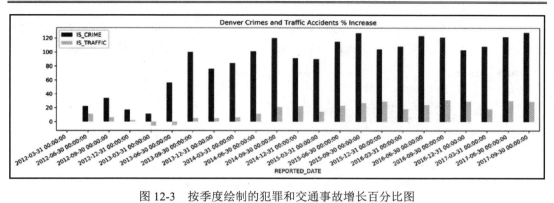

图 12-3　按季度绘制的犯罪和交通事故增长百分比图

12.8　按星期和年份衡量犯罪情况

要同时按星期和按年衡量犯罪情况，需要从 Timestamp 中提取此信息的功能。幸运的是，此功能已内置到具有.dt 属性的任何 Timestamp 列中。

在此秘笈中，我们将使用.dt 属性提供每个犯罪的星期名称和年份，并作为 Series 返回。通过使用这两个 Series 形成的小组来计算所有犯罪情况。最后，对于某些年份只有部分数据的情况进行调整以体现总体比较，并创建犯罪总数的热图。

12.8.1　实战操作

（1）读取丹佛犯罪 hdf5 数据集，将 REPORTED_DATE 保留为一列。

```
>>> crime = pd.read_hdf('data/crime.h5', 'crime')
>>> crime
                    OFFEN/PE_ID     ...   IS_TRAFFIC
0         traffic-accident-dui-duid  ...            1
1         vehicular-eluding-no-chase ...            0
2             disturbing-the-peace   ...            0
3                         curfew    ...            0
4               aggravated-assault   ...            0
...                            ...   ...          ...
460906      burglary-business-by-force ...          0
460907   weapon-unlawful-discharge-of ...            0
460908          traf-habitual-offender ...            0
460909         criminal-mischief-other ...            0
460910                   theft-other   ...            0
```

（2）所有 Timestamp 列均具有特殊属性.dt，该属性可以访问为日期专门设计的各种其他属性和方法。找到每个 REPORTED_DATE 的星期名称，然后计算以下值。

```
>>> (crime
...    ['REPORTED_DATE']
...    .dt.day_name()
...    .value_counts()
... )
Monday       70024
Friday       69621
Wednesday    69538
Thursday     69287
Tuesday      68394
Saturday     58834
Sunday       55213
Name: REPORTED_DATE, dtype: int64
```

（3）可以看到，周末的犯罪和交通事故似乎要少得多。下面按正确的星期顺序排列此数据，并绘制水平条形图。

```
>>> days = ['Monday', 'Tuesday', 'Wednesday', 'Thursday',
...         'Friday', 'Saturday', 'Sunday']
>>> title = 'Denver Crimes and Traffic Accidents per Weekday'
>>> fig, ax = plt.subplots(figsize=(6, 4))
>>> (crime
...    ['REPORTED_DATE']
...    .dt.day_name()
...    .value_counts()
...    .reindex(days)
...    .plot.barh(title=title, ax=ax)
... )
>>> fig.savefig('c12-crimes4.png', dpi=300, bbox_inches='tight')
```

其输出结果如图 12-4 所示。

（4）可以执行非常类似的过程来按年份绘制计数。

```
>>> title = 'Denver Crimes and Traffic Accidents per Year'
>>> fig, ax = plt.subplots(figsize=(6, 4))
>>> (crime
...    ['REPORTED_DATE']
...    .dt.year
...    .value_counts()
...    .sort_index()
```

```
...         .plot.barh(title=title, ax=ax)
...    )
>>> fig.savefig('c12-crimes5.png', dpi=300, bbox_inches='tight')
```

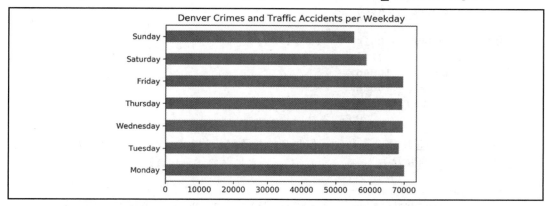

图 12-4　按星期顺序绘制的犯罪情况条形图

其输出结果如图 12-5 所示。

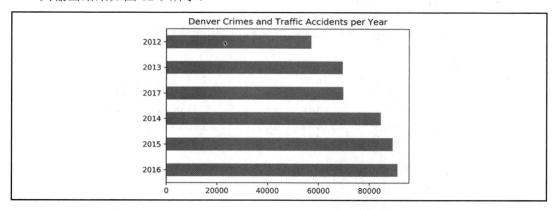

图 12-5　按年度绘制的犯罪情况条形图

（5）我们需要同时按星期和年份分组。一种方法是在 .groupby 方法中使用这些属性。

```
>>> (crime
...     .groupby([crime['REPORTED_DATE'].dt.year.rename('year'),
...               crime['REPORTED_DATE'].dt.day_name().rename('day')])
...     .size()
... )
year  day
2012  Friday      8549
```

```
            Monday     8786
            Saturday   7442
            Sunday     7189
            Thursday   8440
                        ...
    2017    Saturday   8514
            Sunday     8124
            Thursday  10545
            Tuesday   10628
            Wednesday 10576
    Length: 42, dtype: int64
```

（6）现在已经正确聚合了数据，但是该结构不利于进行比较。因此，可以使用.unstack方法来获得一个更具可读性的表。

```
>>> (crime
...     .groupby([crime['REPORTED_DATE'].dt.year.rename('year'),
...               crime['REPORTED_DATE'].dt.day_name().rename('day')])
...     .size()
...     .unstack('day')
... )
day   Friday  Monday  Saturday  Sunday  Thursday  Tuesday
year
2012    8549    8786      7442    7189      8440     8191
2013   10380   10627      8875    8444     10431    10416
2014   12683   12813     10950   10278     12309    12440
2015   13273   13452     11586   10624     13512    13381
2016   14059   13708     11467   10554     14050    13338
2017   10677   10638      8514    8124     10545    10628
```

（7）现在有一个更好的表示形式，并且更易于阅读，但值得注意的是，2017年的数字并不完整。为了更公平地进行比较，可以进行线性外推（linear extrapolation）以估算最终犯罪数量。首先需要找到2017年数据的最后一天。

```
>>> criteria = crime['REPORTED_DATE'].dt.year == 2017
>>> crime.loc[criteria, 'REPORTED_DATE'].dt.dayofyear.max()
272
```

（8）可以看到，2017年目前只有272天的数据。要想获得一整年的数据，最简单的估计是假设全年犯罪率恒定，并将2017年表中的所有值乘以365/272。但是，我们还可以做得更好一些，查看历史数据并计算在各年的前272天发生的犯罪的平均百分比。

```
>>> round(272 / 365, 3)
```

第 12 章 时间序列分析

```
0.745
>>> crime_pct = (crime
...              ['REPORTED_DATE']
...              .dt.dayofyear.le(272)
...              .groupby(crime.REPORTED_DATE.dt.year)
...              .mean()
...              .mul(100)
...              .round(2)
... )

>>> crime_pct
REPORTED_DATE
2012    74.84
2013    72.54
2014    75.06
2015    74.81
2016    75.15
2017   100.00
Name: REPORTED_DATE, dtype: float64

>>> crime_pct.loc[2012:2016].median()
74.84
```

（9）事实证明（也许仅仅是巧合），在各年的前 272 天发生的犯罪百分比几乎与该年过去的天数百分比成正比（272/365=74.52%）。现在可以更新 2017 年的行，并更改列顺序以匹配星期顺序。

```
>>> def update_2017(df_):
...     df_.loc[2017] = (df_
...         .loc[2017]
...         .div(.748)
...         .astype('int')
...     )
...     return df_
>>> (crime
...     .groupby([crime['REPORTED_DATE'].dt.year.rename('year'),
...               crime['REPORTED_DATE'].dt.day_name().rename('day')])
...     .size()
...     .unstack('day')
...     .pipe(update_2017)
...     .reindex(columns=days)
... )
day    Monday   Tuesday  Wednesday  ...  Friday  Saturday  Sunday
```

```
year                              ...
2012     8786      8191      8440 ...     8549      7442      7189
2013    10627     10416     10354 ...    10380      8875      8444
2014    12813     12440     12948 ...    12683     10950     10278
2015    13452     13381     13320 ...    13273     11586     10624
2016    13708     13338     13900 ...    14059     11467     10554
2017    14221     14208     14139 ...    14274     11382     10860
```

（10）现在可以使用获得的数据绘制条形图或折线图，但是此类数据也适于使用热图进行可视化。在 Seaborn 库中包含热图（heatmap）绘制工具。

```
>>> import seaborn as sns
>>> fig, ax = plt.subplots(figsize=(6, 4))
>>> table = (crime
...          .groupby([crime['REPORTED_DATE'].dt.year.rename('year'),
...                    crime['REPORTED_DATE'].dt.day_name().rename('day')])
...          .size()
...          .unstack('day')
...          .pipe(update_2017)
...          .reindex(columns=days)
... )
>>> sns.heatmap(table, cmap='Greys', ax=ax)
>>> fig.savefig('c12-crimes6.png', dpi=300, bbox_inches='tight')
```

其输出结果如图 12-6 所示。

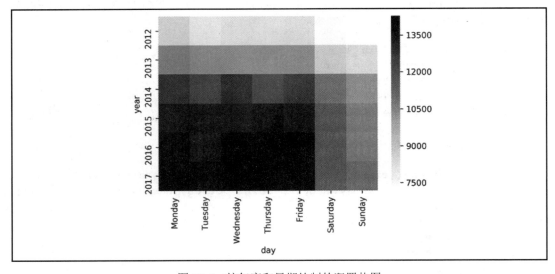

图 12-6　按年度和星期绘制的犯罪热图

(11）犯罪似乎每年都在增加，但这一数据并未说明人口数的增加。现在可以读取丹佛每年人口数的表格。

```
>>> denver_pop = pd.read_csv('data/denver_pop.csv',
...                index_col='Year')
>>> denver_pop
      Population
Year
2017      705000
2016      693000
2015      680000
2014      662000
2013      647000
2012      634000
```

（12）许多犯罪指标都是按每十万居民的比率报告的。因此，在本示例中可以将人口总数除以 100000 得到一个商数值，然后将原始犯罪计数除以该商数值即可得出每 100000 居民的犯罪率。

```
>>> den_100k = denver_pop.div(100_000).squeeze()
>>> normalized = (crime
...     .groupby([crime['REPORTED_DATE'].dt.year.rename('year'),
...               crime['REPORTED_DATE'].dt.day_name() rename('day')])
...     .size()
...     .unstack('day')
...     .pipe(update_2017)
...     .reindex(columns=days)
...     .div(den_100k, axis='index')
...     .astype(int)
... )
>>> normalized
day   Monday  Tuesday  Wednesday  ...  Friday  Saturday  Sunday
2012    1385     1291       1331  ...    1348      1173    1133
2013    1642     1609       1600  ...    1604      1371    1305
2014    1935     1879       1955  ...    1915      1654    1552
2015    1978     1967       1958  ...    1951      1703    1562
2016    1978     1924       2005  ...    2028      1654    1522
2017    2017     2015       2005  ...    2024      1614    1540
```

（13）现在再使用调整了人口增长之后的数据绘制热图。

```
>>> import seaborn as sns
>>> fig, ax = plt.subplots(figsize=(6, 4))
>>> sns.heatmap(normalized, cmap='Greys', ax=ax)
```

```
>>> fig.savefig('c12-crimes7.png', dpi=300, bbox_inches='tight')
```

其输出结果如图 12-7 所示。

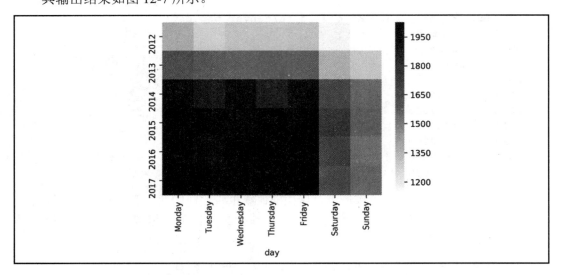

图 12-7　对数据进行标准化处理之后绘制的年度和星期犯罪热图

12.8.2　原理解释

所有包含 Timestamp 的 DataFrame 列都可以使用.dt 属性访问许多其他属性和方法。实际上，.dt 属性中可用的所有这些方法和属性在 Timestamp 对象上也可用。

在步骤（2）中，使用了.dt 属性（仅适用于 Series）提取星期名称并计算发生次数。

在步骤（3）绘图之前，使用了.reindex 方法手动重新排列索引的顺序，在最基本的用例中，该方法接收包含所需顺序的列表。该任务也可以使用.loc 索引器完成，如下所示。

```
>>> (crime
...     ['REPORTED_DATE']
...     .dt.day_name()
...     .value_counts()
...     .loc[days]
... )
Monday       70024
Tuesday      68394
Wednesday    69538
Thursday     69287
Friday       69621
```

```
Saturday        58834
Sunday          55213
Name: REPORTED_DATE, dtype: int64
```

与 .loc 相比，.reindex 方法性能更高，并且在许多情况下都具有更多的参数。

在步骤（4）中，执行了一个非常类似的过程，并再次使用 .dt 属性检索年份，然后使用 .value_counts 方法对出现的次数进行计数。在本示例中，使用了 .sort_index 方法而不是 .reindex 方法，因为年份自然会按所需顺序排序。

此秘笈的目标是将星期和年份进行分组，这已经在步骤（5）中执行。.groupby 方法非常灵活，可以通过多种方式进行分组。在此秘笈中，从 year 和 weekday 列派生出两个 Series。然后，将 .size 方法链接到它，该方法将返回单个值，即每组的长度。

在步骤（5）之后，生成的 Series 就已经很长，只有一列数据，这使得按年和星期进行比较变得很困难。

为了简化可读性，在步骤（6）中使用了 .unstack 方法将星期级别转换为水平列名称。步骤（6）制作了交叉表。在 Pandas 中还可以按以下方式执行此操作。

```
>>> (crime
...     .assign(year=crime.REPORTED_DATE.dt.year,
...             day=crime.REPORTED_DATE.dt.day_name())
...     .pipe(lambda df_: pd.crosstab(df_.year, df_.day))
... )
day     Friday   Monday   ...   Tuesday   Wednesday
year                      ...
2012      8549     8786   ...      8191        8440
2013     10380    10627   ...     10416       10354
2014     12683    12813   ...     12440       12948
2015     13273    13452   ...     13381       13320
2016     14059    13708   ...     13338       13900
2017     10677    10638   ...     10628       10576
```

在步骤（7）中，使用布尔索引仅选择了 2017 年的犯罪记录，然后使用了 .dt 属性中的 .dayofyear 来查找从年初开始经过的总天数。该 Series 的最大值应告诉我们 2017 年有多少天的数据。

步骤（8）非常复杂。首先创建了一个布尔 Series，这是使用以下语句实现的。

```
crime['REPORTED_DATE'].dt.dayofyear.le(272)
```

该语句测试每个犯罪记录是否是在每年第 272 天之前提交的。

在获得该 Series 之后，再次使用 .groupby 方法对返回的 Series 按年份分组，然后使用.mean 方法查找每年第 272 天之前犯罪的百分比。

在步骤（9）中，.loc 属性选择了整个 2017 年的数据行，然后通过除以在步骤（8）中找到的百分比中位数来调整该行数据。

许多犯罪情况的可视化都是通过热图完成的，在步骤（10）中，使用了 Seaborn 库中的工具来绘制热图。cmap 参数采用了数十个可用 matplotlib colormap 的字符串名称。

在步骤（12）中，计算了每 10 万居民的犯罪率（将人口总数除以 100000 得到一个商数值，然后将原始犯罪计数除以该商数值）。这是另一个需要技巧的操作。通常情况下，当你将一个 DataFrame 除以另一个 DataFrame 时，它们会按其列和索引对齐。但是，在此步骤中，没有与 denver_pop 相同的列，因此如果尝试使用它执行除法，则没有值会对齐。要解决此问题，需要使用 squeeze 方法创建 den_100k Series。我们仍然不能让这两个对象相除，因为默认情况下，DataFrame 和 Series 之间的除法会将 DataFrame 的列与 Series 的索引对齐，结果就会像下面这样。

```
>>> (crime
...     .groupby([crime['REPORTED_DATE'].dt.year.rename('year'),
...               crime['REPORTED_DATE'].dt.day_name().rename('day')])
...     .size()
...     .unstack('day')
...     .pipe(update_2017)
...     .reindex(columns=days)
... ) / den_100k
        2012    2013    2014    ...  Thursday  Tuesday  Wednesday
year                             ...
2012    NaN     NaN     NaN     ...  NaN       NaN      NaN
2013    NaN     NaN     NaN     ...  NaN       NaN      NaN
2014    NaN     NaN     NaN     ...  NaN       NaN      NaN
2015    NaN     NaN     NaN     ...  NaN       NaN      NaN
2016    NaN     NaN     NaN     ...  NaN       NaN      NaN
2017    NaN     NaN     NaN     ...  NaN       NaN      NaN
```

我们需要 DataFrame 的索引与 Series 的索引对齐，为此，可以使用.div 方法，该方法允许使用 axis 参数更改对齐方向。

在步骤（13）中绘制了调整数据后犯罪率的热图。

12.8.3 扩展知识

如果要研究特定类型的犯罪，可以执行以下操作。

```
>>> days = ['Monday', 'Tuesday', 'Wednesday', 'Thursday',
...         'Friday', 'Saturday', 'Sunday']
>>> crime_type = 'auto-theft'
>>> normalized = (crime
...    .query('OFFENSE_CATEGORY_ID == @crime_type')
...    .groupby([crime['REPORTED_DATE'].dt.year.rename('year'),
...              crime['REPORTED_DATE'].dt.day_name().rename('day')])
...    .size()
...    .unstack('day')
...    .pipe(update_2017)
...    .reindex(columns=days)
...    .div(den_100k, axis='index')
...    .astype(int)
... )
>>> normalized
day   Monday  Tuesday  Wednesday  ...  Friday  Saturday  Sunday
2012      95       72         72  ...      71        78      76
2013      85       74         74  ...      65        68      67
2014      94       76         72  ...      76        67      67
2015     108      102         89  ...      92        85      78
2016     119      102        100  ...      97        86      85
2017     114      118        111  ...     111        91     102
```

12.9 使用匿名函数进行分组

使用包含 DatetimeIndex 的 DataFrame 为许多新的不同的操作打开了一扇门,本章中的若干个秘笈就展示了这些操作。

此秘笈将对包含 DatetimeIndex 的 DataFrame 使用.groupby 方法,并在此过程中展示该方法的多功能性特点。

12.9.1 实战操作

(1) 读取丹佛犯罪 hdf5 文件,将 REPORTED_DATE 列放在索引中,并对其进行排序。

```
>>> crime = (pd.read_hdf('data/crime.h5', 'crime')
...    .set_index('REPORTED_DATE')
...    .sort_index()
... )
```

（2）DatetimeIndex 具有许多与 Pandas Timestamp 相同的属性和方法。下面查看它们的共同属性。

```
>>> common_attrs = (set(dir(crime.index)) &
...     set(dir(pd.Timestamp)))
>>> [attr for attr in common_attrs if attr[0] != '_']
['tz_convert', 'is_month_start', 'nanosecond', 'day_name',
'microsecond', 'quarter', 'time', 'tzinfo', 'week', 'year',
'to_period', 'freqstr', 'dayofyear', 'is_year_end', 'weekday_name',
'month_name', 'minute', 'hour', 'dayofweek', 'second', 'max', 'min',
'to_numpy', 'tz_localize', 'is_quarter_end', 'to_julian_date',
'strftime', 'day', 'days_in_month', 'weekofyear', 'date', 'daysinmonth',
'month', 'weekday', 'is_year_start', 'is_month_end', 'ceil', 'timetz',
'freq', 'tz', 'is_quarter_start', 'floor', 'normalize', 'resolution',
'is_leap_year', 'round', 'to_pydatetime']
```

（3）我们可以使用 .index 来查找星期名称，这和 12.8 节"按星期和年份衡量犯罪情况"秘笈的步骤（2）中的操作是类似的。

```
>>> crime.index.day_name().value_counts()
Monday       70024
Friday       69621
Wednesday    69538
Thursday     69287
Tuesday      68394
Saturday     58834
Sunday       55213
Name: REPORTED_DATE, dtype: int64
```

（4）.groupby 方法可以接收函数作为参数。该函数将被传递给 .index，并且其返回值将用于形成组。为演示该操作，可以使用函数将 .index 转换为星期名称，然后按组分别计算犯罪和交通事故的总数。

```
>>> (crime
...     .groupby(lambda idx: idx.day_name())
...     [['IS_CRIME', 'IS_TRAFFIC']]
...     .sum()
... )
          IS_CRIME  IS_TRAFFIC
Friday       48833       20814
Monday       52158       17895
Saturday     43363       15516
Sunday       42315       12968
```

Thursday	49470			19845		
Tuesday	49658			18755		
Wednesday	50054			19508		

（5）还可以使用函数列表按天中的小时和按年进行分组，然后重新调整表格的形状以使其更具可读性。

```
>>> funcs = [lambda idx: idx.round('2h').hour, lambda idx: idx.year]
>>> (crime
...     .groupby(funcs)
...     [['IS_CRIME', 'IS_TRAFFIC']]
...     .sum()
...     .unstack()
... )
    IS_CRIME            ...  IS_TRAFFIC
    2012   2013   2014  ...  2015   2016   2017
0   2422   4040   5649  ...  1136   980    782
2   1888   3214   4245  ...  773    718    537
4   1472   2181   2956  ...  471    464    313
6   1067   1365   1750  ...  494    593    462
8   2998   3445   3727  ...  2331   2372   1828
..  ...    ...    ...   ...  ...    ...    ...
14  4266   5698   6708  ...  2840   2763   1990
16  4113   5889   7351  ...  3160   3527   2784
18  3660   5094   6586  ...  3412   3608   2718
20  3521   4895   6130  ...  2071   2184   1491
22  3078   4318   5496  ...  1671   1472   1072
```

（6）如果使用的是 Jupyter，则可以添加.style.highlight_max(color ='lightgrey')语句，以突出显示每一列的最大值。

```
>>> funcs = [lambda idx: idx.round('2h').hour, lambda idx: idx.year]
>>> (crime
...     .groupby(funcs)
...     [['IS_CRIME', 'IS_TRAFFIC']]
...     .sum()
...     .unstack()
...     .style.highlight_max(color='lightgrey')
... )
```

其输出结果如图 12-8 所示。

	IS_CRIME						IS_TRAFFIC					
	2012	2013	2014	2015	2016	2017	2012	2013	2014	2015	2016	2017
0	2422	4040	5649	5649	5377	3811	919	792	978	1136	980	782
2	1888	3214	4245	4050	4091	3041	718	652	779	773	718	537
4	1472	2181	2956	2959	3044	2255	399	378	424	471	464	313
6	1067	1365	1750	2167	2108	1567	411	399	479	494	593	462
8	2998	3445	3727	4161	4488	3251	1957	1955	2210	2331	2372	1828
10	4305	5035	5658	6205	6218	4993	1979	1901	2139	2320	2303	1873
12	4496	5524	6434	6841	7226	5463	2200	2138	2379	2631	2760	1986
14	4266	5698	6708	7218	6896	5396	2241	2245	2630	2840	2763	1990
16	4113	5889	7351	7643	7926	6338	2714	2562	3002	3160	3527	2784
18	3660	5094	6586	7015	7407	6157	3118	2704	3217	3412	3608	2718
20	3521	4895	6130	6360	6963	5272	1787	1806	1994	2071	2184	1491
22	3078	4318	5496	5626	5637	4358	1343	1330	1532	1671	1472	1072

图 12-8　突出显示每一年的犯罪高发时段和交通事故多发时段

12.9.2　原理解释

在步骤（1）中，读取了犯罪数据并将 Timestamp 列放入索引中以创建 DatetimeIndex。

在步骤（2）中，可以看到 DatetimeIndex 具有许多与单个 Timestamp 对象相同的功能。

在步骤（3）中，使用了 DatetimeIndex 的额外功能来提取日期名称。

在步骤（4）中，利用了 .groupby 方法来接收传递给 DatetimeIndex 的函数。匿名函数中的 idx 是 DatetimeIndex，我们用它来检索星期名称。

如步骤（5）所示，可以给 .groupby 方法传递任意数量的自定义函数的列表。在这里，第一个函数使用 .round DatetimeIndex 方法将每个值取整到最接近的 2h 范围，第二个函数返回 .year 属性。在经过分组和聚合之后，再使用 .unstack 方法取消年份的堆叠，使之成为列。

在步骤（6）中，突出显示了每列的最大值。根据已经报告的情况，可见犯罪多半发生在下午 3～5 点，而大多数交通事故则发生在下午 5 点～晚上 7 点。

12.10　按 Timestamp 和其他列分组

.resample 方法无法按时间段之外的其他列进行分组。但是，.groupby 方法则可以同时按时间段和其他列进行分组。

在本秘笈中，我们将演示两种非常相似但又有所不同的方法来按 Timestamp 和其他列进行分组。

12.10.1 实战操作

（1）读取 employee（员工）数据集，并使用 HIRE_DATE（录用日期）列创建 DatetimeIndex。

```
>>> employee = pd.read_csv('data/employee.csv',
...       parse_dates=['JOB_DATE', 'HIRE_DATE'],
...       index_col='HIRE_DATE')
>>> employee
            UNIQUE_ID  ...   JOB_DATE
HIRE_DATE              ...
2006-06-12          0  ... 2012-10-13
2000-07-19          1  ... 2010-09-18
2015-02-03          2  ... 2015-02-03
1982-02-08          3  ... 1991-05-25
1989-06-19          4  ... 1994-10-22
...               ...  ...        ...
2014-06-09       1995  ... 2015-06-09
2003-09-02       1996  ... 2013-10-06
2014-10-13       1997  ... 2015-10-13
2009-01-20       1998  ... 2011-07-02
2009-01-12       1999  ... 2010-07-12
```

（2）首先按性别进行分组，然后找到每个性别的平均工资。

```
>>> (employee
...     .groupby('GENDER')
...     ['BASE_SALARY']
...     .mean()
...     .round(-2)
... )

GENDER
Female    52200.0
Male      57400.0
Name: BASE_SALARY, dtype: float64
```

（3）基于录用日期找到平均薪水，并按录用日期列每 10 年分组。

```
>>> (employee
...     .resample('10AS')
...     ['BASE_SALARY']
...     .mean()
...     .round(-2)
... )
HIRE_DATE
1958-01-01     81200.0
1968-01-01    106500.0
1978-01-01     69600.0
1988-01-01     62300.0
1998-01-01     58200.0
2008-01-01     47200.0
Freq: 10AS-JAN, Name: BASE_SALARY, dtype: float64
```

（4）如果想按性别和 10 年时间跨度分组，则可以在调用.groupby 方法之后调用.resample 方法。

```
>>> (employee
...     .groupby('GENDER')
...     .resample('10AS')
...     ['BASE_SALARY']
...     .mean()
...     .round(-2)
... )
GENDER  HIRE_DATE
Female  1975-01-01     51600.0
        1985-01-01     57600.0
        1995-01-01     55500.0
        2005-01-01     51700.0
        2015-01-01     38600.0
                         ...
Male    1968-01-01    106500.0
        1978-01-01     72300.0
        1988-01-01     64600.0
        1998-01-01     59700.0
        2008-01-01     47200.0
Name: BASE_SALARY, Length: 11, dtype: float64
```

（5）现在我们已经获得了最初想要的结果，但是，每当比较男女工资时，都会遇到一个小问题。这里不妨使用.unstack 方法取消性别级别的堆叠，看看会发生什么。

```
>>> (employee
```

```
...      .groupby('GENDER')
...      .resample('10AS')
...      ['BASE_SALARY']
...      .mean()
...      .round(-2)
...      .unstack('GENDER')
... )
GENDER        Female      Male
HIRE_DATE
1958-0...        NaN   81200.0
1968-0...        NaN  106500.0
1975-0...    51600.0       NaN
1978-0...        NaN   72300.0
1985-0...    57600.0       NaN
...              ...       ...
1995-0...    55500.0       NaN
1998-0...        NaN   59700.0
2005-0...    51700.0       NaN
2008-0...        NaN   47200.0
2015-0...    38600.0       NaN
```

（6）可以看到，男性和女性的 10 年期限不在同一日期开始。发生这种情况的原因是，数据首先按性别分组，然后在每种性别内，根据录用日期组成了更多的组。我们可以验证第一位被录用的男性是在 1958 年，而第一位被录用的女性是在 1975 年。

```
>>> employee[employee['GENDER'] == 'Male'].index.min()
Timestamp('1958-12-29 00:00:00')
>>> employee[employee['GENDER'] == 'Female'].index.min()
Timestamp('1975-06-09 00:00:00')
```

（7）要解决此问题，必须将日期和性别一起分组，这只有使用.groupby 方法才可以实现，具体如下。

```
>>> (employee
...      .groupby(['GENDER', pd.Grouper(freq='10AS')])
...      ['BASE_SALARY']
...      .mean()
...      .round(-2)
... )
GENDER  HIRE_DATE
Female  1968-01-01        NaN
        1978-01-01    57100.0
        1988-01-01    57100.0
        1998-01-01    54700.0
```

```
              2008-01-01     47300.0
                                ...
Male          1968-01-01    106500.0
              1978-01-01     72300.0
              1988-01-01     64600.0
              1998-01-01     59700.0
              2008-01-01     47200.0
Name: BASE_SALARY, Length: 11, dtype: float64
```

（8）现在可以使用.unstack 方法取消 GENDER 列的堆叠并使行完美对齐。

```
>>> (employee
...     .groupby(['GENDER', pd.Grouper(freq='10AS')])
...     ['BASE_SALARY']
...     .mean()
...     .round(-2)
...     .unstack('GENDER')
... )
GENDER          Female      Male
HIRE_DATE
1958-0...          NaN   81200.0
1968-0...          NaN  106500.0
1978-0...      57100.0   72300.0
1988-0...      57100.0   64600.0
1998-0...      54700.0   59700.0
2008-0...      47300.0   47200.0
```

12.10.2 原理解释

步骤（1）中的 read_csv 函数允许将两列都转换为 Timestamp，并同时将它们放入索引中，以创建 DatetimeIndex。

步骤（2）使用单个分组列 gender 执行.groupby 操作。

步骤（3）将.resample 方法与偏移别名 10AS 一起使用，以 10 年的时间增量形成组。A 是年份的别名，而 S 则告诉我们，该期间的开始用作标签。例如，标签 1988-01-01 的数据跨越该日期，直到 1997 年 12 月 31 日为止。

在步骤（4）中，根据最早录用的员工的日期，按性别和 10 年时间跨度分组计算出完全不同的开始日期。

步骤（5）显示了当我们尝试比较男性和女性的工资时，这是如何导致偏差的。不同性别的员工没有相同的 10 年期限。

步骤（6）验证了每种性别最早录用的员工的年份与步骤（4）的输出是匹配的。

要缓解此问题，必须同时按性别和 Timestamp 分组。由于.resample 方法只能按单个 Timestamp 列进行分组，因此只能使用.groupby 方法完成此操作。使用 pd.Grouper 可以复制.resample 的功能。我们将偏移别名传递给 freq 参数，然后将对象放置在一个列表中，该列表包含我们希望分组的所有其他列，如步骤（7）所示。

由于现在男性和女性 10 年跨度的开始日期相同，因此步骤（8）中的重塑数据将针对每种性别进行调整，从而使比较变得更加容易。看起来，随着工作时间的延长，男性的工资往往会更高，尽管在 10 年以下的工作年限中，男性和女性的平均工资是相同的。

12.10.3 扩展知识

从局外人的角度来看，步骤（8）的输出中代表 10 年区间的行并不明显。改善索引标签的方式之一是显示每个时间间隔的开始和结束。这可以通过将当前的索引年份加 9，然后连接到它自身来实现。

```
>>> sal_final = (employee
...     .groupby(['GENDER', pd.Grouper(freq='10AS')])
...     ['BASE_SALARY']
...     .mean()
...     .round(-2)
...     .unstack('GENDER')
... )
>>> years = sal_final.index.year
>>> years_right = years + 9
>>> sal_final.index = years.astype(str) + '-' + years_right.astype(str)
>>> sal_final
GENDER       Female     Male
HIRE_DATE
1958-1967       NaN   81200.0
1968-1977       NaN  106500.0
1978-1987   57100.0   72300.0
1988-1997   57100.0   64600.0
1998-2007   54700.0   59700.0
2008-2017   47300.0   47200.0
```

也可以采用一种完全不同的方法来执行此操作。我们可以使用 cut 函数根据每个员工的受聘年限来创建等宽区间，并从中分组。

```
>>> cuts = pd.cut(employee.index.year, bins=5, precision=0)
>>> cuts.categories.values
```

```
IntervalArray([(1958.0, 1970.0], (1970.0, 1981.0], (1981.0, 1993.0],
               (1993.0, 2004.0], (2004.0, 2016.0]],
              closed='right',
              dtype='interval[float64]')

>>> (employee
...     .groupby([cuts, 'GENDER'])
...     ['BASE_SALARY']
...     .mean()
...     .unstack('GENDER')
...     .round(-2)
... )
GENDER             Female    Male
(1958.0, 1970.0]      NaN  85400.0
(1970.0, 1981.0]  54400.0  72700.0
(1981.0, 1993.0]  55700.0  69300.0
(1993.0, 2004.0]  56500.0  62300.0
(2004.0, 2016.0]  49100.0  49800.0
```

第 13 章 使用 Matplotlib、Pandas 和 Seaborn 进行可视化

本章包含以下秘笈。
- ❑ Matplotlib 入门。
- ❑ Matplotlib 的面向对象指南。
- ❑ 使用 Matplotlib 可视化数据。
- ❑ Pandas 绘图基础。
- ❑ 可视化航班数据集。
- ❑ 使用堆积面积图发现新兴趋势。
- ❑ 了解 Seaborn 和 Pandas 之间的区别。
- ❑ 使用 Seaborn 网格进行多变量分析。
- ❑ 使用 Seaborn 在钻石数据集中发现辛普森悖论。

13.1 介 绍

可视化是探索性数据分析以及 PowerPoint 演示文稿和应用程序中的关键组成部分。

在探索性数据分析过程中,你通常是一个人或在一个小组中工作,需要快速创建绘图以帮助你更好地理解数据。探索性数据分析可以帮助你识别异常值和缺失的数据,也可以引发其他令人感兴趣的问题,这些问题将导致进一步的分析和更直观的显示。这种类型的可视化通常不会从最终用户的角度来考虑问题,严格来说它只是为了帮助你更好地了解当前情况。所以,这种可视化绘图不必是完美的。

在为报表或应用程序准备可视化文件时,必须使用其他方法。你应该注意一些小细节。而且,一般来说,你必须将所有可能的可视化范围缩小到仅最能代表你的数据的少数项目。良好的数据可视化将使观看者享受到提取信息的体验。几乎就像使观众陶醉的电影一样,好的可视化效果将包含大量真正引起人们兴趣的信息。

Python 中主要的数据可视化库是 Matplotlib,旨在模仿 Matlab 的绘图功能。Matplotlib 具有极大的能力来绘制你可以想象的大多数事物,它为用户提供了强大的功能来控制绘图接口的各个方面。

也就是说，对于初学者来说，Matplotlib 不是一个容易掌握的库。值得庆幸的是，Pandas 使我们对数据的可视化变得非常容易，并且通常只需调用一次 plot 方法即可绘制所需的图形。Pandas 没有自己的绘图引擎，它在内部其实是调用 Matplotlib 函数来绘图。

Seaborn 也是一个可视化库，它包装了 Matplotlib，但是自身并不做任何实际的绘图。Seaborn 制作了很漂亮的绘图，并且有许多类型的绘图是无法从 Matplotlib 或 Pandas 中获得的。Seaborn 使用规整（长）数据，而 Pandas 则是使用聚合（宽）数据效果最佳。Seaborn 在其绘图函数中还接收 Pandas DataFrame 对象。

尽管可以在不运行任何 Matplotlib 代码的情况下创建图，但有时还是需要使用它来手动调整更精细的图形细节。对此，本章前两个秘笈将介绍 Matplotlib 的一些基础知识，如果你是 Matplotlib 的初学者，那么它将非常有用。除前两个秘笈外，所有绘图示例都将使用 Pandas 或 Seaborn。

值得一提的是，Python 中的可视化并不是必须依赖于 Matplotlib。除 Matplotlib 外，Bokeh 也已经迅速成为针对 Web 的非常流行的交互式可视化库，并完全独立于 Matplotlib，并且能够生成整个应用。

除 Bokeh 外，还有其他一些绘图库也在发展中，未来的 Pandas 版本也可能会使用除 Matplotlib 外的绘图引擎。

13.2 Matplotlib 入门

对于许多数据科学家而言，他们使用的绝大部分绘图命令都是 Pandas 或 Seaborn，二者均依赖于 Matplotlib 进行绘图。但是，Pandas 和 Seaborn 都无法完全替代 Matplotlib，有时你还需要使用 Matplotlib。因此，本秘笈将简要介绍 Matplotlib 的最关键方面。

如果你是 Jupyter 用户，则需要使用以下指令它。

```
>>> %matplotlib inline
```

该指令告诉 Matplotlib 在 Notebook 中渲染图形。

图 13-1 显示了 Matplotlib 图的分层结构。

Matplotlib 使用了对象的层次结构在输出中显示其所有绘图项。该层次结构是了解有关 Matplotlib 的一切的关键。请注意，以下术语指的是 Matplotlib，而不是具有相同名称的 Pandas 对象（这可能是容易让人搞混的地方）。

Figure（图形）和 Axes（轴）对象是层次结构的两个主要组成部分。Figure 对象位于层次结构的顶部。另外，Figure 对象是将要绘制的所有内容的容器。Figure 中包含一个或多个 Axes 对象。Axes 是使用 Matplotlib 时将与之交互的主要对象，并且可以视为绘图接口。Axes 包含一个 x 轴、一个 y 轴、点、线、标记、标签、图例以及其他绘制的有用项目。

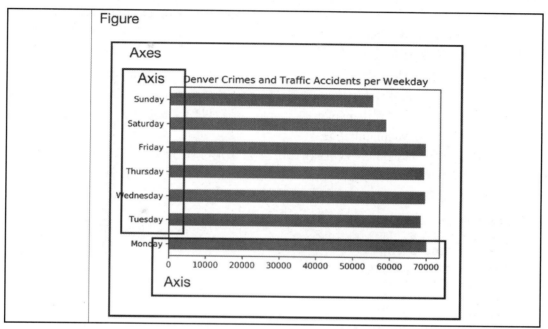

图 13-1　Matplotlib 层次结构

请注意，Axis 也是轴，在 Axes 和 Axis 之间需要进行区分。它们是完全独立的对象。在 Matplotlib 术语中，Axes 对象不是 Axis 的复数，而是如前文所述，该对象创建并控制了大多数有用的绘图元素，而 Axis 则是指绘图的 x 或 y（甚至 z）轴。

由 Axes 对象创建的所有这些有用的绘图元素都被称为 Artist。甚至 Figure 和 Axes 对象本身也是 Artist。对 Artist 的这种区分对本秘笈而言并不重要，但在进行更高级的 Matplotlib 绘图时，尤其是在阅读说明文档时，将很有用。

13.3　Matplotlib 的面向对象指南

Matplotlib 为用户提供了两个不同的接口。有状态接口（stateful interface）使用 pyplot 模块进行所有调用。此接口之所以被称为有状态的，是因为 Matplotlib 在内部跟踪绘图环境的当前状态。每当在有状态接口中创建绘图时，Matplotlib都会找到当前图形或当前轴，并对其进行更改。这种方法固然可以快速绘制一些东西，但是当处理多个图形和轴时可能会变得比较笨拙。

Matplotlib还提供了一个无状态（stateless）或面向对象（object-oriented）的接口，你可以在其中显式使用变量以引用特定绘图对象。每个变量都可以用来更改绘图的某些属性。面向对象的方法是显式的，你始终清楚地知道要修改的对象。

遗憾的是，同时使用这两个选项可能会导致很多混乱，更何况Matplotlib以难以学习而著称。其说明文档提供了使用这两种方法的示例。在实践中，笔者发现将它们结合起来最有用。一方面，可以使用pyplot的subplots函数以创建图形和轴；另一方面，又可以在这些对象上使用Matplotlib方法。

如果你不熟悉Matplotlib，则可能不知道如何识别每种方法之间的区别。在使用有状态接口时，所有命令都是在pyplot模块上调用的函数，通常使用别名plt。

要绘制折线图并在每个轴上添加一些标签，可以按以下方式操作。

```
>>> import matplotlib.pyplot as plt
>>> x = [-3, 5, 7]
>>> y = [10, 2, 5]
>>> fig = plt.figure(figsize=(15,3))
>>> plt.plot(x, y)
>>> plt.xlim(0, 10)
>>> plt.ylim(-3, 8)
>>> plt.xlabel('X Axis')
>>> plt.ylabel('Y axis')
>>> plt.title('Line Plot')
>>> plt.suptitle('Figure Title', size=20, y=1.03)
>>> fig.savefig('c13-fig1.png', dpi=300, bbox_inches='tight')
```

其输出结果如图13-2所示。

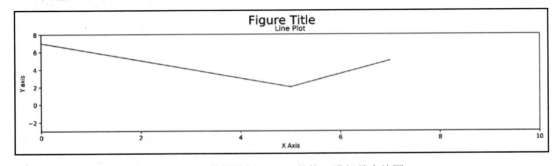

图13-2　使用类似Matlab的接口进行基本绘图

面向对象的方法如下。

```
>>> from matplotlib.figure import Figure
>>> from matplotlib.backends.backend_agg import FigureCanvasAgg as FigureCanvas
>>> from IPython.core.display import display
>>> fig = Figure(figsize=(15, 3))
>>> FigureCanvas(fig)
>>> ax = fig.add_subplot(111)
>>> ax.plot(x, y)
>>> ax.set_xlim(0, 10)
>>> ax.set_ylim(-3, 8)
>>> ax.set_xlabel('X axis')
>>> ax.set_ylabel('Y axis')
>>> ax.set_title('Line Plot')
>>> fig.suptitle('Figure Title', size=20, y=1.03)
>>> display(fig)
>>> fig.savefig('c13-fig2.png', dpi=300, bbox_inches='tight')
```

其输出结果如图 13-3 所示。

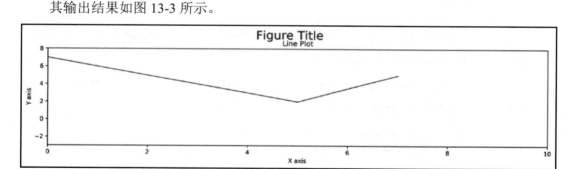

图 13-3　使用面向对象的接口创建基本绘图

在实践中，可以将两种方法结合在一起，示例代码如下。

```
>>> fig, ax = plt.subplots(figsize=(15,3))
>>> ax.plot(x, y)
>>> ax.set(xlim=(0, 10), ylim=(-3, 8),
...        xlabel='X axis', ylabel='Y axis',
...        title='Line Plot')
>>> fig.suptitle('Figure Title', size=20, y=1.03)
>>> fig.savefig('c13-fig3.png', dpi=300, bbox_inches='tight')
```

其输出结果如图 13-4 所示。

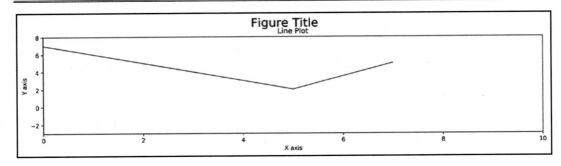

图 13-4　先调用 Matlab 接口创建图形和轴,然后使用方法调用

在此示例中,我们仅使用两个对象(Figure 和 Axes),但是一般来说,图形可以包含数百个对象;每个对象都可以用来以一种非常精细的方式进行修改,而使用有状态接口则不容易实现。在本章中,我们将构建一个空白图形,并使用面向对象的接口修改其一些基本属性。

13.3.1　实战操作

(1)要使用面向对象的方法开始使用 Matplotlib,需要导入 pyplot 模块和别名 plt。

```
>>> import matplotlib.pyplot as plt
```

(2)一般来说,当使用面向对象的方法时,我们将创建一个 Figure,以及一个或多个 Axes 对象。下面使用 subplots 函数创建一个具有单个轴的图形。

```
>>> fig, ax = plt.subplots(nrows=1, ncols=1)
>>> fig.savefig('c13-step2.png', dpi=300)
```

其输出结果如图 13-5 所示。

(3)subplots 函数返回一个包含 Figure 和一个或多个 Axes 对象(本示例只有一个)的两个项目的元组对象,该对象被解包(unpack)到变量 fig 和 ax 中。从现在开始,我们将使用常规的面向对象方法,通过调用方法来使用这些对象。

```
>>> type(fig)
matplotlib.figure.Figure
>>> type(ax)
matplotlib.axes._subplots.AxesSubplot
```

(4)尽管将要调用的 ax 方法比 fig 方法更多,但你可能仍需要与 fig 进行交互。例如,我们可以找到 fig 的大小,然后放大它。

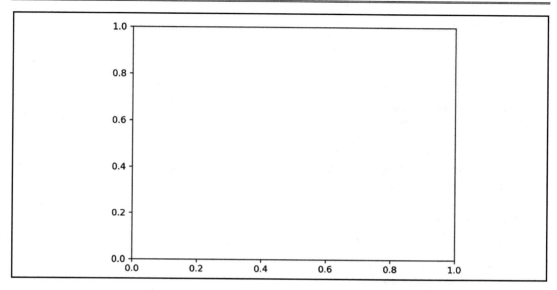

图 13-5　Figure 绘图

```
>>> fig.get_size_inches()
array([ 6., 4.])
>>> fig.set_size_inches(14, 4)
>>> fig.savefig('c13-step4.png', dpi=300)
>>> fig
```

其输出结果如图 13-6 所示。

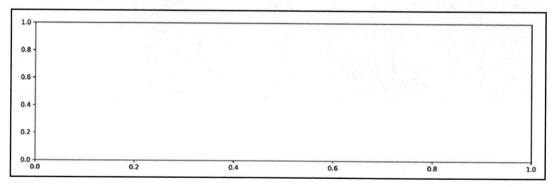

图 13-6　更改图形大小

（5）在开始绘图前，不妨先检查该 matplotlib 的层次结构。可以使用.axes 属性收集图形的所有轴。

```
>>> fig.axes
[<matplotlib.axes._subplots.AxesSubplot at 0x112705ba8>]
```

（6）上一条命令将返回所有 Axes 对象的列表。但是，我们已经将 Axes 对象存储在 ax 变量中。因此可以验证它们是否为同一对象。

```
>>> fig.axes[0] is ax
True
```

（7）为了帮助区分 Figure 和 Axes 对象，可以为每个图形赋予唯一的 facecolor（表面颜色）。Matplotlib 接收各种不同的颜色输入类型。字符串名称支持大约 140 种 HTML 颜色，有关这些颜色的列表，可访问以下网址。

http://bit.ly/2y52UtO

也可以使用包含从 0～1 的浮点数的字符串来表示灰色阴影。

```
>>> fig.set_facecolor('.7')
>>> ax.set_facecolor('.5')
>>> fig.savefig('c13-step7.png', dpi=300, facecolor='.7')
>>> fig
```

其输出结果如图 13-7 所示。

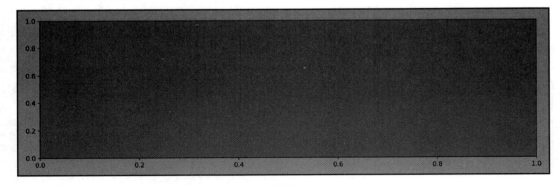

图 13-7　设置 facecolor

（8）现在已经区分了 Figure 和 Axes 对象，可以使用.get_children 方法来查看 Axes 的所有直接子对象。

```
>>> ax_children = ax.get_children()
>>> ax_children
[<matplotlib.spines.Spine at 0x11145b358>,
 <matplotlib.spines.Spine at 0x11145b0f0>,
 <matplotlib.spines.Spine at 0x11145ae80>,
```

```
<matplotlib.spines.Spine at 0x11145ac50>,
<matplotlib.axis.XAxis at 0x11145aa90>,
<matplotlib.axis.YAxis at 0x110fa8d30>,
...]
```

（9）大多数图形都有 4 个 spine（样条）和两个 axis（轴）对象。spine 代表数据边界，它是你看到的与较深的灰色矩形（axes）接触的 4 根物理线。x 和 y 轴对象包含更多的绘图对象，如刻度和它们的标签，以及整个轴的标签。

虽然可以从.get_children 方法的结果中选择 spine，但是使用.spines 属性则可以更轻松地访问它们，示例如下。

```
>>> spines = ax.spines
>>> spines
OrderedDict([ ('left', <matplotlib.spines.Spine at 0x11279e320>),
              ('right', <matplotlib.spines.Spine at 0x11279e0b8>),
              ('bottom', <matplotlib.spines.Spine at 0x11279e048>),
              ('top', <matplotlib.spines.Spine at 0x1127eb5c0>)])
```

（10）spine 包含在有序字典中。我们可以选择左侧的 spine，并更改其位置和宽度，使其更加突出，并使底部的 spine 不可见。

```
>>> spine_left = spines['left']
>>> spine_left.set_position(('outward', -100))
>>> spine_left.set_linewidth(5)
>>> spine_bottom = spines['bottom']
>>> spine_bottom.set_visible(False)
>>> fig.savefig('c13-step10.png', dpi=300, facecolor='.7')
>>> fig
```

其输出结果如图 13-8 所示。

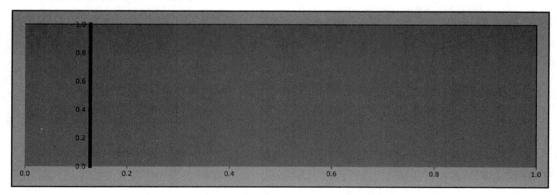

图 13-8　移动或删除 spine

（11）现在来专注讨论轴对象。可以通过.xaxis 和.yaxis 属性访问每个轴。轴对象还可以使用某些 Axes 属性。以下语句使用了两种方式来更改每个轴的某些属性。

```
>>> ax.xaxis.grid(True, which='major', linewidth=2,
...     color='black', linestyle='--')
>>> ax.xaxis.set_ticks([.2, .4, .55, .93])
>>> ax.xaxis.set_label_text('X Axis', family='Verdana',
...     fontsize=15)
>>> ax.set_ylabel('Y Axis', family='Gotham', fontsize=20)
>>> ax.set_yticks([.1, .9])
>>> ax.set_yticklabels(['point 1', 'point 9'], rotation=45)
>>> fig.savefig('c13-step11.png', dpi=300, facecolor='.7')
```

其输出结果如图 13-9 所示。

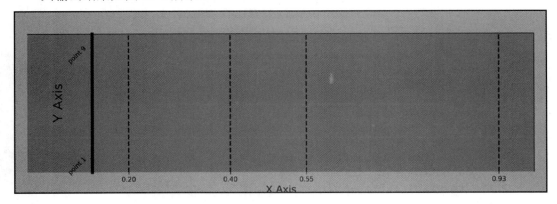

图 13-9　带有标签的轴

13.3.2　原理解释

面向对象方法要掌握的关键思想之一是每个绘图元素都具有 getter 和 setter 方法。

- getter 方法用于取值，都以 get_开头。例如，ax.get_yscale()将检索绘制 y 轴比例类型的字符串（默认为 linear），而 ax.get_xticklabels()则检索 Matplotlib 文本对象的列表（每个对象都有自己的 getter 和 setter 方法）。
- setter 方法用于赋值，可以修改特定的属性或整个对象组。

许多 Matplotlib 都将重点放到特定的绘图元素上，然后通过 getter 和 setter 方法进行检查和修改。

开始使用 Matplotlib 的最简单方法是使用 pyplot 模块，该模块通常使用别名 plt 代指，步骤（1）就是如此。

步骤（2）显示了一种面向对象方法的初始化方式。plt.subplots 函数创建了一个 Figure，以及一个 Axes 对象网格。前两个参数 nrows 和 ncols 定义了一个统一的 Axes 对象网格。例如，plt.subplots(2,4)在一个 Figure 中创建了 8 个相同大小的 Axes 对象。

plt.subplots 返回一个元组。第一个元素是 Figure，第二个元素是 Axes 对象。该元组被解包为两个变量，即 fig 和 ax。如果你不习惯于元组解包，则将步骤（2）编写为如下形式可能会有助于理解。

```
>>> plot_objects = plt.subplots(nrows=1, ncols=1)
>>> type(plot_objects)
tuple
>>> fig = plot_objects[0]
>>> ax = plot_objects[1]
>>> fig.savefig('c13-1-works1.png', dpi=300)
```

其输出结果如图 13-10 所示。

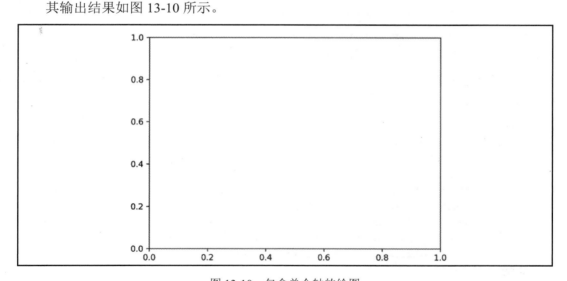

图 13-10　包含单个轴的绘图

如果使用 plt.subplots 创建多个轴，则元组中的第二项是包含所有轴的 NumPy 数组。具体示例如下。

```
>>> fig, axs = plt.subplots(2, 4)
>>> fig.savefig('c13-1-works2.png', dpi=300)
```

其输出结果如图 13-11 所示。

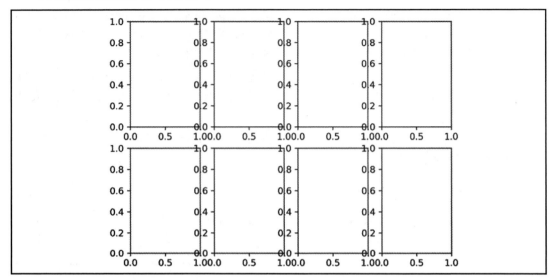

图 13-11　使用轴的网格绘图

axs 变量是一个 NumPy 数组，其中包含 Figure 作为其第一个元素，而 NumPy 数组则作为其第二个元素。

```
>>> axs
array([[<matplotlib.axes._subplots.AxesSubplot object at 0x126820668>,
        <matplotlib.axes._subplots.AxesSubplot object at 0x126844ba8>,
        <matplotlib.axes._subplots.AxesSubplot object at 0x126ad1160>,
        <matplotlib.axes._subplots.AxesSubplot object at 0x126afa6d8>],
       [<matplotlib.axes._subplots.AxesSubplot object at 0x126b21c50>,
        <matplotlib.axes._subplots.AxesSubplot object at 0x126b52208>,
        <matplotlib.axes._subplots.AxesSubplot object at 0x11f695588>,
        <matplotlib.axes._subplots.AxesSubplot object at 0x11f6b3b38>]],
      dtype=object)
```

步骤（3）验证我们确实有通过适当变量引用的 Figure 和 Axes 对象。

在步骤（4）中，我们遇到了第一个 getter 和 setter 方法的示例。Matplotlib 默认将所有 Figure 的宽度设置为 6in 和 4in 高，这不是屏幕上的实际大小，但是如果将 Figure 保存到文件中，则为确切大小（DPI 为 100px/in）。

步骤（5）显示，除 getter 方法外，有时还可以通过其属性访问另一个绘图对象。一一般来说，同时存在属性和 getter 方法来检索同一对象。来看以下示例。

```
>>> ax = axs[0][0]
>>> fig.axes == fig.get_axes()
True
>>> ax.xaxis == ax.get_xaxis()
True
>>> ax.yaxis == ax.get_yaxis()
True
```

许多 Artist 都具有.facecolor 属性，可以将其设置为一种覆盖整个表面的特定颜色，在步骤（7）中就是这样做的。

在步骤（8）中，使用了.get_children 方法以更好地了解对象层次结构。.get_children 方法返回了 Axes 以下所有对象的列表。可以从此列表中选择所有对象，然后使用 setter 方法来修改属性，但一般不这样做。更常见的做法是从属性收集对象或使用 getter 方法。

通常而言，在检索绘图对象时，它们会在列表或字典之类的容器中返回。在步骤（9）中收集 spine 时就是如此。

你必须从它们各自的容器中选择单个对象，以对它们使用 getter 或 setter 方法，在步骤（10）中就是这样做的。也可以使用 for 循环一次遍历每个对象。

步骤（11）以特殊方式添加了网格线。我们希望有一个.get_grid 和.set_grid 方法，但是，目前只有一个.grid 方法，该方法接收布尔值作为第一个参数，以打开和关闭网格线。每个轴都有主要刻度（major tick）和次要刻度（minor tick），但默认情况下，次要刻度是关闭的。which 参数用于选择具有网格线的刻度线类型。

请注意，步骤（11）的前 3 行选择了.xaxis 属性并调用了方法，而后 3 行则从 Axes 对象本身调用了等效方法。第二组方法是 Matplotlib 提供的一种便捷方式，它可以让你少敲入一些代码。

一般来说，大多数对象都只能设置自己的属性，而不能设置其子对象的属性。因此，无法通过 Axes 设置许多 Axis 级的属性，但是在此步骤中，有些属性是可以设置的。两种方法都可以接收。

在步骤（11）的第一行添加了网格线，我们设置了属性.linewidth、.color 和.linestyle。这些都是 Matplotlib 线条（正式称为 Line2D 对象）的所有属性。.set_ticks 方法接收一系列的浮点数，并且仅在指定位置绘制刻度线。使用空列表将完全删除所有刻度。

每个轴都可以被标记一些文本，为此 Matplotlib 使用了一个 Text 对象。在所有可用的文本属性中，这里只修改了少数几个。.set_yticklabels Axes 方法可以采用字符串列表作为每个刻度的标签。你可以设置任意数量的文本属性。

13.3.3 扩展知识

为了帮助找到每个绘图对象的所有可能的属性，可以调用 .properties 方法，该方法将所有这些对象显示为字典。来看 Axis 对象的属性的列表。

```
>>> ax.xaxis.properties()
{'alpha': None,
 'gridlines': <a list of 4 Line2D gridline objects>,
 'label': Text(0.5,22.2,'X Axis'),
 'label_position': 'bottom',
 'label_text': 'X Axis',
 'tick_padding': 3.5,
 'tick_space': 26,
 'ticklabels': <a list of 4 Text major ticklabel objects>,
 'ticklocs': array([ 0.2 , 0.4 , 0.55, 0.93]),
 'ticks_position': 'bottom',
 'visible': True}
```

13.4 使用 Matplotlib 可视化数据

Matplotlib 有几十种绘图方法，几乎可以制作任何你能想象到的类型的绘图。折线图、条形图、直方图、散点图、箱形图、小提琴图、等高线图、饼图以及许多其他绘图都可以用作 Axes 对象上的方法。

从 1.5 版（2015 年发布）之后，Matplotlib 才开始接收来自 Pandas DataFrame 的数据。在此之前，必须将数据从 NumPy 数组或 Python 列表中传递给它。

在此秘笈中，我们将绘制美国犹他州 Alta 滑雪场的年度降雪量。本示例中的绘图受 Trud Antzee 的启发，他创建了类似的挪威降雪量图形。

13.4.1 实战操作

（1）前文已经介绍了如何创建 Axes 并更改其属性，现在就来看如何真正开始可视化数据。

读取犹他州 Alta 滑雪场的降雪数据，并查看每个季节的降雪量。

```
>>> import pandas as pd
>>> import numpy as np
```

第 13 章 使用 Matplotlib、Pandas 和 Seaborn 进行可视化

```
>>> alta = pd.read_csv('data/alta-noaa-1980-2019.csv')
>>> alta
         STATION              NAME       LATITUDE   ...   WT05    WT06    WT11
0        USC00420072    ALTA, UT US      40.5905    ...   NaN     NaN     NaN
1        USC00420072    ALTA, UT US      40.5905    ...   NaN     NaN     NaN
2        USC00420072    ALTA, UT US      40.5905    ...   NaN     NaN     NaN
3        USC00420072    ALTA, UT US      40.5905    ...   NaN     NaN     NaN
4        USC00420072    ALTA, UT US      40.5905    ...   NaN     NaN     NaN
...              ...            ...          ...    ...   ...     ...     ...
14155    USC00420072    ALTA, UT US      40.5905    ...   NaN     NaN     NaN
14156    USC00420072    ALTA, UT US      40.5905    ...   NaN     NaN     NaN
14157    USC00420072    ALTA, UT US      40.5905    ...   NaN     NaN     NaN
14158    USC00420072    ALTA, UT US      40.5905    ...   NaN     NaN     NaN
14159    USC00420072    ALTA, UT US      40.5905    ...   NaN     NaN     NaN
```

（2）获取 2018—2019 年季节的数据。

```
>>> data = (alta
...     .assign(DATE=pd.to_datetime(alta.DATE))
...     .set_index('DATE')
...     .loc['2018-09':'2019-08']
...     .SNWD
... )
>>> data
DATE
2018-09-01    0.0
2018-09-02    0.0
2018-09-03    0.0
2018-09-04    0.0
2018-09-05    0.0
              ...
2019-08-27    0.0
2019-08-28    0.0
2019-08-29    0.0
2019-08-30    0.0
2019-08-31    0.0
Name: SNWD, Length: 364, dtype: float64
```

（3）使用 Matplotlib 可视化此数据。我们可以使用默认绘图，然后调整绘图的外观。请注意：在调用 .savefig 方法时，需要指定 facecolor；否则导出的图像将具有白色 facecolor。

```
>>> blue = '#99ddee'
```

```
>>> white = '#ffffff'
>>> fig, ax = plt.subplots(figsize=(12,4),
...       linewidth=5, facecolor=blue)
>>> ax.set_facecolor(blue)
>>> ax.spines['top'].set_visible(False)
>>> ax.spines['right'].set_visible(False)
>>> ax.spines['bottom'].set_visible(False)
>>> ax.spines['left'].set_visible(False)
>>> ax.tick_params(axis='x', colors=white)
>>> ax.tick_params(axis='y', colors=white)
>>> ax.set_ylabel('Snow Depth (in)', color=white)
>>> ax.set_title('2009-2010', color=white, fontweight='bold')
>>> ax.fill_between(data.index, data, color=white)
>>> fig.savefig('c13-alta1.png', dpi=300, facecolor=blue)
```

其输出结果如图 13-12 所示。

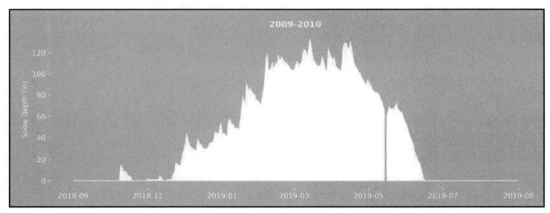

图 13-12　2009—2010 年季节的 Alta 雪位图

（4）在一个 Figure 上可以放置任何数量的绘图。因此，我们可以重构 plot_year 函数并绘制多年数据。

```
>>> import matplotlib.dates as mdt
>>> blue = '#99ddee'
>>> white = '#ffffff'

>>> def plot_year(ax, data, years):
...     ax.set_facecolor(blue)
...     ax.spines['top'].set_visible(False)
```

```
...         ax.spines['right'].set_visible(False)
...         ax.spines['bottom'].set_visible(False)
...         ax.spines['left'].set_visible(False)
...         ax.tick_params(axis='x', colors=white)
...         ax.tick_params(axis='y', colors=white)
...         ax.set_ylabel('Snow Depth (in)', color=white)
...         ax.set_title(years, color=white, fontweight='bold')
...         ax.fill_between(data.index, data, color=white)
>>> years = range(2009, 2019)
>>> fig, axs = plt.subplots(ncols=2, nrows=int(len(years)/2),
...     figsize=(16, 10), linewidth=5, facecolor=blue)
>>> axs = axs.flatten()
>>> max_val = None
>>> max_data = None
>>> max_ax = None
>>> for i,y in enumerate(years):
...     ax = axs[i]
...     data = (alta
...         .assign(DATE=pd.to_datetime(alta.DATE))
...         .set_index('DATE')
...         .loc[f'{y}-09':f'{y+1}-08']
...         .SNWD
...     )
...     if max_val is None or max_val < data.max():
...         max_val = data.max()
...         max_data = data
...         max_ax = ax
...     ax.set_ylim(0, 180)
...     years = f'{y}-{y+1}'
...     plot_year(ax, data, years)
>>> max_ax.annotate(f'Max Snow {max_val}',
...     xy=(mdt.date2num(max_data.idxmax()), max_val),
...     color=white)

>>> fig.suptitle('Alta Snowfall', color=white, fontweight='bold')
>>> fig.tight_layout(rect=[0, 0.03, 1, 0.95])
>>> fig.savefig('c13-alta2.png', dpi=300, facecolor=blue)
```

其输出结果如图 13-13 所示。

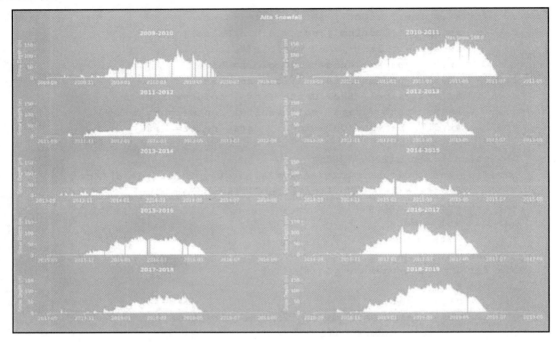

图 13-13　多个季节的 Alta 雪位图

13.4.2　原理解释

在步骤（1）中加载了 NOAA 数据。

在步骤（2）中，我们使用了各种 Pandas 技巧将 DATE 列从字符串转换为日期。然后，将索引设置为 DATE 列，以便可以从 9 月开始切分一年的时间。最后，提取出 SNWD（SNoW depth，降雪深度）列以获得 Pandas Series。

在步骤（3）中，提取出了所有节点数据。使用了 subplots 函数创建 Figure 和 Axes。我们将轴和图形的 facecolor 都设置为浅蓝色。此外还删除了 spine 并将标签颜色设置为白色。最后，使用 .fill_between 绘图函数创建了一个填充的绘图。该绘图（受 Trud 启发）显示了我们想要通过 Matplotlib 强调的内容。在 Matplotlib 中，你几乎可以更改绘图的任何方面。将 Jupyter 与 Matplotlib 结合使用可让你尝试对绘图进行调整。

在步骤（4）中，我们将步骤（3）的操作重构为一个函数，然后在网格中绘制了 10 个图形。在遍历年份数据时，我们还会跟踪最大值。这使我们可以使用 .annotate 方法来注解具有最大显示深度的轴。

13.4.3 扩展知识

人类的大脑并未针对查看数据表进行优化。很多人面对大量繁杂的数据时往往不知所措或茫无头绪。但是，对数据进行可视化处理可以帮助我们深入了解数据。例如，在本示例中，经过可视化后，可以很明显地发现，有些数据是缺失的，因为绘图中产生了空缺。在这种情况下，可以使用.interpolate 方法清理这些空缺。

```
>>> years = range(2009, 2019)
>>> fig, axs = plt.subplots(ncols=2, nrows=int(len(years)/2),
...      figsize=(16, 10), linewidth=5, facecolor=blue)
>>> axs = axs.flatten()
>>> max_val = None
>>> max_data = None
>>> max_ax = None
>>> for i,y in enumerate(years):
...     ax = axs[i]
...     data = (alta.assign(DATE=pd.to_datetime(alta.DATE))
...         .set_index('DATE')
...         .loc[f'{y}-09':f'{y+1}-08']
...         .SNWD
...         .interpolate()
...     )
...     if max_val is None or max_val < data.max():
...         max_val = data.max()
...         max_data = data
...         max_ax = ax
...     ax.set_ylim(0, 180)
...     years = f'{y}-{y+1}'
...     plot_year(ax, data, years)
>>> max_ax.annotate(f'Max Snow {max_val}',
...     xy=(mdt.date2num(max_data.idxmax()), max_val),
...     color=white)

>>> fig.suptitle('Alta Snowfall', color=white, fontweight='bold')
>>> fig.tight_layout(rect=[0, 0.03, 1, 0.95])
>>> fig.savefig('c13-alta3.png', dpi=300, facecolor=blue)
```

其输出结果如图 13-14 所示。

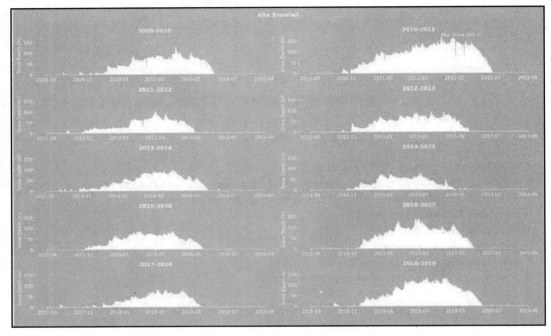

图 13-14 Alta 绘图

即使图 13-14 中的 Alta 绘图也仍然是有问题的。我们可以研究得再深入一些。看起来冬季的降雪量有些时候似乎太多了。下面使用一些 Pandas 语句来查找后续条目之间的绝对差大于某个值（如 50）的地方。

```
>>> (alta
...     .assign(DATE=pd.to_datetime(alta.DATE))
...     .set_index('DATE')
...     .SNWD
...     .to_frame()
...     .assign(next=lambda df_:df_.SNWD.shift(-1),
...             snwd_diff=lambda df_:df_.next-df_.SNWD)
...     .pipe(lambda df_: df_[df_.snwd_diff.abs() > 50])
... )
             SNWD   next  snwd_diff
DATE
1989-11-27   60.0    0.0      -60.0
2007-02-28   87.0    9.0      -78.0
2008-05-22   62.0    0.0      -62.0
2008-05-23    0.0   66.0       66.0
```

2009-01-16	76.0	0.0	-76.0
...	...	...	...
2011-05-18	0.0	136.0	136.0
2012-02-09	58.0	0.0	-58.0
2012-02-10	0.0	56.0	56.0
2013-03-01	75.0	0.0	-75.0
2013-03-02	0.0	78.0	78.0

看起来这些数据似乎有问题。在季节的中间，数据趋于 0（实际上是 0，而不是 np.nan）时，会有一些斑点。因此，可以考虑制作一个 fix_gaps 函数，使用 .pipe 方法来清理它们。

```
>>> def fix_gaps(ser, threshold=50):
...     'Replace values where the shift is > threshold with nan'
...     mask = (ser
...         .to_frame()
...         .assign(next=lambda df_:df_.SNWD.shift(-1),
...                 snwd_diff=lambda df_:df_.next-df_.SNWD)
...         .pipe(lambda df_: df_.snwd_diff.abs() > threshold)
...     )
...     return ser.where(~mask, np.nan)

>>> years = range(2009, 2019)
>>> fig, axs = plt.subplots(ncols=2, nrows=int(len(years)/2),
...     figsize=(16, 10), linewidth=5, facecolor=blue)
>>> axs = axs.flatten()

>>> max_val = None
>>> max_data = None
>>> max_ax = None
>>> for i,y in enumerate(years):
...     ax = axs[i]
...     data = (alta.assign(DATE=pd.to_datetime(alta.DATE))
...         .set_index('DATE')
...         .loc[f'{y}-09':f'{y+1}-08']
...         .SNWD
...         .pipe(fix_gaps)
...         .interpolate()
...     )
...     if max_val is None or max_val < data.max():
...         max_val = data.max()
...         max_data = data
...         max_ax = ax
...     ax.set_ylim(0, 180)
```

```
...         years = f'{y}-{y+1}'
...         plot_year(ax, data, years)
>>> max_ax.annotate(f'Max Snow {max_val}',
...         xy=(mdt.date2num(max_data.idxmax()), max_val),
...         color=white)

>>> fig.suptitle('Alta Snowfall', color=white, fontweight='bold')
>>> fig.tight_layout(rect=[0, 0.03, 1, 0.95])
>>> fig.savefig('c13-alta4.png', dpi=300, facecolor=blue)
```

其输出结果如图 13-15 所示。

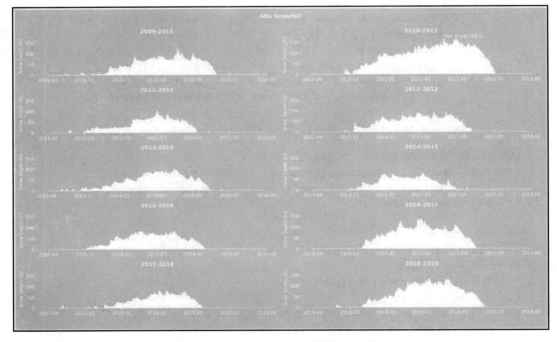

图 13-15 Alta 绘图

13.5 Pandas 绘图基础

Pandas 可以通过自动执行许多步骤让绘图过程变得非常容易。绘图是由 Matplotlib 在内部处理的，可以通过 DataFrame 或 Series 的.plot 属性（它也可以用作方法，但我们将使用该属性进行绘图）公开访问。

在 Pandas 中创建绘图时,将返回 Matplotlib Axes 或 Figure。然后,可以使用 Matplotlib 的全部功能来调整该绘图,直到满意为止。

Pandas 只能生成 Matplotlib 可用图形的一小部分,如折线图、条形图、箱形图和散点图,以及核密度估计(kernel density estimate,KDE)和直方图。Pandas 虽然功能受限,但其优点是绘图更容易,因此 Pandas 接口颇受欢迎,因为它通常只是一行代码。

理解在 Pandas 中绘图的关键之一就是要知道 x 和 y 轴的来源。默认图形(折线图)将在 x 轴上绘制索引,在 y 轴上绘制每一列。对于散点图而言,则需要指定用于 x 和 y 轴的列。直方图、箱形图和 KDE 图将忽略索引,并绘制每列的分布。

此秘笈将展示使用 Pandas 进行绘图的各种示例。

13.5.1 实战操作

(1)创建一个包含有意义索引的小型 DataFrame。

```
>>> df = pd.DataFrame(index=['Atiya', 'Abbas', 'Cornelia',
...     'Stephanie', 'Monte'],
...     data={'Apples':[20, 10, 40, 20, 50],
...     'Oranges':[35, 40, 25, 19, 33]})

>>> df
           Apples   Oranges
Atiya          20        35
Abbas          10        40
Cornelia       40        25
Stephanie      20        19
Monte          50        33
```

(2)条形图使用索引作为 x 轴的标签,而列值则用作条形的高度。可以将 .plot 属性与 .bar 方法一起使用。

```
>>> color = ['.2', '.7']
>>> ax = df.plot.bar(color=color, figsize=(16,4))
>>> ax.get_figure().savefig('c13-pdemo-bar1.png')
```

其输出结果如图 13-16 所示。

(3)KDE 图忽略索引,沿 x 轴使用列名,同时使用列值沿 y 值计算概率密度。

```
>>> ax = df.plot.kde(color=color, figsize=(16,4))
>>> ax.get_figure().savefig('c13-pdemo-kde1.png')
```

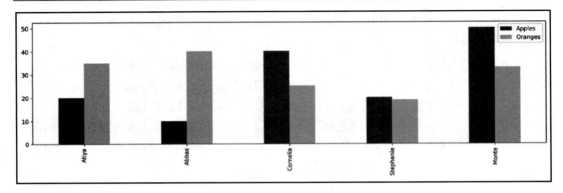

图 13-16　Pandas 条形图

其输出结果如图 13-17 所示。

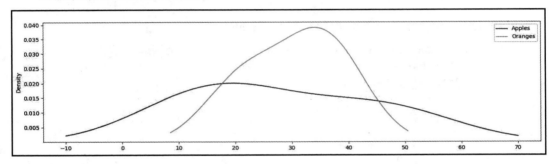

图 13-17　Pandas KDE 图

（4）可以在单个 Figure 中绘制折线图（line plot）、散点图（scatter plot）和条形图（bar plot）。散点图是唯一需要为 x 和 y 值指定列的图形。如果要让散点图使用索引，则必须使用.reset_index 方法将索引变成普通的列。其他两个图（折线图和条形图）则会在 x 轴上使用索引，并为每个数字列创建一组新的折线或条形。

```
>>> fig, (ax1, ax2, ax3) = plt.subplots(1, 3, figsize=(16,4))
>>> fig.suptitle('Two Variable Plots', size=20, y=1.02)
>>> df.plot.line(ax=ax1, title='Line plot')
>>> df.plot.scatter(x='Apples', y='Oranges',
...     ax=ax2, title='Scatterplot')
>>> df.plot.bar(color=color, ax=ax3, title='Bar plot')
>>> fig.savefig('c13-pdemo-scat.png')
```

其输出结果如图 13-18 所示。

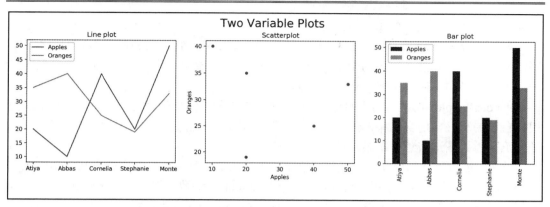

图 13-18　使用 Pandas 在单个 Figure 上绘制多个图表

（5）还可以将 KDE、箱形图（boxplot）和直方图（histogram）放在同一个 Figure 中。这些图用于可视化列的分布。

```
>>> fig, (ax1, ax2, ax3) = plt.subplots(1, 3, figsize=(16,4))
>>> fig.suptitle('One Variable Plots', size=20, y=1.02)
>>> df.plot.kde(color=color, ax=ax1, title='KDE plot')
>>> df.plot.box(ax=ax2, title='Boxplot')
>>> df.plot.hist(color=color, ax=ax3, title='Histogram')
>>> fig.savefig('c13-pdemo-kde2.png')
```

其输出结果如图 13-19 所示。

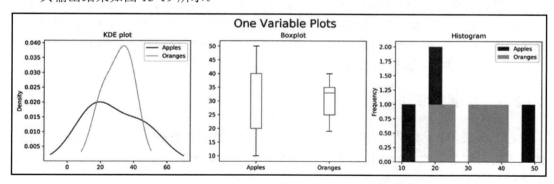

图 13-19　使用 Pandas 绘制 KDE、箱形图和直方图

13.5.2　原理解释

步骤（1）创建了一个很小的 DataFrame 样本，它将帮助我们说明使用 Pandas 进行两

个变量绘图和一个变量绘图之间的差异。

默认情况下，Pandas 将使用 DataFrame 的每个数字列制作一组新的条形、折线、KDE、箱形或直方图，并且当绘图为两个变量的图形时，则将索引用作 x 值。散点图是例外之一，因为它必须明确为 x 和 y 值指定一列。

Pandas 的 .plot 属性具有多种绘图方法，这些方法带有大量参数，允许数据分析人员根据自己的喜好自定义结果。例如，可以设置图形大小、打开和关闭网格线、设置 x 和 y 轴的范围、为图形着色，旋转刻度线等。

此外，还可以使用特定 Matplotlib 绘图方法可用的任何参数。多余的参数将由绘图方法的 **kwds 参数收集，并正确传递给底层的 Matplotlib 函数。例如，在步骤（2）中，我们创建了一个条形图，这意味着我们可以使用 Matplotlib bar 函数中可用的所有参数以及 Pandas 绘图方法中可用的参数。

在步骤（3）中，我们创建了一个单变量 KDE 图，该图将为 DataFrame 中的每个数字列创建一个密度估计。

步骤（4）将所有的两个变量的图形放置在同一个 Figure 中。同样，步骤（5）则是将所有的一个变量的图形放置在一起。

步骤（4）和步骤（5）中的每个步骤都会创建带有 3 个 Axes 对象的 Figure。这里不妨来看以下代码。

```
plt.subplots(1, 3)
```

上述代码创建了一个 Figure，其中 3 个 Axes 分布在 1 行和 3 列上。它返回一个由 Figure 和包含 Axes 的一维 NumPy 数组组成的两个元素的元组。元组的第一个元素被解包到变量 fig 中；元组的第二个元素被解包为另外 3 个变量，每个 Axes 一个。Pandas 的绘图方法带有一个 ax 参数，允许将绘图结果放入 Figure 的特定 Axes 中。

13.5.3 扩展知识

在上述示例中，除散点图外，其他所有图形均不需要指定要使用的列。Pandas 默认情况下会绘制每个数字列，在两个变量的情况下则默认会使用索引。当然，你也可以指定要用于 x 或 y 值的确切列。

```
>>> fig, (ax1, ax2, ax3) = plt.subplots(1, 3, figsize=(16,4))
>>> df.sort_values('Apples').plot.line(x='Apples', y='Oranges',
...      ax=ax1)
>>> df.plot.bar(x='Apples', y='Oranges', ax=ax2)
>>> df.plot.kde(x='Apples', ax=ax3)
```

```
>>> fig.savefig('c13-pdemo-kde3.png')
```

其输出结果如图 13-20 所示。

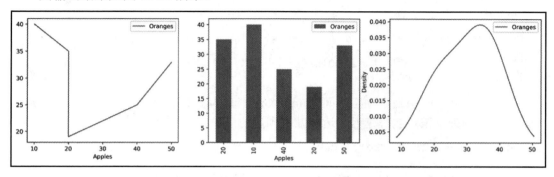

图 13-20　Pandas KDE 图

13.6　可视化航班数据集

数据的可视化操作可以为探索性数据分析提供辅助，而 Pandas 则为轻松快捷地创建它们提供了一个很好的接口。查看新数据集时，一种策略是创建一些单变量图形。这些包括用于分类数据（通常是字符串）的条形图，以及用于连续数据（始终是数字）的直方图、箱形图或 KDE 图。

在本秘笈中，我们将通过使用 Pandas 创建单变量和多变量图形对航班数据集进行一些基本的探索性数据分析。

13.6.1　实战操作

（1）读取航班数据集。

```
>>> flights = pd.read_csv('data/flights.csv')
>>> flights
       MONTH  DAY  WEEKDAY  ...  ARR_DELAY  DIVERTED  CANCELLED
0          1    1        4  ...       65.0         0          0
1          1    1        4  ...      -13.0         0          0
2          1    1        4  ...       35.0         0          0
3          1    1        4  ...       -7.0         0          0
4          1    1        4  ...       39.0         0          0
...      ...  ...      ...  ...        ...       ...        ...
58487     12   31        4  ...      -19.0         0          0
```

58488	12	31	4	...	4.0	0	0
58489	12	31	4	...	-5.0	0	0
58490	12	31	4	...	34.0	0	0
58491	12	31	4	...	-1.0	0	0

（2）在开始绘图之前，不妨先来计算备降（diverted）、取消（canceled）、延迟（delayed）和准时（ontime）航班的数量。我们已经有 DIVERTED（备降）和 CANCELLED（取消）的二进制列。

只要航班到达时间晚于预定时间 15min 或更长时间，即视为航班延误。可以创建两个新的二进制列来跟踪航班延迟到达和准时到达的情况。

```
>>> cols = ['DIVERTED', 'CANCELLED', 'DELAYED']
>>> (flights
...     .assign(DELAYED=flights['ARR_DELAY'].ge(15).astype(int),
...             ON_TIME=lambda df_:1 - df_[cols].any(axis=1))
...     .select_dtypes(int)
...     .sum()
... )
MONTH         363858
DAY           918447
WEEKDAY       229690
SCHED_DEP   81186009
DIST        51057671
SCHED_ARR   90627495
DIVERTED         137
CANCELLED        881
DELAYED        11685
ON_TIME        45789
dtype: int64
```

（3）现在可以在同一 Figure 中为分类列和连续列绘制多个图形。

```
>>> fig, ax_array = plt.subplots(2, 3, figsize=(18,8))
>>> (ax1, ax2, ax3), (ax4, ax5, ax6) = ax_array
>>> fig.suptitle('2015 US Flights - Univariate Summary', size=20)
>>> ac = flights['AIRLINE'].value_counts()
>>> ac.plot.barh(ax=ax1, title='Airline')
>>> (flights
...     ['ORG_AIR']
...     .value_counts()
...     .plot.bar(ax=ax2, rot=0, title='Origin City')
... )
>>> (flights
```

第 13 章 使用 Matplotlib、Pandas 和 Seaborn 进行可视化

```
...        ['DEST_AIR']
...        .value_counts()
...        .head(10)
...        .plot.bar(ax=ax3, rot=0, title='Destination City')
... )
>>> (flights
...     .assign(DELAYED=flights['ARR_DELAY'].ge(15).astype(int),
...             ON_TIME=lambda df_:1 - df_[cols].any(axis=1))
...     [['DIVERTED', 'CANCELLED', 'DELAYED', 'ON_TIME']]
...     .sum()
...     .plot.bar(ax=ax4, rot=0,
...         log=True, title='Flight Status')
... )
>>> flights['DIST'].plot.kde(ax=ax5, xlim=(0, 3000),
...     title='Distance KDE')
>>> flights['ARR_DELAY'].plot.hist(ax=ax6,
...     title='Arrival Delay',
...     range=(0,200)
... )
>>> fig.savefig('c13-uni1.png')
```

其输出结果如图 13-21 所示。

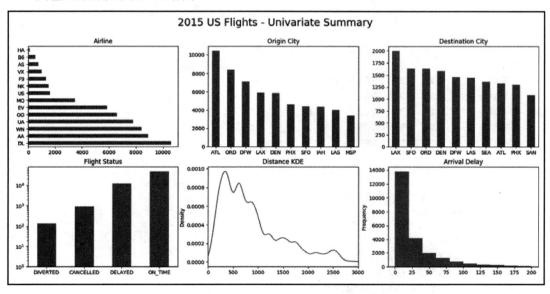

图 13-21　Pandas 单变量图

(4)虽然这不是对所有单变量统计信息的详尽研究,但是也为我们提供了一些变量的大量详细信息。在继续进行多变量图的绘制之前,不妨先绘制出每周的航班次数图。一般来说,在 x 轴上使用带有日期的时间序列进行绘图就应该是这种情况。遗憾的是,本示例在任何列中都没有 Pandas 时间戳,但是我们确实有月份和日期。to_datetime 函数有一个巧妙的技巧,可以识别与 Timestamp 组件匹配的列名。

例如,如果你有一个 DataFrame,其标题明确有 year(年)、month(月)和 day(日)3 列,则将该 DataFrame 传递给 to_datetime 函数时,将返回一系列的时间戳。

在本示例中,要准备当前的 DataFrame,需要为年份添加一列,并使用计划的出发时间来获取小时和分钟。

```
>>> df_date = (flights
...     [['MONTH', 'DAY']]
...     .assign(YEAR=2015,
...             HOUR=flights['SCHED_DEP'] // 100,
...             MINUTE=flights['SCHED_DEP'] % 100)
... )
>>> df_date
       MONTH  DAY  YEAR  HOUR  MINUTE
0          1    1  2015    16      25
1          1    1  2015     8      23
2          1    1  2015    13       5
3          1    1  2015    15      55
4          1    1  2015    17      20
...      ...  ...   ...   ...     ...
58487     12   31  2015     5      15
58488     12   31  2015    19      10
58489     12   31  2015    18      46
58490     12   31  2015     5      25
58491     12   31  2015     8      59
```

(5)可以使用 to_datetime 函数将此 DataFrame 转换为适当的 Timestamp 系列。

```
>>> flight_dep = pd.to_datetime(df_date)
>>> flight_dep
0        2015-01-01 16:25:00
1        2015-01-01 08:23:00
2        2015-01-01 13:05:00
3        2015-01-01 15:55:00
4        2015-01-01 17:20:00
                 ...
58487    2015-12-31 05:15:00
```

```
58488    2015-12-31  19:10:00
58489    2015-12-31  18:46:00
58490    2015-12-31  05:25:00
58491    2015-12-31  08:59:00
Length: 58492, dtype: datetime64[ns]
```

（6）将此结果用作新索引，然后使用.resample方法查找每周的航班计数。

```
>>> flights.index = flight_dep
>>> fc = flights.resample('W').size()
>>> fc.plot.line(figsize=(12,3), title='Flights per Week', grid=True)
>>> fig.savefig('c13-ts1.png')
```

其输出结果如图13-22所示。

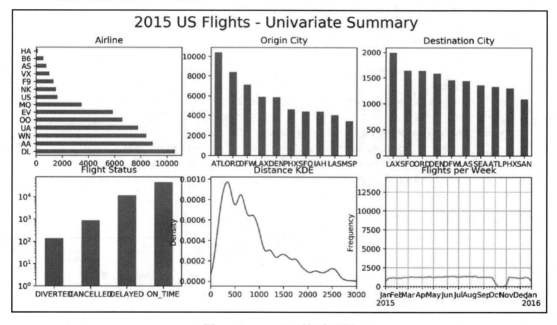

图13-22 Pandas时间序列图

（7）图13-22中的绘图能揭示出一些问题。看起来我们没有10月份的数据。由于缺少这个数据，很难通过可视化分析任何趋势（如果存在趋势的话）。

另外还可以看到，前几周和后几周也低于正常水平，这可能是因为没有整周的数据。

因此，我们可以找到任何少于600个航班的每周数据。然后，可以使用插值方法来填充此缺失的数据。

```
>>> def interp_lt_n(df_, n=600):
...     return (df_
...         .where(df_ > n)
...         .interpolate(limit_direction='both')
... )
>>> fig, ax = plt.subplots(figsize=(16,4))
>>> data = (flights
...     .resample('W')
...     .size()
... )
>>> (data
...     .pipe(interp_lt_n)
...     .iloc[1:-1]
...     .plot.line(color='black', ax=ax)
... )
>>> mask = data<600
>>> (data
...     .pipe(interp_lt_n)
...     [mask]
...     .plot.line(color='.8', linewidth=10)
... )
>>> ax.annotate(xy=(.8, .55), xytext=(.8, .77),
...     xycoords='axes fraction', s='missing data',
...     ha='center', size=20, arrowprops=dict())
>>> ax.set_title('Flights per Week (Interpolated Missing Data)')
>>> fig.savefig('c13-ts2.png')
```

其输出结果如图 13-23 所示。

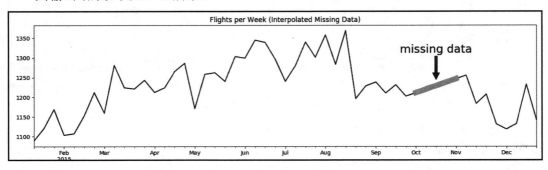

图 13-23　Pandas 时间序列图

原　　文	译　　文
missing data	缺失的数据

（8）现在可以改变方向，专注于多变量绘图。下面查找 10 个机场。
- 入境航班旅行的平均距离最长。
- 至少有 100 个航班。

```
>>> fig, ax = plt.subplots(figsize=(16,4))
>>> (flights
...     .groupby('DEST_AIR')
...     ['DIST']
...     .agg(['mean', 'count'])
...     .query('count > 100')
...     .sort_values('mean')
...     .tail(10)
...     .plot.bar(y='mean', rot=0, legend=False, ax=ax,
...         title='Average Distance per Destination')
... )
>>> fig.savefig('c13-bar1.png')
```

其输出结果如图 13-24 所示。

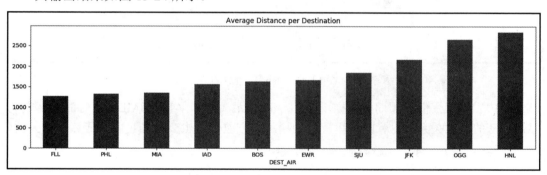

图 13-24　Pandas 条形图

（9）可以看到，前两个目的地机场都在夏威夷，即火奴鲁鲁国际机场（HNL）和卡胡鲁伊机场（OGG），这一点不足为奇。

现在可以同时分析两个变量，例如，对 2000mile 以下的所有航班的距离和飞行时间绘制一幅散点图。

```
>>> fig, ax = plt.subplots(figsize=(8,6))
>>> (flights
...     .reset_index(drop=True)
...     [['DIST', 'AIR_TIME']]
...     .query('DIST <= 2000')
...     .dropna()
```

```
...         .plot.scatter(x='DIST', y='AIR_TIME', ax=ax, alpha=.1, s=1)
... )
>>> fig.savefig('c13-scat1.png')
```

其输出结果如图 13-25 所示。

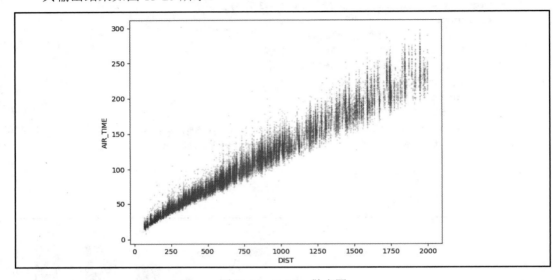

图 13-25 Pandas 散点图

（10）正如我们所预期的那样，DIST（航班距离）和 AIR_TIME（飞行时间）之间存在紧密的线性关系，尽管随着里程数的增加，方差似乎也会增加。可以计算其相关性。

```
flights[['DIST', 'AIR_TIME']].corr()
```

（11）回到绘图问题。有一些航班看起来远离了趋势线。我们可以尝试找到它们。可以使用线性回归模型来识别它们，但是由于 Pandas 不支持线性回归，因此可以采用手动方式。例如，使用 cut 函数将航班距离分为 8 组。

```
>>> (flights
...     .reset_index(drop=True)
...     [['DIST', 'AIR_TIME']]
...     .query('DIST <= 2000')
...     .dropna()
...     .pipe(lambda df_:pd.cut(df_.DIST,
...             bins=range(0, 2001, 250)))
...     .value_counts()
...     .sort_index()
... )
```

```
(0, 250]                6529
(250, 500]             12631
(500, 750]             11506
(750, 1000]             8832
(1000, 1250]            5071
(1250, 1500]            3198
(1500, 1750]            3885
(1750, 2000]            1815
Name: DIST, dtype: int64
```

（12）假设每个组中的所有航班应具有相似的飞行时间，这样就可以为每个航班计算飞行时间偏离该组平均值的标准差的数量。

```
>>> zscore = lambda x: (x - x.mean()) / x.std()
>>> short = (flights
...     [['DIST', 'AIR_TIME']]
...     .query('DIST <= 2000')
...     .dropna()
...     .reset_index(drop=True)
...     .assign(BIN=lambda df_:pd.cut(df_.DIST,
...         bins=range(0, 2001, 250)))
... )

>>> scores = (short
...     .groupby('BIN')
...     ['AIR_TIME']
...     .transform(zscore)
... )

>>> (short.assign(SCORE=scores))
        DIST   AIR_TIME          BIN      SCORE
0        590       94.0   (500, 750]   0.490966
1       1452      154.0  (1250, 1500] -1.267551
2        641       85.0   (500, 750]  -0.296749
3       1192      126.0  (1000, 1250] -1.211020
4       1363      166.0  (1250, 1500] -0.521999
...      ...        ...          ...        ...
53462   1464      166.0  (1250, 1500] -0.521999
53463    414       71.0   (250, 500]   1.376879
53464    262       46.0   (250, 500]  -1.255719
53465    907      124.0   (750, 1000]  0.495005
53466    522       73.0   (500, 750]  -1.347036
```

（13）现在需要一种发现异常值（outlier，也称为离群值）的方法。箱形图提供了检测异常值的视觉效果（超过内部四分位数范围的 1.5 倍即为异常值）。要为每个箱子创建箱形图，需要在列名称中使用箱子名称。可以使用.pivot 方法执行该操作。

```
>>> fig, ax = plt.subplots(figsize=(10,6))
>>> (short.assign(SCORE=scores)
...       .pivot(columns='BIN')
...       ['SCORE']
...       .plot.box(ax=ax)
... )
>>> ax.set_title('Z-Scores for Distance Groups')
>>> fig.savefig('c13-box2.png')
```

其输出结果如图 13-26 所示。

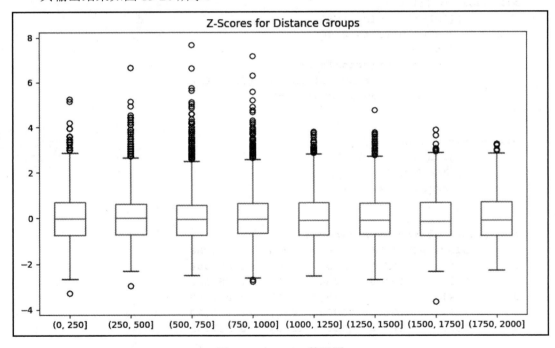

图 13-26　Pandas 箱形图

（14）现在可以来检查距离均值大于 6 个标准偏差的点。因为在步骤（9）中重置了航班 DataFrame 中的索引，所以可以使用它来标识航班 DataFrame 中的每个唯一行。下面创建一个仅包含异常值的单独的 DataFrame。

```
>>> mask = (short
...          .assign(SCORE=scores)
...          .pipe(lambda df_:df_.SCORE.abs() >6)
... )

>>> outliers = (flights
...          [['DIST', 'AIR_TIME']]
...          .query('DIST <= 2000')
...          .dropna()
...          .reset_index(drop=True)
...          [mask]
...          .assign(PLOT_NUM=lambda df_:range(1, len(df_)+1))
... )

>>> outliers
       DIST  AIR_TIME  PLOT_NUM
14972  373   121.0     1
22507  907   199.0     2
40768  643   176.0     3
50141  651   164.0     4
52699  802   210.0     5
```

（15）可以使用此表来识别步骤（9）绘图中的异常值。使用 table 参数时，Pandas 还提供了一种将离群值表格附加到图形底部的方法。

```
>>> fig, ax = plt.subplots(figsize=(8,6))
>>> (short
...    .assign(SCORE=scores)
...    .plot.scatter( x='DIST', y='AIR_TIME',
...                   alpha=.1, s=1, ax=ax,
...                   table=outliers)
... )
>>> outliers.plot.scatter(x='DIST', y='AIR_TIME',
...      s=25, ax=ax, grid=True)
>>> outs = outliers[['AIR_TIME', 'DIST', 'PLOT_NUM']]
>>> for t, d, n in outs.itertuples(index=False):
...      ax.text(d + 5, t + 5, str(n))
>>> plt.setp(ax.get_xticklabels(), y=.1)
>>> plt.setp(ax.get_xticklines(), visible=False)
>>> ax.set_xlabel('')
>>> ax.set_title('Flight Time vs Distance with Outliers')
>>> fig.savefig('c13-scat3.png', dpi=300, bbox_inches='tight')
```

其输出结果如图 13-27 所示。

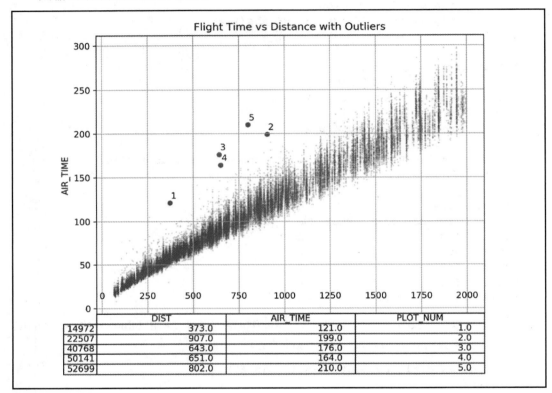

图 13-27　Pandas 散点图

13.6.2　原理解释

在步骤（1）中读取了数据并计算了延迟和按时航班的列之后，即可开始制作单变量绘图（univariate plot）。

步骤（3）中对 subplot 函数的调用将创建一个 2×3 大小相等的 Axes 网格。我们将每个 Axes 解包到其自己的变量中以进行引用。每次对绘图方法的调用都将使用 ax 参数引用 Figure 中的特定 Axes。

.value_counts 方法用于创建 3 个 Series，这些 Series 构成了第一行中的绘图。rot 参数可将刻度标签旋转（rotate）到给定角度。

左下角的图对 y 轴使用了对数标度，这是因为准时航班的数量大约比取消航班的数量大两个数量级。如果不采用对数刻度的话,那么将很难看到左侧的两个条形图。

默认情况下,KDE 图可能会为不可能的值生成正数区域,例如底下一行中的负数英里。因此,我们使用 xlim 参数限制了 x 值的范围。

在右下角创建的延迟到达航班的直方图已传递 range 参数。这并不是 Pandas .plot.hist 方法的方法签名的一部分。相反,此参数由**kwds 参数收集,然后传递给 Matplotlib hist 函数。在这种情况下,无法像前面的绘图一样使用 xlim 参数。该图将被裁剪,而不必为图形的该部分重新计算新箱子的宽度。当然,range 参数不仅限制了 x 轴,而且也将箱子宽度的计算限制到该范围。

步骤(4)创建了一个特殊的额外 DataFrame 来容纳仅包含 datetime 组件的列,以便我们可以在步骤(5)中使用 to_datetime 函数立即将每一行转换为 Timestamp。

在步骤(6)中,使用了 .resample 方法。此方法根据传递的日期偏移别名使用索引来形成组。我们返回了每周航班数(W)作为一个 Series,然后在该 Series 上调用 .plot.line 方法,该方法将索引设置为 x 轴。由于 10 月份的数据缺失,因此出现了一个明显的漏洞。

为了填补这个漏洞,在步骤(7)中,我们使用了 .where 方法,将每周航班数中小于 600 的值设置为缺失。然后,通过线性插值来填充丢失的数据。默认情况下,.interpolate 方法仅在向前方向上进行插值,因此在 DataFrame 开头的所有缺失值都将维持原样。通过将 limit_direction 参数设置为 both,可以确保不会有任何缺失值。

现在可以使用新数据绘图。为了更清楚地显示缺失的数据,我们选择原始数据中缺少的点,并在前一行的相同 Axes 上绘制折线图。通常而言,在注解绘图时,可以使用数据坐标,但是在这种情况下,x 轴的坐标是什么并不明显。要使用 Axes 坐标系——即范围从(0, 0)~(1, 1)的坐标系,可以将 xycoords 参数设置为 axes fraction。现在,新的绘图将错误数据排除在外,使得发现趋势比以前容易得多。通过步骤(7)的绘图结果可以看到,夏季的空中交通流量比一年中的其他时候都要多。

在步骤(8)中,使用了很长的一个方法链对每个目标机场进行分组,并对 DIST(航班距离)列应用了两个函数(mean 和 count)。.query 方法在用于简单过滤的方法中效果很好。当方法链运行到 .plot.bar 方法时,在 DataFrame 中有两列,默认情况下,它将为每一列绘制一个条形图。我们对 count 列不感兴趣,因此仅选择 mean 列来形成条形图。

另外,在使用 DataFrame 进行绘图时,每个列名称都显示在图例(legend)中,这意味着会将 mean 一词放入图例中,而这对于我们来说是无用的,因此可通过将 legend 参数设置为 False 来删除它。

步骤(9)开始研究航班飞行距离与飞行时间之间的关系。由于点的数量众多,因此使用 s 参数缩小了它们的大小。我们还使用 alpha 参数显示了重叠的点。

我们讨论了航班飞行距离与飞行时间之间的相关性,并在步骤(10)中量化了该值。

为了找到平均需要更长的时间到达目的地的航班，在步骤（11）中将每个航班分组为250mile，并在步骤（12）中找到与其组平均值的标准差的数量。

在步骤（13）中，在相同的 Axes 上为 BIN 的每个唯一值创建了一个新的箱形图。

在步骤（14）中，当前 DataFrame short 包含了寻找最慢航班所需的信息，但它不具备我们可能需要进一步研究的所有原始数据。因为在步骤（12）中重置了 short 的索引，所以我们可以使用它来标识原始数据的同一行。此外，我们还为每个包含异常值的行赋予了唯一的整数 PLOT_NUM，以便稍后在绘图时进行标识。

在步骤（15）中，我们从与步骤（9）相同的散点图开始，但是使用了 table 参数将离群值表附加到该图的底部。然后，我们将离群值绘制为顶部的散点图，并确保它们的点更大以便于识别。.itertuples 方法循环遍历每个 DataFrame 行，并以元组形式返回其值。绘图被解包相应的 x 和 y 值，并使用分配给它的编号进行标记。

当表格放置在绘图下方时，它会干扰 x 轴上的绘图对象。因此，我们将刻度标签移动到了轴的内部，并删除了刻度线和轴标签。该表格则提供了有关离群值的信息。

13.7 使用堆积面积图发现新兴趋势

堆积面积图（stacked area chart）是很好的可视化工具，可用于发现新兴趋势，尤其是在市场营销分析中，通常会使用它显示诸如互联网浏览器、手机或车辆之类的产品的市场份额百分比等数据。

在本秘笈中，将使用从颇受欢迎的网站 meetup.com 收集的数据（meetup.com 是一个线下社交平台），并将使用堆积面积图显示 5 个与数据科学相关的聚会组之间的成员分布。

13.7.1 实战操作

（1）读取 meetup（线下聚会）数据集，将 join_date 列转换为 Timestamp，并将其设置为索引。

```
>>> meetup = pd.read_csv('data/meetup_groups.csv',
...       parse_dates=['join_date'],
...       index_col='join_date')
>>> meetup
                              group    ...    country
join_date                                ...
2016-11-18 02:41:29    houston machine learning    ...    us
```

```
2017-05-09 14:16:37     houston machine learning      ...     us
2016-12-30 02:34:16     houston machine learning      ...     us
2016-07-18 00:48:17     houston machine learning      ...     us
2017-05-25 12:58:16     houston machine learning      ...     us
       ...                       ...                  ...    ...
2017-10-07 18:05:24     houston data visualization    ...     us
2017-06-24 14:06:26     houston data visualization    ...     us
2015-10-05 17:08:40     houston data visualization    ...     us
2016-11-04 22:36:24     houston data visualization    ...     us
2016-08-02 17:47:29     houston data visualization    ...     us
```

（2）获取每周加入每个线下聚会小组的人数。

```
>>> (meetup
...     .groupby([pd.Grouper(freq='W'), 'group'])
...     .size()
... )
join_date   group
2010-11-07  houstonr                          5
2010-11-14  houstonr                          11
2010-11-21  houstonr                          2
2010-12-05  houstonr                          1
2011-01-16  houstonr                          2
                                              ..
2017-10-15  houston data science              14
            houston data visualization        13
            houston energy data science       9
            houston machine learning          11
            houstonr                          2
Length: 763, dtype: int64
```

（3）取消 group 小组级别的堆叠，以便每个聚会组都有自己的数据列。

```
>>> (meetup
...     .groupby([pd.Grouper(freq='W'), 'group'])
...     .size()
...     .unstack('group', fill_value=0)
... )
group       houston data science  ...  houstonr
join_date                         ...
2010-11-07                     0  ...         5
2010-11-14                     0  ...        11
2010-11-21                     0  ...         2
2010-12-05                     0  ...         1
```

2011-01-16	0	...	2
...	...	...	...
2017-09-17	16	...	0
2017-09-24	19	...	7
2017-10-01	20	...	1
2017-10-08	22	...	2
2017-10-15	14	...	2

（4）可以看到，此数据表示的是在特定星期加入的成员数量。因此，可以取每列的累加总和来获得成员的总数。

```
>>> (meetup
...     .groupby([pd.Grouper(freq='W'), 'group'])
...     .size()
...     .unstack('group', fill_value=0)
...     .cumsum()
... )
group         houston data science  ...  houstonr
join_date                           ...
2010-11-07                       0  ...         5
2010-11-14                       0  ...        16
2010-11-21                       0  ...        18
2010-12-05                       0  ...        19
2011-01-16                       0  ...        21
...                            ...  ...       ...
2017-09-17                    2105  ...      1056
2017-09-24                    2124  ...      1063
2017-10-01                    2144  ...      1064
2017-10-08                    2166  ...      1066
2017-10-15                    2180  ...      1068
```

（5）许多堆积面积图使用的是总数的百分比，因此每一行总和为1。下面将每一行的值除以行的总值，以找到相对数。

```
>>> (meetup
...     .groupby([pd.Grouper(freq='W'), 'group'])
...     .size()
...     .unstack('group', fill_value=0)
...     .cumsum()
...     .pipe(lambda df_: df_.div(
...         df_.sum(axis='columns'), axis='index'))
... )
group         houston data science  ...  houstonr
```

```
join_date
2010-11-07                       0.000000     ...     1.000000
2010-11-14                       0.000000     ...     1.000000
2010-11-21                       0.000000     ...     1.000000
2010-12-05                       0.000000     ...     1.000000
2011-01-16                       0.000000     ...     1.000000
...                                   ...     ...          ...
2017-09-17                       0.282058     ...     0.141498
2017-09-24                       0.282409     ...     0.141338
2017-10-01                       0.283074     ...     0.140481
2017-10-08                       0.284177     ...     0.139858
2017-10-15                       0.284187     ...     0.139226
```

（6）创建堆积面积图，该图将不断累积列，并堆积在一起。

```python
>>> fig, ax = plt.subplots(figsize=(18,6))
>>> (meetup
...    .groupby([pd.Grouper(freq='W'), 'group'])
...    .size()
...    .unstack('group', fill_value=0)
...    .cumsum()
...    .pipe(lambda df_: df_.div(
...         df_.sum(axis='columns'), axis='index'))
...    .plot.area(ax=ax,
...         cmap='Greys', xlim=('2013-6', None),
...         ylim=(0, 1), legend=False)
... )
>>> ax.figure.suptitle('Houston Meetup Groups', size=25)
>>> ax.set_xlabel('')
>>> ax.yaxis.tick_right()
>>> kwargs = {'xycoords':'axes fraction', 'size':15}
>>> ax.annotate(xy=(.1, .7), s='R Users',
...      color='w', **kwargs)
>>> ax.annotate(xy=(.25, .16), s='Data Visualization',
...      color='k', **kwargs)
>>> ax.annotate(xy=(.5, .55), s='Energy Data Science',
...      color='k', **kwargs)
>>> ax.annotate(xy=(.83, .07), s='Data Science',
...      color='k', **kwargs)
>>> ax.annotate(xy=(.86, .78), s='Machine Learning',
...      color='w', **kwargs)
>>> fig.savefig('c13-stacked1.png')
```

其输出结果如图 13-28 所示。

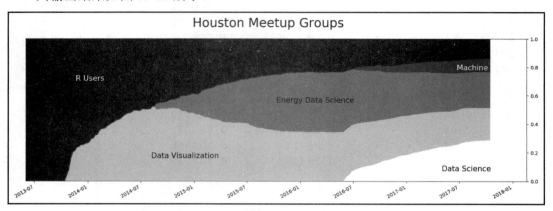

图 13-28　聚会组分布的堆积图

13.7.2　原理解释

我们的目标是确定一段时间内休斯敦 5 个最大的数据科学聚会小组中的成员分布。为此，需要找到每个小组自成立以来每个时间点的总成员数。

在步骤（2）中，我们按每周对线下聚会组进行分组（偏移别名 W），并使用.size 方法返回该周的注册数。

这样获得的 Series 不适合使用 Pandas 绘图。每个聚会组都需要有自己的列，因此可以将 group 索引级别重塑为列。我们还将选项 fill_value 设置为 0，这样，那些在特定星期内没有任何成员加入的组便不会包含缺失值。

我们需要获得每周的会员总数。步骤（4）中的.cumsum 方法提供了此功能。在此步骤之后即可创建堆叠面积图，这是可视化成员总数的好方法。

在步骤（5）中，通过将每个值除以其行总值，可以找到每个组的分布（这是相对于所有组中成员总数的小数）。默认情况下，Pandas 自动按列对齐对象，因此不能使用除法运算符。相反，必须使用.div 方法，并且还要使用带有 index 值的 axis 参数。

在准备好数据后，步骤（6）创建了堆积面积图。值得一提的是，Pandas 允许使用日期时间字符串设置轴限制，但是如果在 Matplotlib 中使用 ax.set_xlim 方法完成此操作，那么这将无法正常工作。另外，该绘图的起始日期提前了几年，是因为 Houston R Users 组的成立要早于其他组。

13.8 了解 Seaborn 和 Pandas 之间的区别

Seaborn 库是一个流行的用于创建可视化的 Python 库。像 Pandas 一样，它本身不执行任何实际的绘图，而是包装使用 Matplotlib。Seaborn 的绘图函数可与 Pandas DataFrame 配合使用，以创建美观漂亮的可视化效果。

尽管 Seaborn 和 Pandas 都减少了 Matplotlib 的开销，但它们处理数据的方式却完全不同。几乎所有的 Seaborn 绘图函数都需要规整（或很长）的数据。

在数据分析过程中，处理规整数据通常会创建聚合（或很宽）的数据。Pandas 适合使用这种格式的数据进行绘图。

在此秘笈中，我们将分别使用 Seaborn 和 Pandas 构建相似的图，以显示它们接收的数据类型（规整数据和宽数据）。

13.8.1 实战操作

（1）读取 employee 员工数据集。

```
>>> employee = pd.read_csv('data/employee.csv',
...      parse_dates=['HIRE_DATE', 'JOB_DATE'])
>>> employee
      UNIQUE_ID    POSITION_TITLE    DEPARTMENT    ... \
0             0        ASSISTAN...    Municipa...  ...
1             1        LIBRARY ...    Library...   ...
2             2        POLICE O...    Houston...   ...
3             3        ENGINEER...    Houston...   ...
4             4        ELECTRICIAN    General...   ...
...         ...               ...          ...     ...
1995       1995        POLICE O...    Houston...   ...
1996       1996        COMMUNIC...    Houston...   ...
1997       1997        POLICE O...    Houston...   ...
1998       1998        POLICE O...    Houston...   ...
1999       1999        FIRE FIG...    Houston...   ...
[2000 rows x 10 columns]
```

（2）导入 Seaborn 库，并使用别名 sns。

```
>>> import seaborn as sns
```

（3）使用 Seaborn 绘制包含各部门计数的条形图。

```
>>> fig, ax = plt.subplots(figsize=(8, 6))
>>> sns.countplot(y='DEPARTMENT', data=employee, ax=ax)
>>> fig.savefig('c13-sns1.png', dpi=300, bbox_inches='tight')
```

其输出结果如图 13-29 所示。

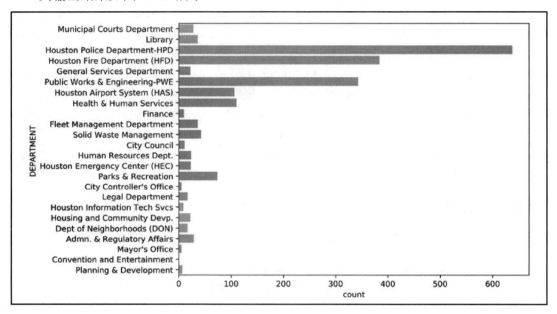

图 13-29　Seaborn 条形图

（4）要使用 Pandas 重现此绘图，需要预先聚合数据。

```
>>> fig, ax = plt.subplots(figsize=(8, 6))
>>> (employee
...     ['DEPARTMENT']
...     .value_counts()
...     .plot.barh(ax=ax)
... )
>>> fig.savefig('c13-sns2.png', dpi=300, bbox_inches='tight')
```

其输出结果如图 13-30 所示。

（5）现在使用 Seaborn 找到每个种族的平均薪水。

```
>>> fig, ax = plt.subplots(figsize=(8, 6))
>>> sns.barplot(y='RACE', x='BASE_SALARY', data=employee, ax=ax)
>>> fig.savefig('c13-sns3.png', dpi=300, bbox_inches='tight')
```

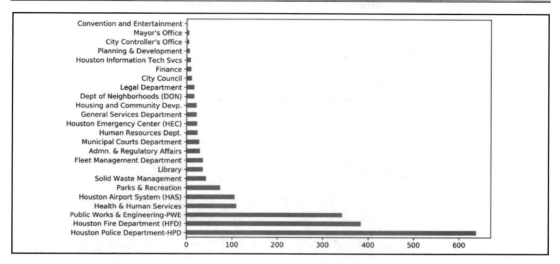

图 13-30　Pandas 条形图

其输出结果如图 13-31 所示。

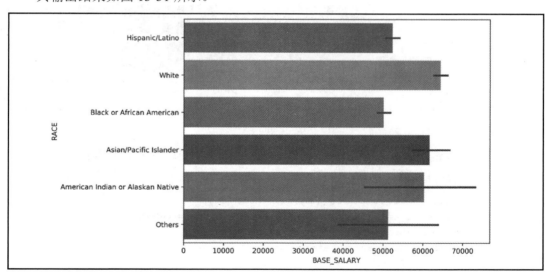

图 13-31　Seaborn 条形图

（6）要使用 Pandas 复制该绘图，需要先按 RACE（种族）列分组。

```
>>> fig, ax = plt.subplots(figsize=(8, 6))
>>> (employee
...     .groupby('RACE', sort=False)
```

```
...         ['BASE_SALARY']
...         .mean()
...         .plot.barh(rot=0, width=.8, ax=ax)
... )
>>> ax.set_xlabel('Mean Salary')
>>> fig.savefig('c13-sns4.png', dpi=300, bbox_inches='tight')
```

其输出结果如图 13-32 所示。

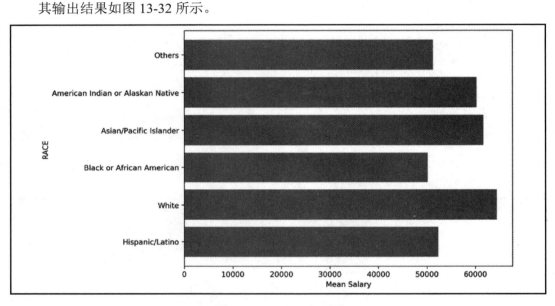

图 13-32　Pandas 条形图

（7）Seaborn 在大多数绘图函数中还具有通过第三个变量 hue 区分数据中各组的能力。例如，我们可以通过 RACE（种族）和 GENDER（性别）列找到平均薪水。

```
>>> fig, ax = plt.subplots(figsize=(18, 6))
>>> sns.barplot(x='RACE', y='BASE_SALARY', hue='GENDER',
...     ax=ax, data=employee, palette='Greys',
...     order=['Hispanic/Latino',
...            'Black or African American',
...            'American Indian or Alaskan Native',
...            'Asian/Pacific Islander', 'Others',
...            'White'])
>>> fig.savefig('c13-sns5.png', dpi=300, bbox_inches='tight')
```

其输出结果如图 13-33 所示。

第 13 章　使用 Matplotlib、Pandas 和 Seaborn 进行可视化

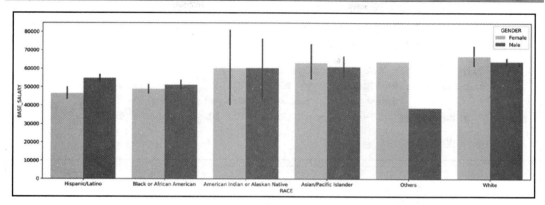

图 13-33　Seaborn 条形图

（8）对于 Pandas 来说，要实现同样的绘图，则必须按 RACE 和 GENDER 进行分组，然后取消 GENDER 的堆叠。

```
>>> fig, ax = plt.subplots(figsize=(18, 6))
>>> (employee
...     .groupby(['RACE', 'GENDER'], sort=False)
...     ['BASE_SALARY']
...     .mean()
...     .unstack('GENDER')
...     .sort_values('Female')
...     .plot.bar(rot=0, ax=ax,
...         width=.8, cmap='viridis')
... )
>>> fig.savefig('c13-sns6.png', dpi=300, bbox_inches='tight')
```

其输出结果如图 13-34 所示。

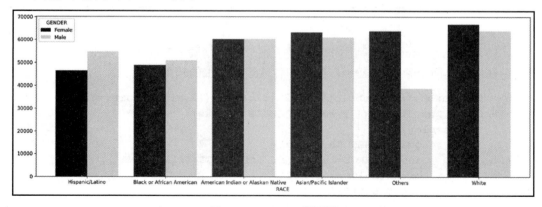

图 13-34　Pandas 条形图

（9）箱形图是 Seaborn 和 Pandas 都可以绘制的另一类图形。首先使用 Seaborn 创建一个按 RACE 和 GENDER 区分的薪水的箱形图。

```
>>> fig, ax = plt.subplots(figsize=(8, 6))
>>> sns.boxplot(x='GENDER', y='BASE_SALARY', data=employee,
...             hue='RACE', palette='Greys', ax=ax)
>>> fig.savefig('c13-sns7.png', dpi=300, bbox_inches='tight')
```

其输出结果如图 13-35 所示。

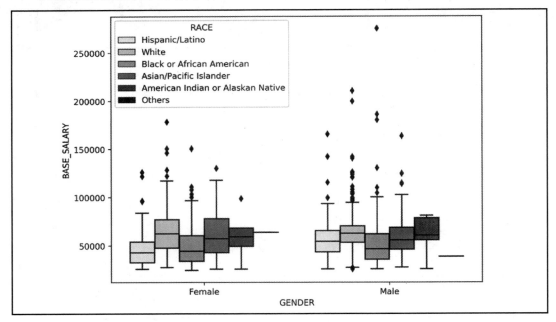

图 13-35　Seaborn 箱形图

（10）要使用 Pandas 绘制图 13-35 中的箱形图的精确复制品不是一件易事。这首先需要为性别创建两个独立的 Axes，然后按种族绘制薪水的箱形图。

```
>>> fig, axs = plt.subplots(1, 2, figsize=(12, 6), sharey=True)
>>> for g, ax in zip(['Female', 'Male'], axs):
...     (employee
...         .query('GENDER == @g')
...         .assign(RACE=lambda df_:df_.RACE.fillna('NA'))
...         .pivot(columns='RACE')
...         ['BASE_SALARY']
...         .plot.box(ax=ax, rot=30)
```

```
...            )
...            ax.set_title(g + ' Salary')
...            ax.set_xlabel('')
>>> fig.savefig('c13-sns8.png', bbox_inches='tight')
```

其输出结果如图 13-36 所示。

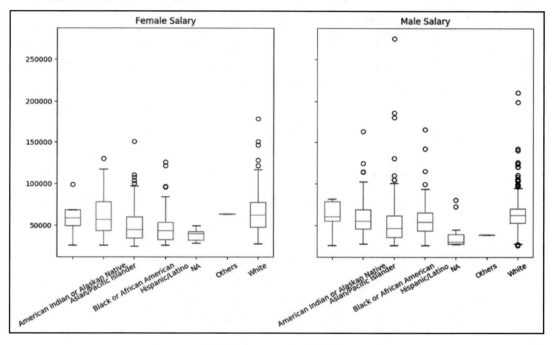

图 13-36　Pandas 箱形图

13.8.2　原理解释

在步骤（2）中导入了 Seaborn，这会更改 Matplotlib 的许多默认属性。在类似字典的对象 plt.rcParams 中可以访问大约 300 个默认绘图参数。要恢复 Matplotlib 的默认值，可调用不带参数的 plt.rcdefaults 函数。

导入 Seaborn 时，Pandas 绘图的样式也会受到影响。本示例中的 employee 数据集满足规整数据的要求，因此适用于几乎所有的 Seaborn 绘图函数。

Seaborn 将进行所有聚合，你只需将 DataFrame 提供给 data 参数，并使用其字符串名称引用列。例如，在步骤（3）中，countplot 函数会毫不费力地对每次出现的 DEPARTMENT 进行计数，以创建条形图。大多数 Seaborn 绘图函数都有 x 和 y 参数。可以通过切换 x

和 y 的值来制作垂直条形图。Pandas 则需要做更多的工作来获得相同的绘图。

在步骤（4）中，必须使用 .value_counts 方法预先计算箱子的高度。

Seaborn 能够使用 barplot 函数进行更复杂的聚合，如步骤（5）和步骤（7）所示。hue 参数进一步在 x 轴上拆分了每个组。在步骤（6）和步骤（8）中，通过对 x 和 hue 变量进行分组，Pandas 几乎都能够复制这些绘图。

在 Seaborn 和 Pandas 中都可以绘制箱形图，并且可以使用规整数据绘制而无须任何聚合。即使不需要聚合，Seaborn 仍然占据上风，因为它可以使用 hue 参数将数据整齐地分为不同的组。如步骤（10）所示，Pandas 无法轻松复制 Seaborn 函数的功能，而是每个组都需要使用 .query 方法进行拆分，并绘制在其自己的 Axes 上。

13.9 使用 Seaborn 网格进行多变量分析

Seaborn 能够在网格中绘制多个图形。Seaborn 中的某些函数在 Matplotlib 的 Axis 级别不起作用，而在 Figure 级别则是有效的，这样的函数包括 catplot、lmplot、pairplot、jointplot 和 clustermap。

在大多数情况下，figure 或 grid 函数使用 axes 函数来构建网格。从 grid 函数返回的最终对象是网格类型，其中有 4 种不同的类型。高级用例需要使用网格类型，但是在绝大多数情况下，你将调用底层 grid 函数来生成实际的网格而不是构造函数本身。

在本秘笈中，我们将按性别和种族研究工作年限与薪水之间的关系。首先使用 Seaborn Axes 函数创建回归图，然后使用 grid 函数为图形添加更多的维度。

13.9.1 实战操作

（1）读取 employee 数据集，并为工作年限创建一列。

```
>>> emp = pd.read_csv('data/employee.csv',
...     parse_dates=['HIRE_DATE', 'JOB_DATE'])
>>> def yrs_exp(df_):
...     days_hired = pd.to_datetime('12-1-2016') - df_.HIRE_DATE
...     return days_hired.dt.days / 365.25

>>> emp = (emp
...     .assign(YEARS_EXPERIENCE=yrs_exp)
... )
```

```
>>> emp[['HIRE_DATE', 'YEARS_EXPERIENCE']]
       HIRE_DATE  YEARS_EXPERIENCE
0      2006-06-12         10.472494
1      2000-07-19         16.369946
2      2015-02-03          1.826184
3      1982-02-08         34.812488
4      1989-06-19         27.452994
...           ...               ...
1995   2014-06-09          2.480544
1996   2003-09-02         13.248732
1997   2014-10-13          2.135567
1998   2009-01-20          7.863269
1999   2009-01-12          7.885172
```

(2) 使用拟合的回归线创建一个散点图,以表示工作年限和薪水之间的关系。

```
>>> fig, ax = plt.subplots(figsize=(8, 6))
>>> sns.regplot(x='YEARS_EXPERIENCE', y='BASE_SALARY',
...     data=emp, ax=ax)
>>> fig.savefig('c13-scat4.png', dpi=300, bbox_inches='tight')
```

其输出结果如图 13-37 所示。

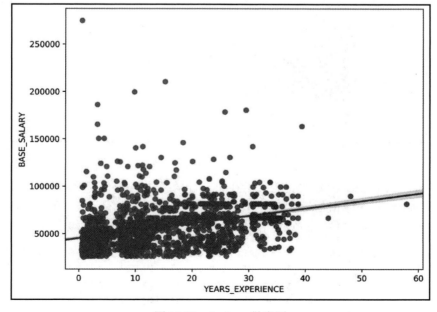

图 13-37　Seaborn 散点图

（3）regplot 函数无法为不同的列绘制多条回归线。下面使用 lmplot 函数绘制一个 Seaborn 网格，为男性和女性添加回归线。

```
>>> grid = sns.lmplot(x='YEARS_EXPERIENCE', y='BASE_SALARY',
...       hue='GENDER', palette='Greys',
...       scatter_kws={'s':10}, data=emp)
>>> grid.fig.set_size_inches(8, 6)
>>> grid.fig.savefig('c13-scat5.png', dpi=300, bbox_inches='tight')
```

其输出结果如图 13-38 所示。

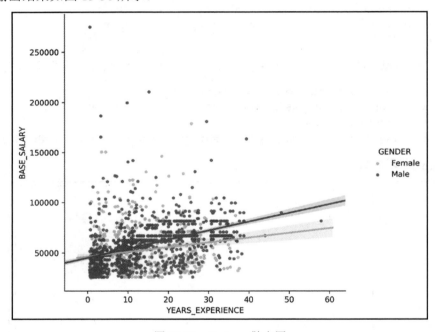

图 13-38　Seaborn 散点图

（4）Seaborn grid 函数的真正力量在于它能够基于另一个变量添加更多的 Axes。lmplot 函数具有 col 和 row 参数，可用于将数据进一步分为不同的组。例如，可以为数据集中的每个唯一种族创建一个单独的图，并且仍然按性别拟合回归线。

```
>>> grid = sns.lmplot( x='YEARS_EXPERIENCE', y='BASE_SALARY',
...                    hue='GENDER', col='RACE', col_wrap=3,
...                    palette='Greys', sharex=False,
...                    line_kws = {'linewidth':5},
...                    data=emp)
```

```
>>> grid.set(ylim=(20000, 120000))
>>> grid.fig.savefig('c13-scat6.png', dpi=300, bbox_inches='tight')
```

其输出结果如图 13-39 所示。

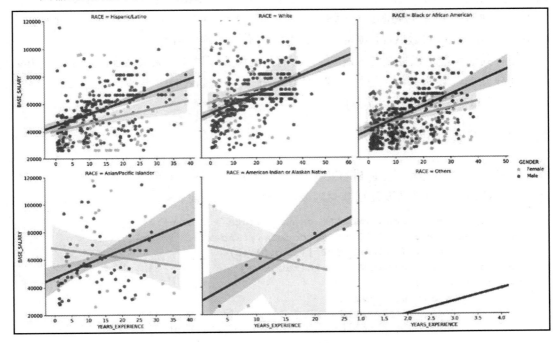

图 13-39　Seaborn 散点图

13.9.2　原理解释

在步骤（1）中，通过使用 Pandas 日期功能创建了另一个连续变量。该数据是 2016 年 12 月 1 日从美国休斯敦市收集的。我们使用该日期来确定每个员工在该市工作了多长时间。当减去日期时（如第 2 行代码所示），将返回一个 Timedelta 对象，其最大单位为天。将该结果的天数除以 365.25，即可计算出工作年限。

步骤（2）使用了 regplot 函数创建散点图（这里使用的是估计的回归线）。regplot 函数返回一个 Matplotlib Axes，可使用它来更改 Figure 的大小。要为每个性别创建两条单独的回归线，必须使用 lmplot 函数，该函数将返回 Seaborn FacetGrid。lmplot 函数有一个 hue 参数，可以为列的每个唯一值叠加一条新的具有不同颜色的回归线。

Seaborn 的 FacetGrid 本质上包装的是 Matplotlib 的 Figure，并提供了一些方便的方法来更改其元素。可以使用.fig 属性访问底层 Matplotlib 的 Figure。

步骤（4）显示了返回 FacetGrid 的 Seaborn 函数的常见用例，这将基于第三个甚至第四个变量创建多个图。我们将 col 参数设置为 RACE，在 RACE 列中为 6 个独特种族中的每个种族创建了 6 个回归图。一般来说，这将返回由一行和六列组成的网格，但是本示例使用了 col_wrap 参数（col_wrap=3），这样就会在 3 列之后换行。

还有其他参数可以控制 Grid 的各个方面。例如，可以使用来自 Matplotlib 底层 line 和 scatter 绘图函数的参数。为此，可以将 scatter_kws 或 line_kws 参数设置为字典，该字典以 Matplotlib 参数为键和值配对。

13.9.3 扩展知识

当具有分类特征时，可以进行类似类型的分析。首先，我们将分类变量 RACE 和 DEPARTMENT 中的级别数分别减少到最常见的前两个和前三个。

```
>>> deps = emp['DEPARTMENT'].value_counts().index[:2]
>>> races = emp['RACE'].value_counts().index[:3]
>>> is_dep = emp['DEPARTMENT'].isin(deps)
>>> is_race = emp['RACE'].isin(races)
>>> emp2 = (emp
...        [is_dep & is_race]
...        .assign(DEPARTMENT=lambda df_:
...                df_['DEPARTMENT'].str.extract('(HPD|HFD)',
...                                             expand=True))
... )

>>> emp2.shape
(968, 11)

>>> emp2['DEPARTMENT'].value_counts()
HPD     591
HFD     377
Name: DEPARTMENT, dtype: int64

>>> emp2['RACE'].value_counts()
White                       478
Hispanic/Latino             250
Black or African American   240
Name: RACE, dtype: int64
```

下面使用一个更简单的 Axes 级函数之一（如 violinplot），以查看按性别划分的工作经验年限分布。

```
>>> common_depts = (emp
...     .groupby('DEPARTMENT')
...     .filter(lambda group: len(group) > 50)
... )

>>> fig, ax = plt.subplots(figsize=(8, 6))
>>> sns.violinplot(x='YEARS_EXPERIENCE', y='GENDER',
...     data=common_depts)
>>> fig.savefig('c13-vio1.png', dpi=300, bbox_inches='tight')
```

其输出结果如图 13-40 所示。

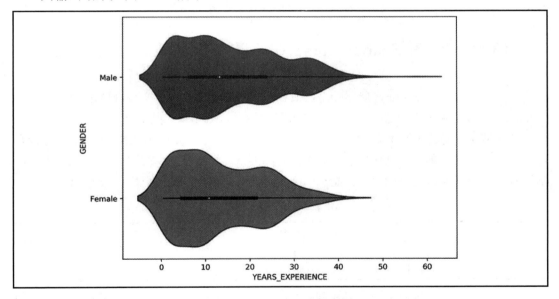

图 13-40　Seaborn 小提琴图

然后，可以使用 catplot 为部门和种族的每个唯一组合添加小提琴图，这需要使用 col 和 row 参数。

```
>>> grid = sns.catplot(x='YEARS_EXPERIENCE', y='GENDER',
...                    col='RACE', row='DEPARTMENT',
...                    height=3, aspect=2,
...                    data=emp2, kind='violin')
>>> grid.fig.savefig('c13-vio2.png', dpi=300, bbox_inches='tight')
```

其输出结果如图 13-41 所示。

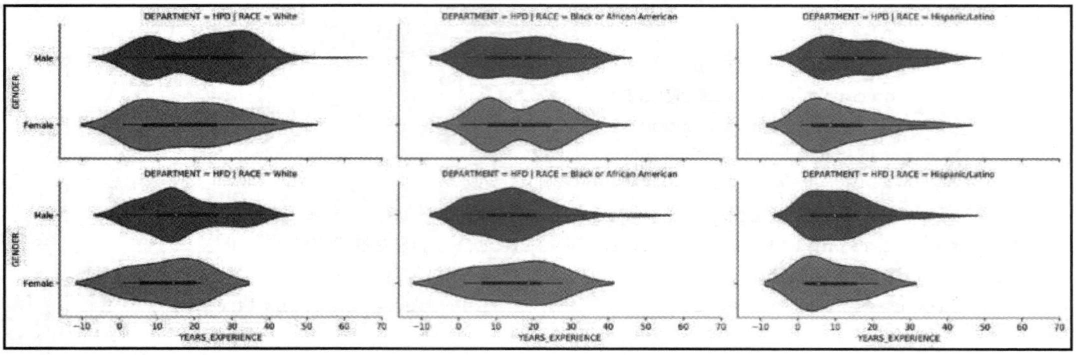

图 13-41　Seaborn 小提琴图

13.10　使用 Seaborn 在钻石数据集中发现辛普森悖论

在进行数据分析时，非常遗憾的是，很容易报告错误的结果。辛普森悖论（Simpson's paradox）是较常见的现象之一。所谓"辛普森悖论"，就是指在某个条件下的两组数据，分别讨论时都会满足某种性质，可是一旦合并考虑，就可能导致相反的结论。例如，假设有两名学生 A 和 B，他们分别接受了 100 个问题的测试。学生 A 回答了 50%正确的问题，而学生 B 则回答了 80%正确的问题。显然，这表明学生 B 的学习成绩更好，如表 13-1 所示。

表 13-1　测试分数和百分比

学　　生	原　始　分　数	正确百分比
A	50/100	50
B	80/100	80

假设两个测试卷非常不同。学生 A 的测试卷中包含 95 道难题，只有 5 道易题，而学生 B 测试卷的难度则完全相反，如表 13-2 所示。

表 13-2　测试卷的难度和比例

学　　生	难　　题	易　　题	困难问题答对百分比	容易问题答对百分比	总　百　分　比
A	45/95	5/5	47	100	50
B	2/5	78/95	40	82	80

考虑到测试卷题目的难易程度后，现在我们看到一个完全不同的结果。

就难题而言,学生 A 在总共 95 道难题中答对了 45 道,答对的百分比为 47%;学生 B 在总共 5 道难题中仅答对了 2 道,答对的百分比为 40%,逊色于学生 A。

就易题而言,学生 A 在总共 5 道易题中答对了 5 道,答对的百分比为 100%;学生 B 在总共 95 道易题中答对了 78 道,答对的百分比为 82%,同样逊色于学生 A。

无论是难题还是易题,学生 A 都领先于学生 B,但是,总体上看,学生 A 的分数却比学生 B 的分数低得多,这就是辛普森悖论的典型例子。汇总的整体显示了与每个单独部分完全相反的情况。

在此秘笈中,我们将首先得出一个令人不可思议的结果,该结果似乎表明,高品质的钻石比低品质的钻石更有价值。然后通过对数据进行更细粒度的分析来揭示辛普森悖论,表明事实恰恰相反。

13.10.1 实战操作

(1)读取 diamonds(钻石)数据集。

```
>>> dia = pd.read_csv('data/diamonds.csv')
>>> dia
       carat        cut  color  ...     x     y     z
0       0.23      Ideal      E  ...  3.95  3.98  2.43
1       0.21    Premium      E  ...  3.89  3.84  2.31
2       0.23       Good      E  ...  4.05  4.07  2.31
3       0.29    Premium      I  ...  4.20  4.23  2.63
4       0.31       Good      J  ...  4.34  4.35  2.75
...      ...        ...    ...  ...   ...   ...   ...
53935   0.72      Ideal      D  ...  5.75  5.76  3.50
53936   0.72       Good      D  ...  5.69  5.75  3.61
53937   0.70  Very Good      D  ...  5.66  5.68  3.56
53938   0.86    Premium      H  ...  6.15  6.12  3.74
53939   0.75      Ideal      D  ...  5.83  5.87  3.64
```

(2)在开始分析之前,可以将 cut(切割)、color(颜色)和 clarity(净度)列更改为有序的分类变量。

```
>>> cut_cats = ['Fair', 'Good', 'Very Good', 'Premium', 'Ideal']
>>> color_cats = ['J', 'I', 'H', 'G', 'F', 'E', 'D']
>>> clarity_cats = ['I1', 'SI2', 'SI1', 'VS2',
...                 'VS1', 'VVS2', 'VVS1', 'IF']
>>> dia2 = (dia
```

```
...         .assign(cut=pd.Categorical(dia['cut'],
...                     categories=cut_cats,
...                     ordered=True),
...                 color=pd.Categorical(dia['color'],
...                     categories=color_cats,
...                     ordered=True),
...                 clarity=pd.Categorical(dia['clarity'],
...                     categories=clarity_cats,
...                     ordered=True))
... )

>>> dia2
       carat        cut  color  ...     x     y     z
0       0.23      Ideal      E  ...  3.95  3.98  2.43
1       0.21    Premium      E  ...  3.89  3.84  2.31
2       0.23       Good      E  ...  4.05  4.07  2.31
3       0.29    Premium      I  ...  4.20  4.23  2.63
4       0.31       Good      J  ...  4.34  4.35  2.75
...      ...        ...    ...  ...   ...   ...   ...
53935   0.72      Ideal      D  ...  5.75  5.76  3.50
53936   0.72       Good      D  ...  5.69  5.75  3.61
53937   0.70  Very Good      D  ...  5.66  5.68  3.56
53938   0.86    Premium      H  ...  6.15  6.12  3.74
53939   0.75      Ideal      D  ...  5.83  5.87  3.64
```

（3）Seaborn 为其绘图使用分类顺序。现在可以对 cut、color 和 clarity 列的每个级别的平均价格绘制条形图。

```
>>> import seaborn as sns
>>> fig, (ax1, ax2, ax3) = plt.subplots(1, 3, figsize=(14,4))
>>> sns.barplot(x='color', y='price', data=dia2, ax=ax1)
>>> sns.barplot(x='cut', y='price', data=dia2, ax=ax2)
>>> sns.barplot(x='clarity', y='price', data=dia2, ax=ax3)
>>> fig.suptitle('Price Decreasing with Increasing Quality?')
>>> fig.savefig('c13-bar4.png', dpi=300, bbox_inches='tight')
```

其输出结果如图 13-42 所示。

（4）看起来颜色和价格似乎呈下降趋势。最高质量的切割和净度水平反而价格低廉。这是怎么回事？接下来我们进行更深入的挖掘，并再次绘制每种钻石颜色的价格，但为 clarity 列的每个级别绘制一个新图。

```
>>> grid = sns.catplot(x='color', y='price', col='clarity',
...     col_wrap=4, data=dia2, kind='bar')
>>> grid.fig.savefig('c13-bar5.png', dpi=300, bbox_inches='tight')
```

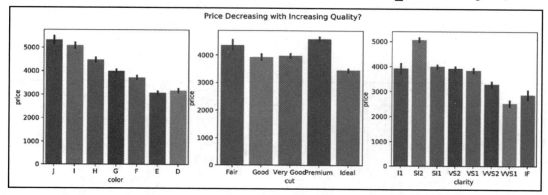

图 13-42　Seaborn 条形图

其输出结果如图 13-43 所示。

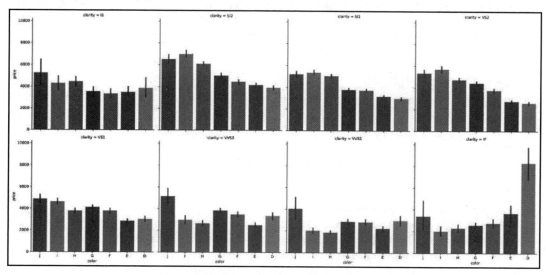

图 13-43　Seaborn 条形图

（5）图 13-43 中的条形图更具启发性。尽管价格似乎随着颜色质量的提高而下降，但当净度达到最高水平时，价格却没有下降，而是大幅度上扬。

前面仅仅关注了钻石的价格，而没有关注其大小。现在让我们从步骤（3）重新创建图，但使用克拉（carat）大小代替价格。

```
>>> fig, (ax1, ax2, ax3) = plt.subplots(1, 3, figsize=(14,4))
>>> sns.barplot(x='color', y='carat', data=dia2, ax=ax1)
>>> sns.barplot(x='cut', y='carat', data=dia2, ax=ax2)
>>> sns.barplot(x='clarity', y='carat', data=dia2, ax=ax3)
>>> fig.suptitle('Diamond size decreases with quality')
>>> fig.savefig('c13-bar6.png', dpi=300, bbox_inches='tight')
```

其输出结果如图 13-44 所示。

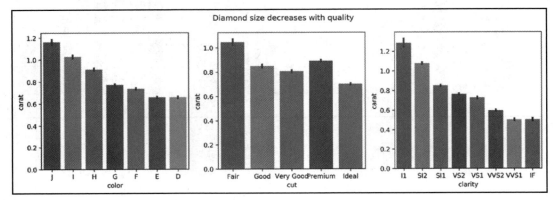

图 13-44　Seaborn 条形图

（6）图 13-44 中的可视化结果有点意思，高质量的钻石看起来尺寸较小，这在直觉上是有道理的。下面创建一个新变量，将克拉值分为 5 个不同的部分，然后创建一个点图。接下来的图形显示，实际上当将钻石按尺寸细分时，钻石质量越高则价格越贵。

```
>>> dia2 = (dia2
...     .assign(carat_category=pd.qcut(dia2.carat, 5))
...    )

>>> from matplotlib.cm import Greys
>>> greys = Greys(np.arange(50,250,40))
>>> grid = sns.catplot(x='clarity', y='price', data=dia2,
...     hue='carat_category', col='color',
...     col_wrap=4, kind='point', palette=greys)
>>> grid.fig.suptitle('Diamond price by size, color and clarity',
...     y=1.02, size=20)
>>> grid.fig.savefig('c13-bar7.png', dpi=300, bbox_inches='tight')
```

其输出结果如图 13-45 所示。

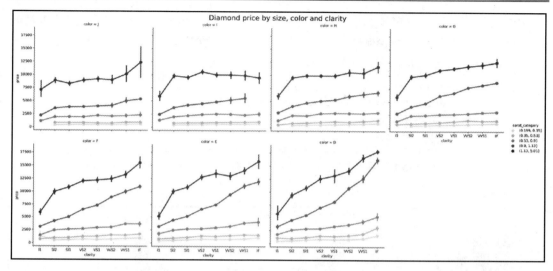

图 13-45　Seaborn 点图

13.10.2　原理解释

在此秘笈中，创建分类列非常重要，因为可以对其进行排序。Seaborn 使用此顺序将标签放置在图形上。步骤（3）和步骤（4）显示了随着钻石质量的提高却出现了价格下降的趋势，这就是引人关注的辛普森悖论——整体的汇总结果在未考虑其他变量时往往可能得出与事实相反的结论。

揭示这一矛盾的关键在于关注克拉的大小。步骤（5）告诉我们，克拉的大小也会随着质量的提高而减小。考虑到这一事实，我们使用 qcut 函数将钻石尺寸划分成 5 个相等大小的箱子（bin）。默认情况下，qcut 函数会根据给定的分位数将变量分为离散分类。通过将整数传递给它（carat_category=pd.qcut(dia2.carat, 5)），可以创建等距的分位数。当然，你也可以选择将一系列显式的非规则分位数传递给它。

使用这个新变量，即可按每个钻石尺寸的分组绘制其平均价格的点图。步骤（6）就是这样做的。Seaborn 中的点图将创建一条线连接每个分类的均值。每个点上的竖线是该组的标准偏差。绘图结果证实，只要将克拉的大小保持不变，钻石的确就会随着其品质的提高而变得更加昂贵。

13.10.3　扩展知识

步骤（3）和步骤（5）中的条形图可以使用更高级的 Seaborn PairGrid 构造函数创建，

该构造函数可以绘制双变量关系。使用 PairGrid 构造函数是一个两步过程：第一步是调用 PairGrid 构造函数，并告诉该构造函数哪个变量用于 x 值，哪个变量用于 y 值；第二步则是调用.map 方法将图形应用于 x 和 y 列的所有组合。

```
>>> g = sns.PairGrid(dia2, height=5,
...     x_vars=["color", "cut", "clarity"],
...     y_vars=["price"])
>>> g.map(sns.barplot)
>>> g.fig.suptitle('Replication of Step 3 with PairGrid', y=1.02)
>>> g.fig.savefig('c13-bar8.png', dpi=300, bbox_inches='tight')
```

其输出结果如图 13-46 所示。

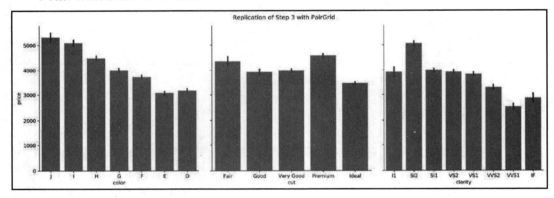

图 13-46　Seaborn 条形图

第 14 章　调试和测试

本章包含以下秘笈。
- 转换数据。
- 测试 .apply 方法的性能。
- 使用 Dask、Pandarell 和 Swifter 等提高 .apply 方法的性能。
- 检查代码。
- 在 Jupyter 中进行调试。
- 管理数据的完整性。
- 结合使用 pytest 和 Pandas。
- 使用 Hypothesis 库生成测试。

14.1　转换数据

Kaggle 是一个流行的数据科学竞赛平台，可以为数据科学家和开发人员提供机器学习竞赛、托管数据库、编写和分享代码。数据分析人员如果能够在 Kaggle 比赛中获得较高排名，将是一项很好的个人能力证明。本章将研究一些 Kaggle 在 2018 年所做的分析调查数据的代码。该调查向 Kaggle 用户询问了有关社会经济方面的信息。

本节将介绍调查数据以及一些分析它的代码。该数据的副标题是 the most comprehensive dataset available on the state of machine learning and data science（有关机器学习和数据科学状态的最全面的数据集）。下面我们深入研究该数据，看看它包含什么内容。该数据可从以下网址中下载。

https://www.kaggle.com/kaggle/kaggle-survey-2018

14.1.1　实战操作

（1）将数据加载到 DataFrame 中。

```
>>> import pandas as pd
>>> import numpy as np
>>> import zipfile
```

```
>>> url = 'data/kaggle-survey-2018.zip'

>>> with zipfile.ZipFile(url) as z:
...     print(z.namelist())
...     kag = pd.read_csv(z.open('multipleChoiceResponses.csv'))
...     df = kag.iloc[1:]
['multipleChoiceResponses.csv', 'freeFormResponses.csv',
'SurveySchema.csv']
```

（2）查看数据和数据类型。

```
>>> df.T
                    1           2           3      ...    23857
Time from...       710         434         718     ...     370
Q1              Female        Male      Female     ...    Male
Q1_OTHER_...        -1          -1          -1     ...      -1
Q2               45-49       30-34       30-34     ...   22-24
Q3          United S...   Indonesia  United S...   ...  Turkey
...                ...         ...         ...     ...     ...
Q50_Part_5         NaN         NaN         NaN     ...     NaN
Q50_Part_6         NaN         NaN         NaN     ...     NaN
Q50_Part_7         NaN         NaN         NaN     ...     NaN
Q50_Part_8         NaN         NaN         NaN     ...     NaN
Q50_OTHER...        -1          -1          -1     ...      -1

>>> df.dtypes
Time from Start to Finish (seconds)     object
Q1                                      object
Q1_OTHER_TEXT                           object
Q2                                      object
Q3                                      object
                                         ...
Q50_Part_5                              object
Q50_Part_6                              object
Q50_Part_7                              object
Q50_Part_8                              object
Q50_OTHER_TEXT                          object
Length: 395, dtype: object
```

（3）很明显，大多数调查数据都是从应答选项中选择的。我们看到所有列的类型都是object。可以使用.value_counts方法来完成探索这些值的标准过程。

```
>>> df.Q1.value_counts(dropna=False)
Male                              19430
```

```
Female                      4010
Prefer not to say            340
Prefer to self-describe       79
Name: Q1, dtype: int64
```

（4）现在我们将感兴趣的每一列都提取出来作为一个 Series。可以将大多数值过滤为有数量限制的值。

在本示例中，可以使用 Series.rename 方法为列指定更有意义的名称。

一些值（如 Q2、Q8 和 Q9 等）都具有范围答案。以 Q2 为例，它询问的是年龄，其中包含了诸如 45～49 和 22～24 之类的值。可以使用.str.slice 方法提取前两个字符，并将类型从字符串转换为整数。

在 Q4 列中，调查的是教育背景，将该值可以转换为序数。

最后，在将要处理的许多列转换为数字并清除了其他一些我们不感兴趣的列之后，使用 pd.concat 将所有 Series 放回 DataFrame 中。

我们将所有这些处理代码放入一个名为 tweak_kag 的函数中。

```
>>> def tweak_kag(df):
...     na_mask = df.Q9.isna()
...     hide_mask = df.Q9.str.startswith('I do not').fillna(False)
...     df = df[~na_mask & ~hide_mask]
...
...     q1 = (df.Q1
...         .replace({ 'Prefer not to say': 'Another',
...                    'Prefer to self-describe': 'Another'})
...         .rename('Gender')
...     )
...     q2 = df.Q2.str.slice(0,2).astype(int).rename('Age')
...     def limit_countries(val):
...         if val in {'United States of America', 'India', 'China'}:
...             return val
...         return 'Another'
...     q3 = df.Q3.apply(limit_countries).rename('Country')
...
...     q4 = (df.Q4
...         .replace({'Master\'s degree': 18,
...         'Bachelor\'s degree': 16,
...         'Doctoral degree': 20,
...         'Some college/university study without earning a bachelor\'s degree': 13,
...         'Professional degree': 19,
```

```
...         'I prefer not to answer': None,
...         'No formal education past high school': 12})
...        .fillna(11)
...        .rename('Edu')
...    )
...
...    def only_cs_stat_val(val):
...        if val not in {'cs', 'eng', 'stat'}:
...            return 'another'
...        return val
...
...    q5 = (df.Q5
...          .replace({
...              'Computer science (software engineering, etc.)': 'cs',
...              'Engineering (non-computer focused)': 'eng',
...              'Mathematics or statistics': 'stat'})
...          .apply(only_cs_stat_val)
...          .rename('Studies'))
...    def limit_occupation(val):
...        if val in {'Student', 'Data Scientist', 'Software Engineer', 'Not employed',
...                   'Data Engineer'}:
...            return val
...        return 'Another'
...
...    q6 = df.Q6.apply(limit_occupation).rename('Occupation')
...
...    q8 = (df.Q8
...        .str.replace('+', '')
...        .str.split('-', expand=True)
...        .iloc[:,0]
...        .fillna(-1)
...        .astype(int)
...        .rename('Experience')
...    )
...
...    q9 = (df.Q9
...        .str.replace('+','')
...        .str.replace(',','')
...        .str.replace('500000', '500')
...        .str.replace('I do not wish to disclose my approximate yearly compensation','')
```

```
...             .str.split('-', expand=True)
...             .iloc[:,0]
...             .astype(int)
...             .mul(1000)
...             .rename('Salary'))
...    return pd.concat([q1, q2, q3, q4, q5, q6, q8, q9], axis=1)

>>> tweak_kag(df)
       Gender    Age      Country  ...   Occupation  Experience
2        Male     30      Another  ...      Another           5
3      Female     30    United S...  ...  Data Sci...          0
5        Male     22        India  ...      Another           0
7        Male     35      Another  ...      Another          10
8        Male     18        India  ...      Another           0
...       ...    ...          ...  ...          ...         ...
23844    Male     30      Another  ...    Software...         10
23845    Male     22      Another  ...      Student           0
23854    Male     30      Another  ...      Another           5
23855    Male     45      Another  ...      Another           5
23857    Male     22      Another  ...    Software...          0

>>> tweak_kag(df).dtypes
Gender         object
Age             int64
Country        object
Edu           float64
Studies        object
Occupation     object
Experience      int64
Salary    int64 dtype: object
```

14.1.2 原理解释

在调查数据中包含丰富的信息，但是由于所有列都是 object 类型，因此很难分析该调查数据。我们编写的 tweak_kag 函数过滤了未提供薪水信息的受访者，还将诸如 Age（年龄）、Edu（受教育程度）、Experience（工作经验）和 Salary（薪水）之类的列转换为数值，以便量化分析。其余分类列将被删减以降低基数。

清洗数据将使后续分析变得更容易。例如，现在我们可以轻松地按国家/地区分组并关联薪水和工作经验。

```
>>> kag = tweak_kag(df)
```

```
>>> (kag
...    .groupby('Country')
...    .apply(lambda g: g.Salary.corr(g.Experience))
... )
Country
Another                     0.289827
China                       0.252974
India                       0.167335
United States of America    0.354125
dtype: float64
```

14.2 测试.apply方法的性能

Series 和 DataFrame 的.apply 方法是 Pandas 中最慢的操作之一。在本秘笈中，我们将探讨其速度，并查看是否可以比较正在发生的事情。

14.2.1 实战操作

（1）现在可以使用 Jupyter 中的%% timeit 单元格魔术（cell magic）命令来对.apply 方法的应用进行计时。

以下是来自 tweak_kag 函数的代码，该代码限制了 country 列（Q3）的基数。

```
>>> %%timeit
>>> def limit_countries(val):
...     if val in {'United States of America', 'India', 'China'}:
...         return val
...     return 'Another'

>>> q3 = df.Q3.apply(limit_countries).rename('Country')
6.42 ms ± 1.22 ms per loop (mean ± std. dev. of 7 runs, 100 loops each)
```

（2）现在改为使用.replace 方法而不是.apply 方法，看看是否可以提高性能。

```
>>> %%timeit
>>> other_values = df.Q3.value_counts().iloc[3:].index
>>> q3_2 = df.Q3.replace(other_values, 'Another')
27.7 ms ± 535 µs per loop (mean ± std. dev. of 7 runs, 10 loops each)
```

（3）很遗憾，这比.apply 方法还要慢得多。可尝试其他方式，如果使用.isin 和.where 方法相结合，则其运行速度约为.apply 方法的两倍。

```
>>> %%timeit
>>> values = {'United States of America', 'India', 'China'}
>>> q3_3 = df.Q3.where(df.Q3.isin(values), 'Another')
3.39 ms ± 570 µs per loop (mean ± std. dev. of 7 runs, 100 loops each)
```

（4）不妨来尝试 np.where 函数。虽然 np.where 函数不是 Pandas 的一部分，但是 Pandas 通常可以与 NumPy 函数一起使用。

```
>>> %%timeit
>>> values = {'United States of America', 'India', 'China'}
>>> q3_4 = pd.Series(np.where(df.Q3.isin(values), df.Q3, 'Another'),
...       index=df.index)
2.75 ms ± 345 µs per loop (mean ± std. dev. of 7 runs, 100 loops each)
```

（5）由此可见，np.where 函数是速度最快的。
别忘记检查它们的结果是否相同。

```
>>> q3.equals(q3_2)
True
>>> q3.equals(q3_3)
True
>>> q3.equals(q3_4)
True
```

14.2.2 原理解释

此秘笈对.apply、.replace 和.where 方法进行了基准测试。在这 3 个方法中，.where 方法的速度最快。最后，我们还尝试了 NumPy 的 where 函数，该函数的速度比 Pandas 还快。但是，如果使用 NumPy 函数，则需要将结果转换回 Series（并为其提供与原始 DataFrame 相同的索引）。

14.2.3 扩展知识

.apply 方法的说明文档指出：如果传入 NumPy 函数，那么它将运行一条快速路径，并将整个 Series 传递给该函数；但是，如果传入 Python 函数，则会为 Series 中的每个值调用该函数。这可能会造成混淆，因为.apply 方法的行为取决于传递给它的参数。

如果你发现自己将函数传递给了.apply 方法（或已经执行了 groupby 操作并正在调用.agg、.transform 或其他将函数作为参数的方法），并且不记得有哪些参数将被传递给函数时，可以使用以下代码来帮忙调试。当然，你也可以查看说明文档，甚至可以查看.apply 方法的代码。

```
>>> def limit_countries(val):
...     if val in  {'United States of America', 'India', 'China'}:
...         return val
...     return 'Another'

>>> q3 = df.Q3.apply(limit_countries).rename('Country')

>>> def debug(something):
...     # what is something? A cell, series, dataframe?
...     print(type(something), something)
...     1/0

>>> q3.apply(debug)
<class 'str'> United States of America
Traceback (most recent call last)
...
ZeroDivisionError: division by zero
```

上面的输出显示，传递到 debug 函数中的是字符串（来自 Series q3 的标量值）。如果你不想抛出异常，则可以设置一个全局变量来保存传递到函数中的参数。

```
>>> the_item = None
>>> def debug(something):
...     global the_item
...     the_item = something
...     return something

>>> _ = q3.apply(debug)

>>> the_item
'Another'
```

需要注意的是，传递给.apply 方法的函数将会通过 Series 中的每个项目调用一次。对单个项目进行操作是一条很缓慢的路径，如果可能的话，应尽量避免这样做。下一个秘笈将显示加快调用.apply 的另一个选项。

14.3 使用 Dask、Pandarell 和 Swifter 等提高.apply 方法的性能

.apply 方法有时候是很方便的。很多库都可以使此类操作并行化，有多种机制可以做

到这一点。最简单的方法就是尝试利用向量化（vectorization）。数学运算在 Pandas 中就是矢量化的，如果你向数字 Series 中添加一个数字（如 5），则 Pandas 不会让每个值加 5，相反，它将利用现代 CPU 的功能一次执行该操作。

如果不能进行向量化（如在使用前面的 limit_countries 函数的情况下），那么你还可以有其他选择。本节将介绍其中一些方法。

请注意，你将需要安装本节介绍的库，因为这些库都不包含在 Pandas 中。

本节示例在调查数据的 Country 列中显示了一些限定值。

14.3.1　实战操作

（1）导入并初始化 pandarallel 库。pandarallel 库尝试在所有可用 CPU 上并行化 Pandas 操作。请注意，pandarallel 库在 Linux 和 Mac 系统上运行良好。由于 pandarallel 库利用了共享内存技术，因此除非在适用于 Linux 的 Windows 子系统上执行 Python；否则该库将无法在 Windows 上运行。

```
>>> from pandarallel import pandarallel
>>> pandarallel.initialize()
```

（2）pandarallel 库增强了 DataFrame，以添加一些额外的方法。

可以按以下方式使用.parallel_apply 方法。

```
>>> def limit_countries(val):
...     if val in  {'United States of America', 'India', 'China'}:
...         return val
...     return 'Another'

>>> %%timeit
>>> res_p = df.Q3.parallel_apply(limit_countries).rename('Country')
133 ms ± 11.1 ms per loop (mean ± std. dev. of 7 runs, 10 loops each)
```

（3）现在可以尝试另一个库。导入 swifter 库。

```
>>> import swifter
```

（4）swifter 库同样增强了 DataFrame 以添加.swifter 访问器。

可以按以下方式使用 swifter 库。

```
>>> %%timeit
>>> res_s = df.Q3.swifter.apply(limit_countries).rename('Country')
187 ms ± 31.4 ms per loop (mean ± std. dev. of 7 runs, 10 loops each)
```

（5）导入 dask 库。

```
>>> import dask
```

（6）使用 dask .map_partitions 函数。

```
>>> %%timeit
>>> res_d = (dask.dataframe.from_pandas(
...         df, npartitions=4)
...     .map_partitions(lambda df: df.Q3.apply(limit_countries))
...     .rename('Countries')
... )
29.1 s ± 1.75 s per loop (mean ± std. dev. of 7 runs, 1 loop each)
```

（7）使用 np.vectorize 函数。

```
>>> np_fn = np.vectorize(limit_countries)

>>> %%timeit
>>> res_v = df.Q3.apply(np_fn).rename('Country')
643 ms ± 86.8 ms per loop (mean ± std. dev. of 7 runs, 1 loop each)
```

（8）导入 numba 库并使用 jit 装饰器装饰 numba 函数。

```
>>> from numba import jit
>>> @jit
... def limit_countries2(val):
...     if val in   ['United States of America', 'India', 'China']:
...         return val
...     return 'Another'
```

（9）使用装饰之后的 numba 函数。

```
>>> %%timeit
>>> res_n = df.Q3.apply(limit_countries2).rename('Country')
158 ms ± 16.1 ms per loop (mean ± std. dev. of 7 runs, 10 loops each)
```

14.3.2 原理解释

请注意，并行化代码本身也是有开销的。在上面的示例中，全部都是在使用普通的 Pandas 代码以串行方式运行时速度更快。仅当数据量非常大，大到串行处理的成本远超过并行代码的开销时，并行处理才是有意义的。pandarallel 库的示例至少有一百万个样本，而我们的数据集比该数据集要小得多，因此在我们的示例中 .apply 方法更快。

在步骤（1）和步骤（2）中使用了 pandarallel 库。pandarallel 库利用了标准库中的

multiprocessing 库来尝试并行运行计算。初始化 pandarallel 库时，可以指定 nb_workers 参数，以指示要使用的 CPU 数（默认情况下，它将使用所有 CPU）。该示例显示了如何使用.parallel_apply 方法，该方法类似于 Pandas 中的.apply 方法。pandarallel 库还可以与 groupby 对象和 Series 对象一起使用。

步骤（3）和步骤（4）显示了 swifter 库的用法。swifter 库将.swifter 属性添加到 DataFrame 和 Series 中。另外，swifter 库采用了不同的方法来加速代码。它将尝试查看操作是否可以向量化。如果不行，则它将查看 Pandas 将花费多长时间（通过运行一个小样本），然后确定是利用 dask 库还是坚持使用 Pandas。同样，确定要使用哪个路径的逻辑本身也是有开销的，因此盲目使用此库可能并不会导致最高效的代码。

Swifter 网站上有一个 Notebook，其中比较了 swifter、np.vectorize、dask 和 Pandas 的执行效率。它对不同类型的函数进行了广泛的基准测试。对于所谓的非向量化函数（例如前面示例中的 limit_countries 就是这样的函数，因为它具有正常的 Python 逻辑），需要数据量达到一百万行，Pandas .apply 方法才会失去优势。

在步骤（5）和步骤（6）中，演示了 dask 库的应用。同样，加载数据和利用库提供的并行化会产生一些开销。dask 的许多用户完全放弃了 Pandas 而只使用 dask，这是因为它实现了类似的功能，但又可以将处理扩展到大数据（并在集群上运行）。

在步骤（7）中，尝试使用了 NumPy 的 vectorize 函数。它从任意 Python 函数创建 NumPy ufunc（在 NumPy 数组上运行的通用函数）。本示例尝试利用了 NumPy 广播规则。在这种情况下，使用它并不会提高性能。

步骤（8）和步骤（9）演示了如何使用 numba 库。我们利用 jit 装饰器创建了一个新函数 limit_countries2。该装饰器将 Python 函数转换为本地代码。同样，此函数也不适合通过此装饰器提高执行速度。

再强调一下，上述许多选项在处理大数据的情况下可能会提高性能。但在我们的示例中，盲目应用它们只会降低代码执行速度。

14.4　检查代码

Jupyter 环境具有一个扩展，使你可以快速获取类、方法或函数的说明文档或源代码。我们强烈建议你习惯使用这些功能。如果你可以留在 Jupyter 环境中响应可能出现的问题，则可以提高生产率。

本节将演示如何查看.apply 方法的源代码。分别直接在 DataFrame 或 Series 对象上查看 DataFrame 或 Series 方法的文档是最容易的。本书一直在建议对 Pandas 对象进行链接操作。遗憾的是，Jupyter（和其他编辑器环境）无法执行代码补全或查找从方法链调用返回的中间对象的文档。因此，建议直接在未链接的方法上执行查找操作。

14.4.1 实战操作

（1）加载调查数据。

```
>>> import zipfile
>>> url = 'data/kaggle-survey-2018.zip'

>>> with zipfile.ZipFile(url) as z:
...     kag = pd.read_csv(z.open('multipleChoiceResponses.csv'))
...     df = kag.iloc[1:]
```

（2）使用 Jupyter ? 扩展查找 .apply 的说明文档。当然，你也可以在 Jupyter 中按 Shift + Tab 快捷键 4 次以获得它。

```
>>> df.Q3.apply?
Signature: df.Q3.apply(func, convert_dtype=True, args=(), **kwds)
Docstring:
Invoke function on values of Series.

Can be ufunc (a NumPy function that applies to the entire Series)
or a Python function that only works on single values.

Parameters
----------
func : function
    Python function or NumPy ufunc to apply.
convert_dtype : bool, default True
    Try to find better dtype for elementwise function results. If
    False, leave as dtype=object.
args : tuple
    Positional arguments passed to func after the series value.**kwds
    Additional keyword arguments passed to func.

Returns
-------
Series or DataFrame
    If func returns a Series object the result will be a DataFrame.

See Also
```

Series.map: For element-wise operations.
Series.agg: Only perform aggregating type operations.
Series.transform: Only perform transforming type operations.

Examples

    ...

File: ~/.env/364/lib/python3.6/site-packages/pandas/core/series.py
Type: method

（3）使用??可以查看源代码（没有Shift+Tab快捷键可以获取代码）。

```
>>> df.Q3.apply??
Signature: df.Q3.apply(func, convert_dtype=True, args=(), **kwds)
Source:
    def apply(self, func, convert_dtype=True, args=(), **kwds):
        ...

        if len(self) == 0:
            return self._constructor(dtype=self.dtype, index=self.index).__finalize__(
                self
            )

        # dispatch to agg
        if isinstance(func, (list, dict)):
            return self.aggregate(func, *args, **kwds)

        # if we are a string, try to dispatch
        if isinstance(func, str):
            return self._try_aggregate_string_function(func, *args, **kwds)

        # handle ufuncs and lambdas
        if kwds or args and not isinstance(func, np.ufunc):

            def f(x):
                return func(x, *args, **kwds)

        else:
```

```
            f = func

        with np.errstate(all="ignore"):
            if isinstance(f, np.ufunc):
                return f(self)

            # row-wise access
            if is_extension_type(self.dtype):
                mapped = self._values.map(f)
            else:
                values = self.astype(object).values
                mapped = lib.map_infer(values, f, convert=convert_dtype)

        if len(mapped) and isinstance(mapped[0], Series):
            # GH 25959 use pd.array instead of tolist
            # so extension arrays can be used
            return self._constructor_expanddim(pd.array(mapped), index=self.index)
        else:
            return self._constructor(mapped, index=self.index).__finalize__(self)
File:      ~/.env/364/lib/python3.6/site-packages/pandas/core/series.py
Type:      method
```

（4）可以看到，该方法试图找出合适的代码来调用。如果所有这些都失败，则最终它将计算 mapped 变量。下面我们尝试找出 lib.map_infer 的作用。

```
>>> import pandas.core.series
>>> pandas.core.series.lib
<module 'pandas._libs.lib' from '.env/364/lib/python3.6/site-packages/pandas/_libs/lib.cpython-36m-darwin.so'>

>>> pandas.core.series.lib.map_infer??
Docstring:
Substitute for np.vectorize with pandas-friendly dtype inference

Parameters
----------
arr : ndarray
f : function

Returns
```

```
-------
mapped : ndarray
Type: builtin_function_or_method
```

14.4.2 原理解释

Jupyter 能够检查 Python 对象的文档字符串和源代码。标准的 Python REPL 可以利用内置的 help 函数来查看文档字符串，但是无法显示源代码。

但是，Jupyter 还藏着一些妙招。如果在函数或方法后面加上一个问号（?），那么它将显示该代码的说明文档。请注意，这不是有效的 Python 语法，它是 Jupyter 的功能。如果添加两个问号（??），则 Jupyter 将显示该函数或方法的源代码。

此秘笈显示了如何跟踪源代码，以了解 Pandas 中的.apply 方法在后台的工作方式。

在步骤（3）中，如果没有结果，则可以看到一个快捷方式。

我们还可以看到字符串函数（即传递字符串字面意思）的工作方式。例如，getattr 函数就可以从 DataFrame 中提取相应的方法。

接下来，代码检查它应用的是否是 NumPy 函数。如果它是 np.ufunc 的实例，那么将调用该函数，或者将对底层._values 属性调用.map 方法；否则将调用 lib.map_infer。

在步骤（4）中，我们尝试检查了 lib.map_infer，但发现它是一个 so 文件（在 Windows 上为 pyd）。这是一个编译文件，通常是用 C 编写 Python 或使用 Cython 的结果。Jupyter 无法向我们显示编译文件的源。

14.4.3 扩展知识

当你查看函数或方法的源代码时，Jupyter 将在窗格底部显示其所属的文件。如果真的需要深入研究源代码，则可以在 Jupyter 之外的编辑器中将其打开。然后，可以使用编辑器浏览该代码和任何相应的代码（大多数编辑器比 Jupyter 具有更好的代码导航功能）。

14.5 在 Jupyter 中进行调试

前面的秘笈展示了如何理解 Pandas 代码以及如何通过 Jupyter 对其进行检查。本节将研究在 Jupyter 中使用 IPython 调试器（IPython debugger，ipdb）。

本节将创建一个函数，当尝试将其与 Series .apply 方法一起使用时，函数将抛出错误，然后我们将使用 ipdb 对其进行调试。

14.5.1 实战操作

（1）加载调查数据。

```
>>> import zipfile
>>> url = 'data/kaggle-survey-2018.zip'

>>> with zipfile.ZipFile(url) as z:
...     kag = pd.read_csv(z.open('multipleChoiceResponses.csv'))
...     df = kag.iloc[1:]
```

（2）尝试编写并运行一个函数以给一个 Series 加 1。

```
>>> def add1(x):
...     return x + 1

>>> df.Q3.apply(add1)
---------------------------------------------------------------
TypeError                                 Traceback (most recent call last)
<ipython-input-9-6ce28d2fea57> in <module>
      2     return x + 1
      3
----> 4 df.Q3.apply(add1)

~/.env/364/lib/python3.6/site-packages/pandas/core/series.py in apply(self, func, convert_dtype, args, **kwds)
   4043             else:
   4044                 values = self.astype(object).values
-> 4045                 mapped = lib.map_infer(values, f, convert=convert_dtype)
   4046
   4047             if len(mapped) and isinstance(mapped[0], Series):

pandas/_libs/lib.pyx in pandas._libs.lib.map_infer()

<ipython-input-9-6ce28d2fea57> in add1(x)
      1 def add1(x):
----> 2     return x + 1
      3
      4 df.Q3.apply(add1)
```

第 14 章　调试和测试

TypeError: must be str, not int

（3）在出现异常后，可立即使用%debug 单元格魔术命令来进入调试窗口。这会将调试器打开到引发异常的位置。

可以使用调试器命令在堆栈中导航。按 U 键会将堆栈弹出到调用当前行的函数中。可以使用打印命令（p）检查对象，如图 14-1 所示。

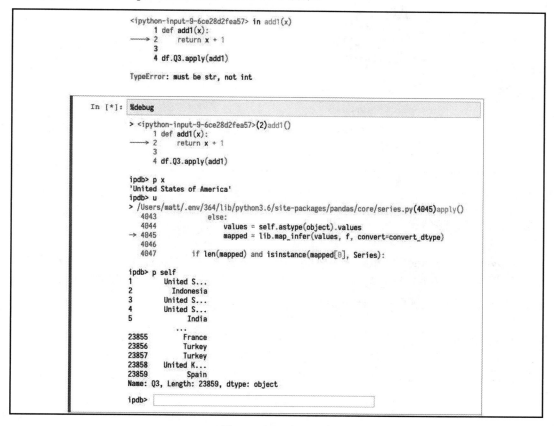

图 14-1　Jupyter 调试

（4）如果要进入代码而不需要抛出异常，则可以使用 IPython 调试器中的 set_trace 函数，这将使你立即进入调试器并找到该行。

>>> from IPython.core.debugger import set_trace

>>> def add1(x):

```
...        set_trace()
...        return x + 1

>>> df.Q3.apply(add1)
```

此时的显示如图 14-2 所示。

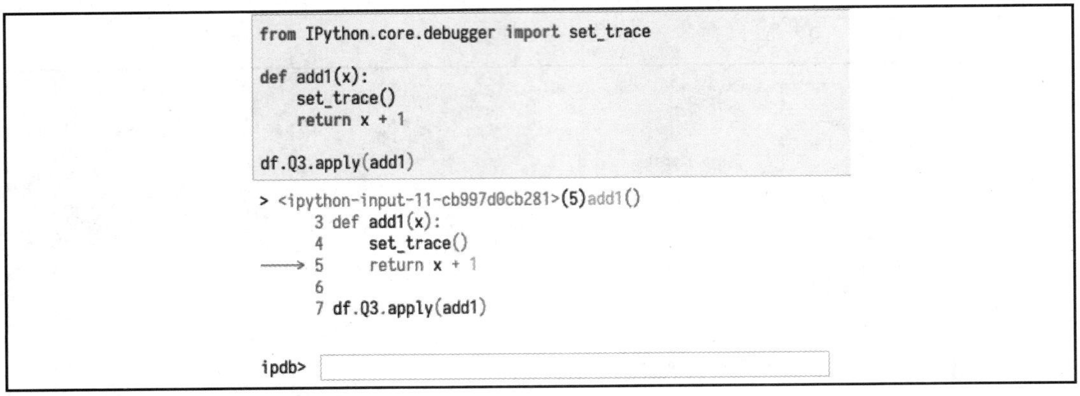

```
from IPython.core.debugger import set_trace

def add1(x):
    set_trace()
    return x + 1

df.Q3.apply(add1)
> <ipython-input-11-cb997d0cb281>(5)add1()
      3 def add1(x):
      4     set_trace()
----> 5     return x + 1
      6
      7 df.Q3.apply(add1)

ipdb>
```

图 14-2　Jupyter 调试

14.5.2　原理解释

Jupyter（从 IPython 派生）随 IPython 调试器一起提供。这将复制标准库中 pdb 模块的功能，但具有语法高亮之类的优点。它也具有按 Tab 键即自动补全代码的功能，但这仅在 IPython 控制台中有效，在 Jupyter 中则不起作用。

14.5.3　扩展知识

如果你不熟悉使用调试器，则还可以使用以下救急方式：命令 h 将打印出你可以从调试器运行的所有命令。

```
ipdb> h

Documented commands (type help <topic>):
========================================
EOF    cl         disable  interact  next   psource  rv      unt
a      clear      display  j         p      q        s       until
alias  commands   down     jump      pdef   quit     source  up
args   condition  enable   l         pdoc   r        step    w
```

```
b         cont       exit      list       pfile     restart   tbreak    whatis
break     continue   h         ll         pinfo     return    u         where
bt        d          help      longlist   pinfo2    retval    unalias
c         debug      ignore    n          pp        run       undisplay
```

我最常用的命令是 s、n、l、u、d 和 c。如果你想知道 s 的作用，则可以将其输入。

```
ipdb> h s
s(tep)
        Execute the current line, stop at the first possible occasion
        (either in a function that is called or in the current function).
```

上述命令是在告诉调试器打印出 step（s）的帮助 help（h）文档。

因为在 Jupyter 中进行的编码通常都是小规模的，所以调试器显得有点牛刀杀鸡。当然，掌握其用法毕竟是一件好事，特别是如果你想深入研究 Pandas 源代码并了解正在发生的事情的话，调试器还是很好用的。

14.6 管理数据的完整性

Great Expectations 是一种第三方工具，可让你捕获和定义数据集的属性。我们可以保存这些属性，然后使用它们来验证将来的数据以确保数据完整性。这在构建机器学习模型时非常有用，因为新的分类数据值和数字离群值往往会导致模型性能不佳或出现错误。本节将研究 Kaggle 数据集，并建立一个 Expectation Suite 来测试和验证数据。

14.6.1 实战操作

（1）使用先前定义的 tweak_kag 函数读取数据。

```
>>> kag = tweak_kag(df)
```

（2）使用 Great Expectations from_pandas 函数读取 Great Expectations DataFrame（DataFrame 的子类，其中包含一些额外的方法）。

```
>>> import great_expectations as ge
>>> kag_ge = ge.from_pandas(kag)
```

（3）检查 DataFrame 上的其他方法。

```
>>> sorted([x for x in set(dir(kag_ge)) - set(dir(kag))
...         if not x.startswith('_')])
['autoinspect',
```

```
'batch_fingerprint',
'batch_id',
'batch_kwargs',
'column_aggregate_expectation',
'column_map_expectation',
'column_pair_map_expectation',
'discard_failing_expectations',
'edit_expectation_suite',
'expect_column_bootstrapped_ks_test_p_value_to_be_greater_than',
'expect_column_chisquare_test_p_value_to_be_greater_than',
'expect_column_distinct_values_to_be_in_set',
'expect_column_distinct_values_to_contain_set',
'expect_column_distinct_values_to_equal_set',
'expect_column_kl_divergence_to_be_less_than',
'expect_column_max_to_be_between',
'expect_column_mean_to_be_between',
'expect_column_median_to_be_between',
'expect_column_min_to_be_between',
'expect_column_most_common_value_to_be_in_set',
'expect_column_pair_values_A_to_be_greater_than_B',
'expect_column_pair_values_to_be_equal',
'expect_column_pair_values_to_be_in_set',
'expect_column_parameterized_distribution_ks_test_p_value_to_be_greater_than',
'expect_column_proportion_of_unique_values_to_be_between',
'expect_column_quantile_values_to_be_between',
'expect_column_stdev_to_be_between',
'expect_column_sum_to_be_between',
'expect_column_to_exist',
'expect_column_unique_value_count_to_be_between',
'expect_column_value_lengths_to_be_between',
'expect_column_value_lengths_to_equal',
'expect_column_values_to_be_between',
'expect_column_values_to_be_dateutil_parseable',
'expect_column_values_to_be_decreasing',
'expect_column_values_to_be_in_set',
'expect_column_values_to_be_in_type_list',
'expect_column_values_to_be_increasing',
'expect_column_values_to_be_json_parseable',
'expect_column_values_to_be_null',
'expect_column_values_to_be_of_type',
'expect_column_values_to_be_unique',
'expect_column_values_to_match_json_schema',
```

```
'expect_column_values_to_match_regex',
'expect_column_values_to_match_regex_list',
'expect_column_values_to_match_strftime_format',
'expect_column_values_to_not_be_in_set',
'expect_column_values_to_not_be_null',
'expect_column_values_to_not_match_regex',
'expect_column_values_ to_not_match_regex_list',
'expect_multicolumn_values_to_be_unique',
'expect_table_column_count_to_be_between',
'expect_table_column_count_to_equal',
'expect_table_columns_to_match_ordered_list',
'expect_table_row_count_to_be_between',
'expect_table_row_count_to_equal',
'expectation',
'find_expectation_indexes',
'find_expectations',
'from_dataset',
'get_column_count',
'get_column_count_in_range',
'get_column_hist',
'get_column_max',
'get_column_mean',
'get_column_median',
'get_column_min',
'get_column_modes',
'get_column_nonnull_count',
'get_column_partition',
'get_column_quantiles',
'get_column_stdev',
'get_column_sum',
'get_column_unique_count',
'get_column_value_counts',
'get_config_value',
'get_data_asset_name',
'get_default_expectation_arguments',
'get_evaluation_parameter',
'get_expectation_suite',
'get_expectation_suite_name',
'get_expectations_config',
'get_row_count',
'get_table_columns',
'hashable_getters',
'multicolumn_map_expectation',
```

```
'profile',
'remove_expectation',
'save_expectation_suite',
'save_expectation_suite_name',
'set_config_value',
'set_data_asset_name',
'set_default_expectation_argument',
'set_evaluation_parameter',
'test_column_aggregate_expectation_function',
'test_column_map_expectation_function',
'test_expectation_function',
'validate']
```

（4）Great Expectations 对表的形状、缺失值、类型、范围、字符串、日期、聚合函数、列对（column pair）、分布和文件属性等均有期望（expectation）。这些期望都是可以使用的。库将跟踪我们使用的期望，然后将它们保存为 Expectation Suite。

```
>>> kag_ge.expect_column_to_exist('Salary')
{'success': True}

>>> kag_ge.expect_column_mean_to_be_between(
...     'Salary', min_value=10_000, max_value=100_000)
{'success': True,
'result': {'observed_value': 43869.66102793441,
'element_count': 15429,
'missing_count': 0,
'missing_percent': 0.0}}

>>> kag_ge.expect_column_values_to_be_between(
...     'Salary', min_value=0, max_value=500_000)
{'success': True,
'result': {'element_count': 15429,
'missing_count': 0,
'missing_percent': 0.0,
'unexpected_count': 0,
'unexpected_percent': 0.0,
'unexpected_percent_nonmissing': 0.0,
'partial_unexpected_list': []}}

>>> kag_ge.expect_column_values_to_not_be_null('Salary')
{'success': True,
'result': {'element_count': 15429,
'unexpected_count': 0,
```

```
'unexpected_percent': 0.0,
'partial_unexpected_list': []}}

>>> kag_ge.expect_column_values_to_match_regex(
...         'Country', r'America|India|Another|China')
{'success': True,
'result': {'element_count': 15429,
'missing_count': 0,
'missing_percent': 0.0,
'unexpected_count': 0,
'unexpected_percent': 0.0,
'unexpected_percent_nonmissing': 0.0,
'partial_unexpected_list': []}}

>>> kag_ge.expect_column_values_to_be_of_type(
...         'Salary', type_='int')
{'success': True, 'result': {'observed_value': 'int64'}}
```

（5）将期望保存到文件中。Great Expectations 使用 JSON 来指定这些期望。

```
>>> kag_ge.save_expectation_suite('kaggle_expectations.json')
```

该文件应如下所示。

```
{
    "data_asset_name": null,
    "expectation_suite_name": "default",
    "meta": {
        "great_expectations. version ": "0.8.6"
    },
    "expectations": [
        {
            "expectation_type": "expect_column_to_exist",
            "kwargs": {
                "column": "Salary"
            }
        },
        {
            "expectation_type": "expect_column_mean_to_be_between",
            "kwargs": {
                "column": "Salary",
                "min_value": 10000,
                "max_value": 100000
            }
```

```
        },
        {
            "expectation_type": "expect_column_values_to_be_between",
            "kwargs": {
                "column": "Salary",
                "min_value": 0,
                "max_value": 500000
            }
        },
        {
            "expectation_type": "expect_column_values_to_not_be_null",
            "kwargs": {
                "column": "Salary"
            }
        },
        {
            "expectation_type": "expect_column_values_to_match_regex",
            "kwargs": {
                "column": "Country",
                "regex": "America|India|Another|China"
            }
        },
        {
            "expectation_type": "expect_column_values_to_be_of_type",
            "kwargs": {
                "column": "Salary",
                "type_": "int"
            }
        }
    ],
    "data_asset_type": "Dataset"
}
```

（6）使用 Expectation Suite 评估在 CSV 文件中找到的数据。我们会将 Kaggle 数据保存到 CSV 文件中，并进行测试以确保其仍然可以通过。

```
>>> kag_ge.to_csv('kag.csv')
>>> import json
>>> ge.validate(ge.read_csv('kag.csv'),
...     expectation_suite=json.load(
...         open('kaggle_expectations.json')))
{'results': [{'success': True,
```

```
         'expectation_config': {'expectation_type': 'expect_column_to_ exist',
         'kwargs': {'column': 'Salary'}},
         'exception_info': {'raised_exception': False,
         'exception_message': None,
         'exception_traceback': None}},
{'success': True,
         'result': {'observed_value': 43869.66102793441,
         'element_count': 15429,
         'missing_count': 0,
         'missing_percent': 0.0},
         'expectation_config': {'expectation_type': 'expect_column_mean_
to_be_between',
         'kwargs': {'column': 'Salary', 'min_value': 10000, 'max_value':
100000}},
         'exception_info': {'raised_exception': False,
         'exception_message': None,
         'exception_traceback': None}},
{'success': True,
         'result': {'element_count': 15429,
         'missing_count': 0,
         'missing_percent': 0.0,
         'unexpected_count': 0,
         'unexpected_percent': 0.0,
         'unexpected_percent_nonmissing': 0.0,
         'partial_unexpected_list': []},
         'expectation_config': {'expectation_type': 'expect_column_
values_to_be_between',
         'kwargs': {'column': 'Salary', 'min_value': 0, 'max_value': 500000}},
         'exception_info': {'raised_exception': False,
         'exception_message': None,
         'exception_traceback': None}},
{'success': True,
         'result': {'element_count': 15429,
         'unexpected_count': 0,
         'unexpected_percent': 0.0,
         'partial_unexpected_list': []},
         'expectation_config': {'expectation_type': 'expect_column_
values_to_not_be_null',
         'kwargs': {'column': 'Salary'}},
         'exception_info': {'raised_exception': False,
         'exception_message': None,
         'exception_traceback': None}},
```

```
{'success': True,
    'result': {'observed_value': 'int64'},
    'expectation_config': {'expectation_type': 'expect_column_
values_to_be_of_type',
    'kwargs': {'column': 'Salary', 'type_': 'int'}},
    'exception_info': {'raised_exception': False,
    'exception_message': None,
    'exception_traceback': None}},
{'success': True,
    'result': {'element_count': 15429,
    'missing_count': 0,
    'missing_percent': 0.0,
    'unexpected_count': 0,
    'unexpected_percent': 0.0,
    'unexpected_percent_nonmissing': 0.0,
    'partial_unexpected_list': []},
    'expectation_config': {'expectation_type': 'expect_column_
values_to_match_regex',
    'kwargs': {'column': 'Country', 'regex': 'America|India|Another|
China'}},
    'exception_info': {'raised_exception': False,
    'exception_message': None,
    'exception_traceback': None}}],
'success': True,
'statistics': {'evaluated_expectations': 6,
'successful_expectations': 6,
'unsuccessful_expectations': 0,
'success_percent': 100.0},
'meta': {'great_expectations. version ': '0.8.6',
'data_asset_name': None,
'expectation_suite_name': 'default',
'run_id': '2020-01-08T214957.098199Z'}}
```

14.6.2 原理解释

Great Expectations 库扩展了 Pandas DataFrame。可以使用它来验证原始数据或通过 Pandas 调整过的数据。在我们的示例中，展示了如何为 DataFrame 创建期望。

在步骤（3）中列出了许多内置的期望。可以利用这些期望，或者根据需要建立自定义期望。验证数据的结果是带有 success 条目的 JSON 对象。可以将它们集成到测试套件中，以确保数据处理管道可以使用新数据。

14.7 结合使用 pytest 和 Pandas

本节将演示如何测试 Pandas 代码。我们将通过测试工件来做到这一点。本节测试使用的是第三方库 pytest。

本秘笈将不使用 Jupyter，而是使用命令行。

14.7.1 实战操作

（1）创建项目数据布局。pytest 库支持以若干种不同样式布局的项目。我们将创建一个如下所示的文件夹结构。

```
kag-demo-pytest/
├── data
│   └── kaggle-survey-2018.zip
├── kag.py
└── test
    └── test_kag.py
```

kag.py 文件包含用于加载原始数据和对其进行调整的代码。它看起来应该如下所示。

```python
import pandas as pd

import zipfile


def load_raw(zip_fname):
    with zipfile.ZipFile(zip_fname) as z:
        kag = pd.read_csv(z.open('multipleChoiceResponses.csv'))
        df = kag.iloc[1:]
    return df


def tweak_kag(df):
    na_mask = df.Q9.isna()
    hide_mask = df.Q9.str.startswith('I do not').fillna(False)
    df = df[~na_mask & ~hide_mask]

    q1 = (df.Q1
        .replace({ 'Prefer not to say': 'Another',
```

```python
                        'Prefer to self-describe': 'Another'})
            .rename('Gender')
    )
    q2 = df.Q2.str.slice(0,2).astype(int).rename('Age')
    def limit_countries(val):
        if val in  {'United States of America', 'India', 'China'}:
            return val
        return 'Another'
    q3 = df.Q3.apply(limit_countries).rename('Country')

    q4 = (df.Q4
     .replace({'Master\'s degree': 18,
     'Bachelor\'s degree': 16,
     'Doctoral degree': 20,
     'Some college/university study without earning a bachelor's
degree': 13,
     'Professional degree': 19,
     'I prefer not to answer': None,
     'No formal education past high school': 12})
     .fillna(11)
     .rename('Edu')
    )

    def only_cs_stat_val(val):
        if val not in {'cs', 'eng', 'stat'}:
            return 'another'
        return val

    q5 = (df.Q5
            .replace({
                'Computer science (software engineering, etc.)': 'cs',
                'Engineering (non-computer focused)': 'eng',
                'Mathematics or statistics': 'stat'})
            .apply(only_cs_stat_val)
            .rename('Studies'))
    def limit_occupation(val):
        if val in {'Student', 'Data Scientist', 'Software Engineer',
'Not employed',
                    'Data Engineer'}:
            return val
        return 'Another'

    q6 = df.Q6.apply(limit_occupation).rename('Occupation')
```

第 14 章 调试和测试

```python
    q8 = (df.Q8
        .str.replace('+', '')
        .str.split('-', expand=True)
        .iloc[:,0]
        .fillna(-1)
        .astype(int)
        .rename('Experience')
    )

    q9 = (df.Q9
        .str.replace('+','')
        .str.replace(',','')
        .str.replace('500000', '500')
        .str.replace('I do not wish to disclose my approximate yearly compensation','')
        .str.split('-', expand=True)
        .iloc[:,0]
        .astype(int)
        .mul(1000)
        .rename('Salary'))
    return pd.concat([q1, q2, q3, q4, q5, q6, q8, q9], axis=1)
```

test_kag.py 文件看起来如下所示。

```python
import pytest

import kag


@pytest.fixture(scope='session')
def df():
    df = kag.load_raw('data/kaggle-survey-2018.zip')
    return kag.tweak_kag(df)


def test_salary_mean(df):
    assert 10_000 < df.Salary.mean() < 100_000

def test_salary_between(df):
    assert df.Salary.min() >= 0
    assert df.Salary.max() <= 500_000

def test_salary_not_null(df):
```

```
        assert not df.Salary.isna().any()

def test_country_values(df):
    assert set(df.Country.unique()) == {'Another', 'United States
of America', 'India', 'China'}

def test_salary_dtype(df):
    assert df.Salary.dtype == int
```

（2）从 kag-demo 目录运行测试。如果已经安装了 pytest 库，将有一个 pytest 可执行文件。如果尝试运行该命令，则会收到错误消息。

```
(env)$ pytest
=================== test session starts ===================
platform darwin -- Python 3.6.4, pytest-3.10.1, py-1.7.0,
pluggy-0.8.0
rootdir: /Users/matt/pandas-cookbook/kag-demo, inifile:
plugins: asyncio-0.10.0
collected 0 items / 1 errors

========================= ERRORS =========================
_____ERROR collecting test/test_kag.py_____
ImportError while importing test module '/Users/matt/pandas-cookbook/kag-
demo/test/test_kag.py'.
Hint: make sure your test modules/packages have valid Python names.
Traceback:
test/test_kag.py:3: in <module>
    import kag
E   ModuleNotFoundError: No module named 'kag'
!!!!!!!! Interrupted: 1 errors during collection !!!!!!!!
================ 1 error in 0.15 seconds ================
```

上述错误是因为 pytest 想要使用已安装的代码来运行测试。因为笔者没有使用 pip（或其他机制）来安装 kag.py，所以 pytest 抱怨说它无法在安装代码的位置找到该模块。

（3）帮助 pytest 找到 kag.py 文件的解决方法是将 pytest 作为模块调用。所以，可改为运行以下命令。

```
$ python -m pytest
==================== test session starts ====================
platform darwin -- Python 3.6.4, pytest-3.10.1, py-1.7.0, pluggy-0.8.0
rootdir: /Users/matt/pandas-cookbook/kag-demo, inifile:
collected 5 items
```

```
test/test_kag.py .....
[100%]

============== 5 passed, 1 warnings in 3.51 seconds ===============
```

以这种方式调用 pytest 会将当前目录添加到 PYTHONPATH 中，并且 kag 模块的导入现在已经成功了。

14.7.2 原理解释

对于 pytest 库的完整介绍超出了本书的讨论范围。当然，本示例中的 test_kag.py 文件已经包含指定的测试，pytest 完全能够理解它们。任何以 test_ 开头的函数名称都将被识别为测试。这些测试函数的参数 df 被称为 fixture。fixture 是 pytest 框架的精髓，它可以为可靠的和可重复执行的测试提供固定的基线（你可以将它理解为测试的固定配置，使不同范围的测试都能够获得统一的配置）。

可以看到，在 test_kag.py 文件的顶层附近，指定了一个名为 df 的函数，该函数使用了以下装饰。

```
@pytest.fixture(scope='session')
```

测试会话开始时，将调用此函数一次。任何测试函数只要它包含了名为 df 的参数，都将获得 df 函数的输出。其作用域被指定为 session（会话）作用域，因此数据仅被加载一次（用于整个测试会话）。如果没有指定该作用域，那么该 fixture 的作用域将处于函数级别（默认）。在函数级作用域内，该 fixture 将会为每个使用它作为参数的测试函数运行一次，这会使得测试需要运行 12s（而不是像在我的机器上那样仅运行 3s 多）。

14.7.3 扩展知识

也可以从 pytest 运行 Great Expectations 测试。例如，将以下函数添加到 test_kag.py（你也许需要更改自己的期望套件的路径）。

```
def test_ge(df):
    import json
    import great_expectations as ge
    res = ge.validate(ge.from_pandas(df),
        expectation_suite=json.load(open('kaggle_expectations.json')))
    failures = []
    for exp in res['results']:
```

```
        if not exp['success']:
            failures.append(json.dumps(exp, indent=2))
    if failures:
        assert False, '\n'.join(failures)
    else:
        assert True
```

14.8 使用 Hypothesis 库生成测试

Hypothesis 库是用于生成测试或执行基于属性的测试（property-based test）的第三方库。Hypothesis 的本意是"假设"，你可以创建一个策略（strategy），也就是一个生成数据样本的对象，然后运行代码以测试该策略生成的输出。总之，Hypothesis 可以帮助省去很多生成数据的麻烦，通过更少的工作在代码中发现更多的错误。在本节中，我们将使用 Hypothesis 生成的数据来测试不变量是否成立。

对于这种类型的测试，完全可以单独编写一本厚厚的著作，因此，为简便起见，本节将仅演示一个使用该库的示例。

我们将演示如何生成 Kaggle 调查数据，然后使用生成的调查数据，让 tweak_kag 函数运行它，以验证该函数在新数据上是有效的。

我们将利用 14.7 节"结合使用 pytest 和 Pandas"中的测试代码。Hypothesis 库可以与 pytest 一起使用，因此本示例也可以使用相同的布局。

14.8.1 实战操作

（1）创建项目数据布局。如果你已经有了 14.7 节"结合使用 pytest 和 Pandas"的代码，则只要添加一个 test_hypot.py 文件和一个 conftest.py 文件即可。

```
kag-demo-hypo/
├── data
│   └── kaggle-survey-2018.zip
├── kag.py
└── test
    ├── conftest.py
    ├── test_hypot.py
    └── test_kag.py
```

（2）将共享的 fixture 放到 conftest.py 文件中（如前文所述，你可以将 fixture 理解为固定参数配置）。conftest.py 文件是 pytest 在尝试查找 fixture 时寻找的特殊文件。我们不需要导入它，但是其中定义的任何 fixture 都可以被其他测试文件使用。

将 fixture 代码从 test_kag.py 文件移至 conftest.py 文件中，以使其具有以下代码。我们还需要做一些重构，以创建一个 raw_ 函数，该函数不是可以在测试之外调用的 fixture。

```python
import pytest

import kag

@pytest.fixture(scope='session')
def raw():
    return raw_()

def raw_():
    return kag.load_raw('data/kaggle-survey-2018.zip')

@pytest.fixture(scope='session')
def df(raw):
    return kag.tweak_kag(raw)
```

将以下代码放入 test_hypot.py 文件中。

```python
from hypothesis import given, strategies
from hypothesis.extra.pandas import column, data_frames

from conftest import raw_

import kag

def hypot_df_generator():
    df = raw_()
    cols = []
    for col in ['Q1', 'Q2', 'Q3', 'Q4', 'Q5', 'Q6', 'Q8', 'Q9']:
        cols.append(column(col, elements=strategies.sampled_from(df[col].unique())))
    return data_frames(columns=cols)


@given(hypot_df_generator())
def test_countries(gen_df):
    if gen_df.shape[0] == 0:
        return
    kag_ = kag.tweak_kag(gen_df)
    assert len(kag_.Country.unique()) <= 4
```

函数 hypot_df_generator 构造了一个 Hypothesis 搜索策略。该搜索策略可以生成不同类型的数据。我们可以手动创建这些策略。在本示例中，使用了现有的 CSV 文件填充不同的值。对于我们感兴趣的列来说，这些值都是可能值。

函数 test_countries 是一个 pytest 测试，它使用了 @given(hypot_df_generator()) 装饰器进行装饰。该装饰将把 gen_df 对象传递给测试函数。该对象是一个符合搜索策略规范的 DataFrame。

现在可以针对该 DataFrame 测试我们的不变量（invariant）。在本示例中，将运行 tweak_kag 函数，并确保 Country 列中的唯一国家/地区数小于或等于 4。

（3）转到 kag_demo 目录并运行测试。以下是仅运行 test_countries 测试的命令。

```
$ python -m pytest -k test_countries
The output looks like this:
======================= test session starts =======================
platform darwin -- Python 3.6.4, pytest-5.3.2, py-1.7.0, pluggy-0.13.1
rootdir: /Users/matt/kag-demo
plugins: asyncio-0.10.0, hypothesis-5.1.2
collected 6 items / 5 deselected / 1 selected

test/test_hypot.py F                                         [100%]

============================ FAILURES =============================
_____test_countries_____

    @given(hypot_df_generator())
>   def test_countries(gen_df):

test/test_hypot.py:19:
_ _ _ _ _ _ _ _ _ _ _ _ _ _ _ _ _ _ _ _ _ _ _ _ _ _ _ _ _ _ _ _ _ _
test/test_hypot.py:23: in test_countries
    kag_ = kag.tweak_kag(gen_df)
kag.py:63: in tweak_kag
    q8 = (df.Q8
/Users/matt/.env/364/lib/python3.6/site-packages/pandas/core/generic.py:5175: in __getattr__
    return object.__getattribute__(self, name)
/Users/matt/.env/364/lib/python3.6/site-packages/pandas/core/accessor.py:175: in __get__
    accessor_obj = self._accessor(obj)
```

```
/Users/matt/.env/364/lib/python3.6/site-packages/pandas/core/
strings.py:1917: in __init__
    self._inferred_dtype = self._validate(data)
_ _ _ _ _ _ _ _ _ _ _ _ _ _ _ _ _ _ _ _ _ _ _ _ _ _ _ _ _ _

data = Series([], Name: Q8, dtype: float64)

    @staticmethod
    def _validate(data):
        """
        Auxiliary function for StringMethods, infers and checks
dtype of data.
        This is a "first line of defence" at the creation of the
StringMethods-
        object (see _make_accessor), and just checks that the
dtype is in the
        *union* of the allowed types over all string methods
below; this
        restriction is then refined on a per-method basis using
the decorator
        @forbid_nonstring_types (more info in the corresponding
docstring).

        This really should exclude all series/index with any non-
string values,
        but that isn't practical for performance reasons until we
have a str
        dtype (GH 9343 / 13877)

        Parameters
        ----------
        data : The content of the Series

        Returns
        -------
        dtype : inferred dtype of data
        """
        if isinstance(data, ABCMultiIndex):
            raise AttributeError(
                "Can only use .str accessor with Index, " "not
MultiIndex"
            )
```

```
        # see _libs/lib.pyx for list of inferred types
        allowed_types = ["string", "empty", "bytes", "mixed",
"mixed-integer"]

        values = getattr(data, "values", data) # Series / Index
        values = getattr(values, "categories", values) #
categorical / normal

        try:
            inferred_dtype = lib.infer_dtype(values, skipna=True)
        except ValueError:
            # GH#27571 mostly occurs with ExtensionArray
            inferred_dtype = None

        if inferred_dtype not in allowed_types:
>           raise AttributeError("Can only use .str accessor with
string " "values!")
E           AttributeError: Can only use .str accessor with string
values!

/Users/matt/.env/364/lib/python3.6/site-packages/pandas/core/
strings.py:1967: AttributeError
------------------------- Hypothesis -------------------------
Falsifying example: test_countries(
    gen_df=      Q1       Q2                       Q3    ...
Q6   Q8   Q9
    0    Female  45-49    United States of America    ...  Consultant
NaN  NaN

    [1 rows x 8 columns],
)
=========== 1 failed, 5 deselected, 1 warning in 2.23s ===========
```

可以看到，上述输出中有很多噪声，但是如果对其进行扫描，那么你会发现其实是处理列 Q8 的代码出现了问题。这是因为它为 Q8 生成了一个带有 NaN 条目的单行。如果使用此 DataFrame 运行 tweak_kag，则 Pandas 会推断 Q8 列具有浮点类型，这样在尝试使用 .str 访问器时就会出错。

这是一个错误吗？对此很难给出确切的答案。但这至少表明，如果原始数据中仅包含缺失值，那么我们的代码将无法正常工作。

14.8.2 原理解释

Hypothesis 库会尝试生成符合规范的数据范围。数据分析人员可以使用此生成的数据来测试不变量是否成立。在本示例中,我们看到调查数据包含了缺失值。当生成带有单行缺失数据的 DataFrame 时,tweak_kag 函数无法正常工作。.str 访问器仅在一列中至少有一个字符串值时才正常有效。

我们可以解决这些问题,并继续测试其他不变量。传统的测试方法一般是创建测试用例、指定输入、指定预期的输出、运行代码,然后判断输出是否符合预期;而 Hypothesis 测试方法则是描述输入数据、描述输出属性、自动生成大量测试用例,以判断输出属性是否违背。虽然流程看起来类似,但是传统测试方法每次都只能指定一个测试用例,容易出现"一叶障目"的情况,而 Hypothesis 测试方法则是描述指定一组测试用例,测试覆盖的情况会大大增加,发现错误的可能性自然也大大增加。这就好比我们站在高处,"一览众山小",自然能够看到整个森林,因此使用 Hypothesis 库是解决"一叶障目"问题的好方法。